RURAL HUNTERDON

RURAL HUNTERDON

AN AGRICULTURAL HISTORY

By Hubert G. Schmidt

GREENWOOD PRESS, PUBLISHERS
WESTPORT, CONNECTICUT

The Library of Congress has catalogued this publication as follows:

Library of Congress Cataloging in Publication Data

Schmidt, Hubert Glasgow, 1905-
Rural Hunterdon.

Bibliography: p.
1. Hunterdon Co., N. J.--Economic conditions.
I. Title.
[HC107.N52H97 1972] 330.9749'71 77-139149
ISBN 0-8371-5765-X

Originally published in 1945
by Rutgers University Press, New Brunswick

Reprinted with the permission
of Rutgers University Press

First Greenwood Reprinting 1972

Library of Congress Catalogue Card Number 77-139149

ISBN 0-8371-5765-X

Printed in the United States of America

To the memory of

JACOB GOODALE LIPMAN

who, as dean of the College of Agriculture at Rutgers University and as director of the New Jersey Agricultural Experiment Station, contributed richly to the agricultural scholarship of the State, of the nation —and of the world.

Contents

Foreword

THE IDEA of an agricultural or economic history of Hunterdon County, New Jersey, received its inception many years ago as the result of conversations between the late Dean Jacob G. Lipman of Rutgers University and H. E. Deats of Hunterdon County. The former, equally well-known as a soil chemist and as an administrator, had a genuine love for agricultural history; the latter, a prominent farmer and secretary of the Hunterdon County Historical Society, had for several decades assiduously collected every kind of written and printed material in any way related to the county's development. The hopes of these two men were realized in 1938 when the Hunterdon County Board of Agriculture established a two-year fellowship at Rutgers University, the holder to do the necessary research and write a history of the economic development of the county. The work which I did as recipient of that fellowship and in my spare time since has resulted in the present book.

It was apparent from the beginning that the historian of Hunterdon County would have plenty of grist for his mill. By good fortune a great deal of valuable historical material for the region has been preserved, particularly for the period since the American Revolution. Official records of all kinds are unusually complete, and the same is true for the files of local newspapers. The account books of farmers, artisans, and business men have been retained in surprising numbers; and many families have kept correspondence, diaries, and other personal papers, often through several generations. The reminiscences written by older men, accounts written by local historians, and traditions committed to paper at one time or another have also proved numerous. A great many of these records which might otherwise have been lost have been gathered together and preserved by a very active local historical society, and others have gravitated to Rutgers University Library at New Brunswick and to the New Jersey Historical Society Library at Newark. It is doubtful whether many other rural counties of the United States have preserved as many valuable source materials relating to their past.

All of these sources are listed in the bibliography at the end of this book, and many will be referred to in the text, but a few of them merit special mention here. The most important are the Capner Papers, which include the correspondence and other papers of several English families who settled in the county between 1786 and 1800. The letters from Hunterdon County, which were fortunately brought back to America by later arrivals, portray in great detail the problems involved in immigration, the adjustments which newcomers must make in a new land, and the effect of the impact of European methods on American agriculture. Those written from England are not pertinent to this study, but give an excellent account of conditions in Leicestershire for the same period. The Capners and their relatives and friends were acute and unprejudiced observers, and their hundreds of letters have proved a treasure house. The collection has an application beyond Hunterdon County alone and should some day be carefully edited and put into printed form.

Other important material worthy of publication, at least in part, includes two diaries and the papers of John Emley. The diary of James H. Blackwell of present-day Flemington Junction shows the day-by-day activities of a progressive general farmer between 1843 and 1866. That of Elmer E. Bonnell gives a similar account by a farmer and peach orchardist located near Clinton from 1878 to 1895, during the heyday of peach growing in this area. The papers of John Emley, which cover approximately the period from 1750 to 1770, include a portion of his records and correspondence as surveyor and land agent for large absentee landowners, including the West Jersey Society.

Additional sources less special in nature also have proved valuable for this particular work. The account books of Janeway and Broughton, owners of a general store probably at Bound Brook in Somerset County, run from 1735 to 1747 and show much about the types of commodities, the farm produce of the period, and the buying habits of the people. This store drew customers from a large territory, including a part of Hunterdon County. The accounts of a general store in the Whitehouse region just before the Revolution are less complete but are of use for similar reasons, as are the numerous day books and ledgers of nineteenth-century stores to be cited later. Of these the most valuable for purposes of comparison belonged to Solomon Holcombe. He was the proprietor of a general store at Mount Airy for over fifty years, and his day books from 1824 to 1832 and from 1868 to 1871, at the beginning and near the end of his career, are complete.

The best and most comprehensive farm accounts found were those of Dr. John Bowne of present-day Bowne Station, running from 1801 to 1834. They include his records both as a physician and as a farmer. Other valuable farm accounts which were used were those kept by Thomas Lauderdale of West Amwell Township between 1886 and 1898, and those by his son William between 1914 and 1940. Of some use, too, were the accounts of James H. Blackwell during the same period as his diary. A very complete set of accounts over a long period were kept at the Case tannery near Flemington. Day books, ledgers, hide books, and other records in complete form date from 1783 to 1851. The machine shop and foundry accounts of Hiram Deats near Quakertown from 1831 to 1840 and of Deats, Case, and Company at Pittstown from 1869 to 1874 were also very informative.

Official records of various kinds, too, have proved invaluable in this study. Especially outstanding were the Hunterdon County portion of the New Jersey Wills at the Office of the Secretary of State at Trenton. The title is a misnomer, as not only wills but other probate papers are included. The collection begins with the earliest settlers of the county and comes up to the present. The inventories of estates in particular, and especially those of the Colonial period, gave information which could have been learned from no other source as to how the people lived and worked. For example, these papers showed that the grain cradle was in fairly common use considerably earlier than is usually supposed.

Other official records of note are the Justice of Peace dockets of Peter Brunner of Readington Township from 1780 to 1788 and of his successor, Ezekial Cole, from 1790 to 1804. Both men seem to have acted as fatherly advisers for a sizable neighborhood, and their records are much more comprehensive than was usually the case for such dockets. Even more important were the schedules of the local census enumerators in 1850. The records for the whole county, bound in a single volume, are now at the Rutgers University Library. Together they give vital information about every person, every farm, and every place of business in Hunterdon County in that year or for the year previous. The official records of West Amwell Township, kindly lent to me by the authorities there, also gave some facts which could never have been found elsewhere.

The source materials discussed thus far are all in manuscript form. The files of the local newspapers constitute a very important printed source, the only one which will be mentioned here. The earliest paper of the county as now constituted was the *Hunterdon Gazette,* established at Flemington, the county seat, in 1825. It was ably edited and was a powerful force for many years; a Whig organ, it became a political orphan with the disappearance of that party. After 1838 the *Gazette* had a competitor in the *Hunterdon Democrat,* a paper with powerful political backing, and in 1856 another paper with obvious political outlook, the *Hunterdon Republican,* entered the field. After a decade of three-cornered cutthroat competition, the *Gazette* and *Democrat* were combined into the *Hunterdon County Democrat.* Files of this newspaper and its two predecessors in nearly complete form, as well as part of the files of the *Democrat-Advertiser,* a competitor between 1881 and 1926, are extant. The files of the *Republican* are not complete for the whole period, but provide a valuable check for information found in other publications. Scattered issues and clippings ot other local papers were used, and the files of the *Federalist and New-Jersey Gazette* from 1796 to 1802 and those of the *Lambertville Press* from 1858 to 1861 proved to be of real merit.

After years of delving into these and other sources relating primarily to Hunterdon County, it would be understandable if the research student lost his perspective and fell into the error of treating developments in Hunterdon County as if they were something unique. It has been necessary, therefore, to keep in mind continually the various forces which swept through the Middle Atlantic region and through rural America generally. I have tried at all times to be conscious of the general developments of which those in Hunterdon County were a small part, and I hope that the reader will do the same. Nevertheless, it has been necessary to treat the county as a unit of development and has seemed best to go into considerable detail at the risk of being damned as an antiquarian. Other regions have had a similar development, it is true, but with enough variations so that the history of Hunterdon County cannot tell their complete story. That should be told in each case by other students. When enough thorough studies of local areas, not necessarily counties, have been made, I am sure that history in general will benefit a great deal.

In telling the story of Hunterdon County I have been faced with the problem of knowing just when to stop. The research for this work yielded data which could very well serve as the basis for an interesting history of the social and institutional development within the county. However, it has been found advisable to hold generally to the original scope and to bring in only those social and other aspects most closely related to economic growth. I have found it difficult to keep within these bounds, and have occasionally become so interested in pursuing a line of thought as to disregard them. Some day, fate willing, I hope to write for publication the story of the religious, educational, cultural, and institutional growth of the county.

In the present work there are a few points of style which should be explained. Quotations are given verbatim for the most part; and original spelling, punctuation, and capitalization are usually retained. Only when the error might be laid at my door have I introduced the expression *sic*. Occasionally a period or a comma has been inserted to clarify the meaning, particularly if the spacing used by the original writer indicates that a break in the thought was intended. Words or parts of words in brackets are also occasionally introduced for clarity.

Another point of considerable importance is that of documentation. Footnotes are used in general only to make positive identifications, to give credit for quotations, and to give the sources for statistical material and for statements which might well provoke doubt in the reader. The omission of the greater part of the documentation will doubtless please the average reader but may be frowned upon by some scholars. To them let me hasten to say that typed copies of this work, changed but little as to text and fully annotated, are to be found in Rutgers University Library. Furthermore, I shall be glad to answer any queries relating to sources of information which are addressed to me in care of the Rutgers University Press. In the footnotes which are retained, information which would duplicate that given in the bibliography has been omitted.

In conclusion I wish to acknowledge my deep gratitude to a number of persons who have given important aid and assistance. Mr. H. E. Deats, whom I mentioned in the first paragraph of this preface, has given unsparingly of his time and energy in locating important source materials among the large collection of the Hunterdon County Historical Society. Several staff members of the Rutgers University Library have also willingly cooperated from the beginning in making available their large collection of New Jersey materials. One of them, Mr. Donald Sinclair, was a most helpful critic; and I regret that his absence abroad in the service of his country prevents my further use of his technical knowledge before bringing this book to publication. Other libraries which have proven helpful are the New Jersey Historical Society Library at Newark, the Princeton University Library, and the Bucks County Historical Society Library at Doylestown, Pennsylvania. Horace M. Mann, curator of the Agricultural Museum at the latter institution gave much assistance in choosing and furnishing photographs as illustrations.

Persons who have in their official capacity furnished important information for this study are D. M. Babbitt, Hunterdon County Agricultural Agent, B. F. Ramsburg, Hunterdon County Club Agent, Harry A. Moore, Hunterdon County Superintendent of Schools, and Howard Mason of the United States Soil Conservation Service. Others who have given answers to specific queries or have reminisced concerning earlier days are C. B. Andrews, the late Sylvanus Apgar, John H. Bessick, J. Hart Case, Miss Amydell Case, Dr. Barclay Fuhrmann, Miss Julia Haver, Mrs. L. Higgins, William Lauderdale, Paul Loeser, William F. Reagan, Eugene Shirer, and Fred Totten. I owe a special debt to D. Howard Moreau, editor of the *Hunterdon County Democrat,* for lending me his deservedly treasured newspaper files and for help in selecting illustrations.

The text of this book has been read and criticized in whole or in part by Dr. Robert T. Thompson, Dr. Irving S. Kull, Professor Frank G. Helyar, Dr. L. Ethan Ellis, and Dr. Willard C. Thompson, all of Rutgers University. I am sincerely grateful for the constant guidance and many helpful suggestions given by them. Any mistakes made will be in spite of their help and should be laid at my door alone. Dr. Carl W. Woodward, President of Rhode Island State College, and Dr. Howard W. Beers, of the staff of the University of Kentucky, also gave helpful suggestions during the early years of my research when they were connected with Rutgers University. My typist, Mrs. Beulah Van Riper, deserves much credit for her patience and perseverance.

Permission for the use of photographs for illustrations was granted by the Historical Society and the Agricultural Museum of Bucks County, Pennsylvania, for Figs. I–IV, VI, VIII, IX, XVII, and the lower portion of VII; by the *Hunterdon County Democrat* for Figs. X, XIII, XIV, XVIII, XX, upper portion of VII, and the title page vignette; by Hiram E. Deats for Figs. XII, XVI, XIX, XXI, and XXII; by Alvah Haver for Fig. XI; and by the county engineer for Fig. XV.

HUBERT G. SCHMIDT

New Brunswick
New Jersey
June, 1945

RURAL HUNTERDON

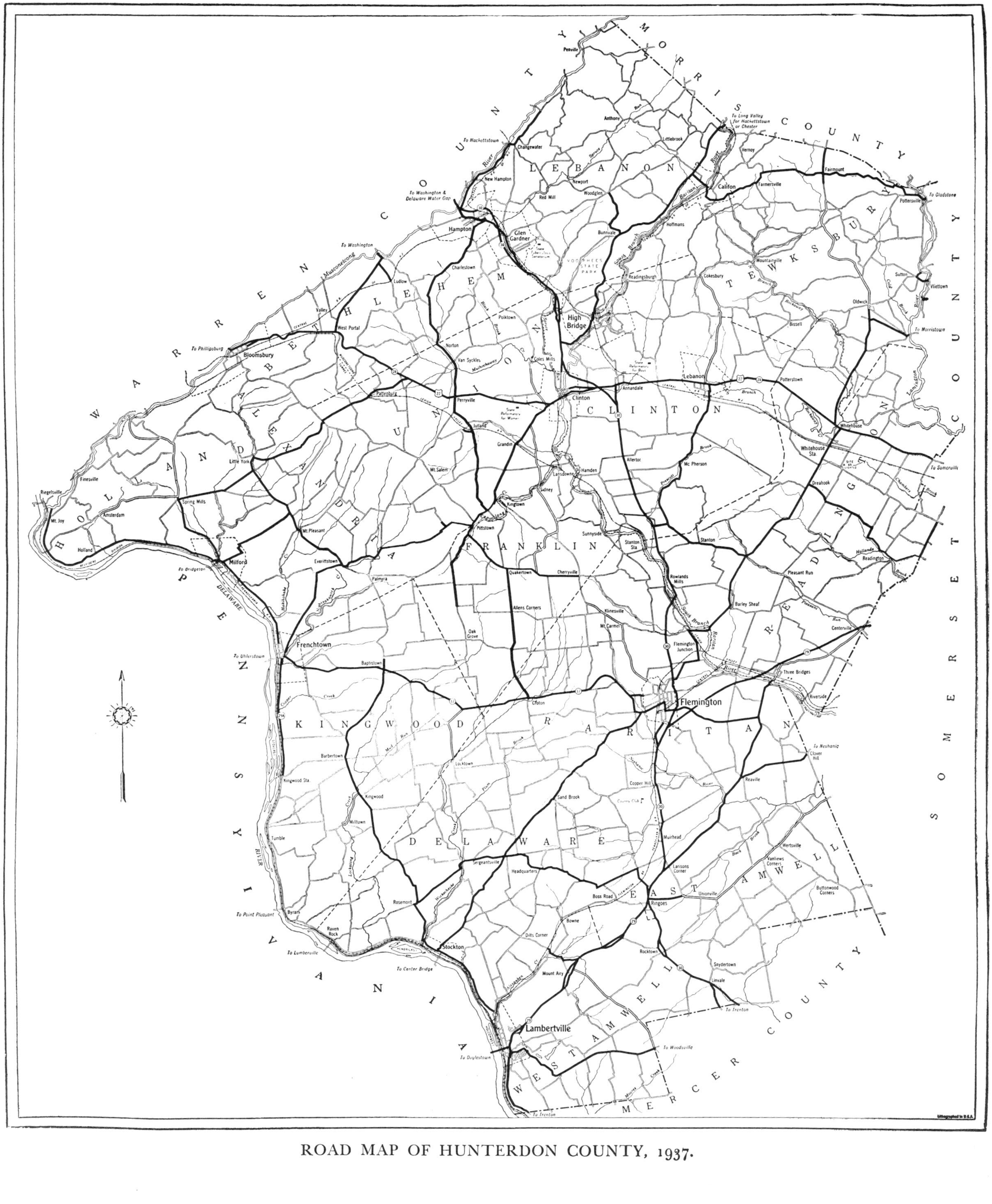

ROAD MAP OF HUNTERDON COUNTY, 1937.

CHAPTER I

Predominantly Rural

IT has been nearly two and one-half centuries since the first pioneers settled in the region which is now Hunterdon County, New Jersey. During that long period most of the inhabitants of the county have been primarily interested in agriculture or in occupations supplemental to it. Their welfare, economic and otherwise, has depended to a large degree to the present day on their ability to till the soil and to raise domestic animals. Trade, transportation, and, until recently, manufacturing have been largely concerned with farm products or with goods needed by farmers. The inhabitants of the numerous villages which have existed since early days have been tied in one way or another to the economic growth of the countryside about them. Only as manufacturing has been divorced from agriculture in the present century has the growth of the village population been partially independent of rural growth. The total population of the county during its entire history has ebbed or flowed as local agriculture has flourished or declined.

The story of the economic development of the county is, therefore, largely that of the basic industry, agriculture. Furthermore, the social organization and cultural growth have been those of a people engrossed in gaining a livelihood from the soil, and the political outlook has been always typically rural. The most important leaders of the county, from a long range point of view, have been the outstanding farmers, though politicians and others have at times taken the limelight. Therefore, it is appropriate that this study be devoted principally to farmers and farming.

Agriculture here, as elsewhere, has gone through many changes, some drastic, others of small importance. Sometimes these changes have come in fairly rapidly, for example, the production of sorghum during the Civil War, the revival of flax growing under the stimulus of bounties in the 1880's, the decline of peach growing about 1900, and the adoption of the soybean during the 1930's. Other developments have, on the other hand, been slow in coming, *e. g.*, the spread of the potato as a garden and field crop in the second half of the eighteenth century, the growth of poultry raising to commercial importance after 1875, and the various changes in dairying since the Civil War. Agriculture has gone through continual modification, and for that reason there have been none of the dramatic turning points which are sometimes to be found in political history. Therefore, a topical arrangement which carries each of the aspects of the story through the whole period of the county's history has seemed most logical. Perhaps some of the obvious disadvantages of such a plan can be avoided by first giving a brief summary of events in chronological order.

The first true farmers came into southern Hunterdon about 1700, and during the next half century most parts of the present county were opened up to settlement. The pioneering families were of diverse nationality, for the most part English, Scotch-Irish, Dutch-American, Rhineland German, and French. The effect of the wilderness on each group was much the same, however, and all of them adopted a similar type of general agriculture. They found a region of fertile, well-drained soils which had been utilized to a small degree only by the native Indians. Before they could exploit these soils, it was necessary to eradicate the stubborn growth of timber from the fields which they wished to till. Fortunately the woods and streams provided much of the family food supply and grazing of a kind for the live stock until such time as crops might be raised.

A pioneer bringing his family into the wilderness was faced with a multitude of tasks, which he attacked in the order of prime necessity. While his branded, hobbled, and belled animals roamed the woods, he and members of his family built a cabin and constructed the rude furniture which was necessary. The small clearing about the cabin became a garden, protected by a rail or picket fence. The brush and smaller trees of a nearby plot were cleared and burned and the larger trees girdled in order to provide a corn field for the following spring. Rails split from straight tree trunks were used to construct a worm fence around the field, since all growing crops must be guarded from the depredations of roaming cattle, horses, sheep, and swine. This field might be used for wheat, flax, and buckwheat in following years and other fields carved from the wilderness and planted to corn. Over a period of years the pioneer would clear sufficient land for all of his crops and for pasturage.

The pioneer did not give his livestock much attention. He had no feed for them during the first winter, and if he provided any shelter it was only a hastily constructed open shed. During the second winter he usually had some corn fodder to give them during the bitterest weather, and in later years they were fed oat and grass hay. The animals which survived such treatment were a hardy lot, but not very productive by the standards of today. Cows gave no milk during the winter and not a great deal at any time. The beef from animals which were killed was of poor quality, even if some corn had been fed prior to butchering. The rangy, half-wild hogs took on added weight from feeding on acorns and nuts in the fall, but the pork was tough and stringy. The sheep produced short fleeces and poor mutton. The horses, except for an occasional one which was given better care, were undersized and emaciated, not capable of sustained hard work.

During the period in which the pioneer was enlarging his crop land and increasing his flocks and herds, he sometimes found it difficult to provide sufficient food and clothing to sustain his rapidly growing family. However, if he were a good hunter and trapper and his wife a skillful gardener, they could manage to get by until a crop of corn was grown and some hogs were ready to butcher, and in another year or two some wheat and buckwheat products might be added to the restricted diet. As for clothing, that brought from the older settlements was supplemented with buckskin until the sheep could provide sufficient wool and a crop of flax could be grown. After that the women of the family by arduous labor provided the family with clothes of wool, linsey-woolsey, and tow. Since the small income from furs, corn whiskey, and other frontier products must be spent on such necessities as powder and shot, salt, nails, and hardware, the family had to produce nearly everything for itself. The first family in a locality had to rely entirely on its own efforts. Later comers might receive help from neighbors in raising a cabin and perhaps in other ways, but as long as frontier conditions existed they, too, must be self-reliant and independent.

Much of the northern portion of the county and considerable blocks of land elsewhere were settled slowly, and true pioneers were to be found within its boundaries as late as the Revolution; but in most of the region roads were soon opened to the outside, and settlers poured in rapidly. Because of this the transition period from frontier to settled countryside was rather short for many communities. By 1770 the farmers were able to market their surplus wheat, beef, and pork, and to use their new purchasing power to buy the necessities and luxury goods which were brought in by traders and merchants. In a relatively short time the settlement lost its frontier aspects and acquired the characteristics of an older, more settled community. Log cabins were replaced by frame dwellings, and frame barns were provided for livestock, which now had greater value than formerly. The farm family was still independent in many ways, but its old total self-sufficiency was gone.

More or less rapidly the pattern of existence for the nineteenth century was set. No longer isolated, the farm family became part of a community and shared in the economic, social, religious, and political life of its neighbors. Equally important, the settlement was connected with the outside world by roads over which the wagoners of the day were constantly carting out surplus farm products and carting in commodities needed or desired by the rural population. The farm family of Hunterdon County had thus become part of a larger world and was affected by the same forces which operated in other rural portions of the Middle Atlantic Colonies.

This remained a region of general farming for two hundred years, with little specialization until the latter part of the nineteenth century. Nearly every farmer raised all of the crops and types of livestock common to the area, and was chiefly concerned with the production of food and fiber for his own family. Nevertheless, the opening of roads to outside markets resulted in some adjustments in farming methods. The numbers of beef cattle and hogs increased, and their feed was supplemented by corn and other grain; likewise draft animals increased, and their care improved. However, domestic animals generally were a poor lot, chiefly because of indiscriminate breeding and unbalanced feeding. The need for more winter feed led to a greater acreage of timothy and other "English grasses." The number of apple orchards increased markedly because of the demand for brandy. The acreage of wheat, the chief cash grain crop, and of corn, used for fattening meat animals, was expanded considerably. The enlarged operations unfortunately led to the destruction of timber reserves and the farming of slopes which eroded easily. The increased production of corn and wheat combined with lack of knowledge of the nature of the soil led to land exhaustion. By the end of the eighteenth century the dire effects of poor crop practices were very noticeable.

Production for a market not only enabled the farmer to purchase goods from the outside but also made it possible for him to hire various types of services. The pioneer had been a jack-of-all-trades and had manufactured nearly everything used in the home or on the farm. Farmers now turned increasingly to the artisans and mechanics who had settled in their midst and were making a living by manufacturing goods needed on the farm or by processing things produced on the farm. Although some home manufactures such as spinning were given up tardily, the farmer and his family were gradually emancipated from many tasks and given more time for farming operations as such. The increasing scale of operations often made it necessary for a farmer to hire or in other ways acquire a supplementary labor force.

By 1800 the bad effects of a century of exploitation forced changes in agricultural practices. Most of the arable land had been cleared, and already many fields were abandoned or converted into pasture. The farmers who did not move to new frontiers were faced with the problem of serious soil depletion. Through necessity they came to utilize all manure, straw, and other soil building materials produced on the farm, and some of them learned to turn under green crops. Many were using crushed gypsum or land plaster, and most farmers a little later turned to calcined lime as a cure-all. Unfortunately, the lime in many cases was incompletely slaked and, despite increased yields for a year or two after application, did the soil more harm than good. By the middle of the century farmers were turning to phosphate and other chemical products, and the cost of fertilizers soon became an important item in the budget of every farmer.

The decline in land fertility had led naturally to a smaller acreage of wheat, and many farmers had to look elsewhere for a cash crop. Where the land was suitable, some of them turned to the growing of apples and the production of brandy; others imported better breeds of sheep and produced wool; still others, because of the increased amounts of pasture and meadow land, became breeders or feeders of cattle. A new interest in selective breeding developed, feeding practices were improved, and better care was given to animals. Beef and butter were soon among the most important market products of Hunterdon County. With new sources of cash income and lowered production and marketing costs brought by canal and rail transportation, the outlook for agriculture in this area seemed bright enough in 1850. Farming was as diversified as it had been a century earlier, though there had been a change of emphasis. From the point of view of the welfare of the land, the increase in livestock raising was decidedly a change for the better.

Paralleling these changes in farming practices was a further decline in home industries. The discontinuance of the spinning of flax, one of the last important industries to give way, led to the ultimate disappearance of a crop formerly grown generally. It was a flourishing period for artisans and mechanics. More than ever before they made the home furnishings, clothing, tools, and other articles for the rural population, which despite some setbacks was still prosperous.

Although things went well enough for a short time after 1850, the second half of the nineteenth century was not a happy period for agriculture in Hunterdon County. Railroad transportation brought competition from newly developed regions where production costs were lower. Noticeable enough before, this competition became ruinous after the Civil War. One by one the old staples failed, though their production did not cease. The Middle West could produce wheat, beef, pork, and wool more cheaply than farmers here, and a change in the public appetite had caused a decrease in the demand for apple brandy. It was natural that this region and others like it near metropolitan markets should turn to a more perishable product. It is little wonder, therefore, that farmers all over the county tried commercial peach production when it had proved profitable for a few. It was a good crop only in localities with the right conditions of soil and air drainage, but there it remained a very important though not always profitable crop for nearly a half century. About 1900 peach growing declined, primarily because of the competition of other regions. A smaller development of other fruit growing came late in the century, but did not flourish for long.

Most farmers continued a general type of agriculture, but were increasingly at a disadvantage, especially after 1870. The invention of the refrigerator car was a particularly hard blow to the many who depended on the sale of fat cattle to bring in a cash income. Faced with disaster, they turned to dairy cattle and increased their milk production in response to an increased demand for butter. The making of butter became an important home industry and remained so until the rapid construction of neighborhood creameries during the last two decades of the century gave the farmer a market for cream. A slow development, but important for the future, was the sale of whole milk. Farmers or dealers ran wagon routes through the local villages, and an increasing amount of milk was shipped to the metropolitan area by rail. Another consequence of changing economic conditions was the elevation of the humble chicken to a new place in Hunterdon County agriculture. The rapid growth of city markets for eggs and meat and the invention of the incubator and other labor-saving equipment led more and more farmers to turn to poultry. The commercial production of truck crops, particularly tomatoes for canneries and city markets, also expanded. Despite these developments agriculture continued to decline until about 1900.

The total valuation of farm products fell, land valuations spiraled downward, and the agricultural population declined steadily.

It was this half century from 1850 to 1900 which saw the greatest changes in agricultural techniques. Many simpler labor-saving machines were used before 1850, and most of the later developments were foreshadowed by that time. But it was after 1850 that successful machines came into use which saved manpower to a substantial degree and made the individual farmer much more productive. The change was not immediately to the advantage of Hunterdon County and other areas like it. Eastern farmers were more conservative and less prosperous than those of the Middle West, and their fields were usually smaller, stonier, and rougher. Therefore, they were slower to accept the new developments, and to take advantage of lowered costs of production. The farmers who were to survive eventually adopted the new machines in order to eliminate expensive farm labor as much as possible. As machines were evolved to fit his operations, he became increasingly efficient and much of the drudgery was taken out of farming. This fact, combined with the development of new specialized types of farming, accounts for the rapid comeback made by agriculture in Hunterdon County after 1900.

Although Hunterdon County agriculture was still groping about in a changing world, searching for a new basis for prosperity, it had by 1900 taken on many of the aspects of the commercialized era which lay ahead. Much of the farming was still general in the sense that farmers grew a number of crops and kept livestock of several kinds, but they no longer produced most of the things needed on the farm. House furnishings, clothing, tools, farm equipment, commercial fertilizer, even much of the family food supply and the feed for livestock were purchased. Agriculture was still diversified, but far from self-sufficing. This meant that the cash income of the farmer had to be far larger than in an earlier period. Dealing with middlemen twice he sold unprocessed products for a distant market and bought processed commodities made elsewhere. The farmer was becoming increasingly a business man, and business acumen was as important for success as willingness to work hard.

Another marked economic change in Hunterdon County during the second half of the nineteenth century was the almost complete separation of industry from agriculture, due in large degree to the improvement in transportation. Artisan and mechanic groups, whose place in an ordered society had seemed so secure in 1850, had been rapidly absorbed into the factory system, so that the number of craftsmen left to supply the needs of the rural population was very small. Furthermore, farm products were seldom processed in small nearby mills and shops, as had once been the case. Some factories had grown up to supply local farm markets and to process local farm products, but they, too, were eventually driven to the wall or absorbed by larger industrial concentrations. Certain large scale industries did survive and others came in, but in most cases they bought their raw products elsewhere and produced for a national or international market. The trends here mentioned continued into, and became even more marked in, the period after 1900. The gulf between the farming and laboring groups of the county has become steadily wider.

After the turn of the present century agriculture in Hunterdon County recovered rapidly from its long decline. This was due in large degree to almost revolutionary changes in production. The growth of dairying and poultry raising of the previous period was accelerated greatly, and the raising of certain vegetables was considerably increased. The development of huge urban markets, now brought very close by motor transportation and good roads, caused a great demand and rising prices. The number of dairy cattle expanded rapidly and the production of the individual cow was greatly increased by selective breeding and scientific feeding. Likewise there was an enormous increase in the poultry industry, leading to specialization in the production of eggs, the raising of meat chickens, and the hatching of baby chicks. The expansion in vegetable production has been most marked in recent years. The new agricultural prosperity of the county was reflected by the fact that farm valuations doubled during the first decade of the century and have risen steadily since.

Paralleling this specialization and commercialization was a rapid and complete mechanization of agriculture. Machines were developed for nearly every farm operation, and horse power was rapidly replaced by motor power. Hunterdon County farmers, faced with high labor costs, were now quick to adopt each new improvement, and the per capita production of the average worker was greatly increased. The many changes in the farm plant, the expansion of operations, and the acquisition of new machinery resulted in a great increase in the farmer's investment. As never before he was a capitalist and business man, with a correspondingly greater premium put on his managerial ability.

In the twentieth century it became increasingly important to utilize the wealth of agricultural knowledge resulting from private experimentation and especially from the work done by State Experiment Stations. Much of this information was available in governmental publications and farm magazines, but a more practical way of getting immediate help for farming problems as they arose was to utilize the New Jersey Agricultural Extension Service. Hunterdon County was slow to take full advantage of this opportunity, but eventually employed county representatives of that organization. High school courses in vocational agriculture and short and regular courses in agriculture at Rutgers University have given more specialized training to ambitious farm boys, and a considerable number of boys and girls have benefited from membership in local 4-H clubs. In the past two decades farmers of the county have also learned the benefits of economic cooperation, particularly in selling farm produce and in buying farm supplies.

The history of Hunterdon County could be written in terms of its religious, educational, cultural, and institutional growth; but this would not tell the fundamental story. In order to understand the development of Hunterdon, it is necessary to know the detailed history of its agriculture and the detailed history of its economy, which has been based upon its agriculture, for Hunterdon is rural.

CHAPTER II

Before the White Man

THE COUNTY OF HUNTERDON is situated on the western side of the state of New Jersey somewhat above its narrow middle portion. The Delaware and Musconetcong rivers give it natural limits on its southwestern and northwestern borders, but its remaining boundaries are very irregular and portions are so rambling and erratic as to lend color to a local tradition that the original surveyors were not teetotalers. The county, forming a compact block of some 279,680 acres, today constitutes one of the largest rural areas of New Jersey. The white settlers who came into this region soon after the year 1700 found it an almost untouched wilderness. They and their successors have transformed it into a settled countryside with an agricultural and industrial economy in harmony with modern conditions. Before telling that story, however, it may be well to say something about the geography of the area.

The hills of Hunterdon have long been the pride of local residents and on occasion have furnished themes for song and story—and for real estate prospectuses. An outcropping of the Appalachians, known as the Highlands, penetrates to a depth of several miles in the north, while the remainder of the county is part of the Piedmont belt. In both areas the prominent topographical features have resulted from stream erosion on rocks of varying degrees of hardness. In the Highlands veins of granite and gneiss have resisted the elements, while those of limestone have been eroded away. The result has been the formation of a series of rounded or flat-topped ridges, four to six hundred feet above the intervening valleys. The ridges run generally in a northeast-southwest direction, becoming more broken near the Delaware. Their summits are from eight hundred to a little over a thousand feet above sea level.

In the red shale areas of the Piedmont a rolling plain from two hundred to four hundred feet above sea level has been created. Rock masses more resistant than red shale have been left protruding. Sourland Mountain, in the extreme south, has a backbone of diabase or trap rock, as do Round Mountain and Cushetunk near the center of the county. The latter is in the shape of a horseshoe and encloses a red shale area known as Round Valley. Beds of conglomerate stone full of quartzite pebbles form the basis of the high hills south of Pattenburg known as the Barrens. West of Flemington is the Hunterdon Plateau, a level upland from five hundred to seven hundred feet above sea level. The basic rock of this tableland is argillite or "blue jingler," a claystone of very fine grain. A number of other elevations scattered over the Piedmont owe their existence to the relative toughness of their stony structure.

Because of the sweeping view from Cherryville, on the edge of the Hunterdon Plateau, it has long been a popular belief that that village is the highest point in the county. The local residents, five hundred feet above Flemington in the valley below, fancy that they can see over the tops of the heights to the north of them. Actually the village is only 702 feet above sea level, while Cushetunk is 834 feet and the ridges of the Highlands are even higher. Two ridges, one near West Portal and the other near Anthony, rise to a height of 1073 feet, and are the highest points of the county. Sourland Mountain is somewhat lower than the Hunterdon Plateau. The lowest place in the county is in the extreme south. At normal flow the surface of the Delaware as it leaves the county is only thirty feet above sea level.

The drainage of the county is divided nearly equally between the river systems of the Delaware and the Raritan. A series of parallel streams flowing in a generally south-southwest direction drain the western half. Beginning at the north, the largest of these creeks bear the euphonious names of Musconetcong, Queequakomisikunk, Hakihohake, Nashisakawick, Lackatong, Wickecheoke, and Alexauken. The Musconetcong is now often classed as a river. Most of the flow to the Raritan from Hunterdon County is by way of the South Branch, which enters the county at the northeast and flows southwest as far as High Bridge. At that point it breaks through the ridges and turns to the southeast across the Piedmont. Spruce Run, a tributary, parallels the South Branch to Glen Gardner, then turns to join it at Clinton. The Mulhockaway and Capoolong creeks, tributaries respectively of Spruce Run and the South Branch, together with several less important streams contest the drainage of the central part of the county with the feeders of the Delaware. The eastern edge of the county slopes toward the south and southeast, and its streams eventually join either the South Branch or the North Branch outside of Hunterdon County. The largest of these streams are the Neshanic River in the southeast and the Rockaway Creek in the northwest. The latter stream is a tributary of the Lamington River, which forms the boundary of the county for several miles.

Most of the streams are still busily wearing down their valleys and cutting deeper into their stream beds. The Delaware has cut a channel in places over four hundred feet in depth, and its tributaries in their lower courses have done nearly as well. The streams of the Highlands have narrow valleys between high parallel ridges, while those of the Piedmont usually have wide flood plains. For the most part the county is well drained. An exception is the Hunterdon Plateau, particularly the part about Croton as a center, known in earlier times as "the Great Swamp." A letter written by Thomas Capner in 1787 mentions that there were places where a pole might be driven down into the muck to a depth of twenty feet. Stream erosion since the clearing of the forests has made for better drainage, but even yet the upper reaches of the Wickecheoke, Lockatong, and Capoolong creeks are sluggish. This region was the only area of any size in the county in which the pioneers found insufficient stream fall to permit the use of water power.

Soil surveys, described in two pamphlets written by A. L. Patrick and others, show that the soils of the county are as varied as the stones from which they were created. In the Highlands the Chester soils are most common. These brown soils are the result of the decomposition of granite and gneiss. The Chester loams and gravelly loams are good soils, adapted to the growing of grain, forage crops, and fruit. The stony loams, thin and gritty, cover the slopes and summits of the ridges and are better suited for the growing of timber and for pasturage than for the growing of crops. The deterioration of the limestone veins between the ridges has resulted in the formation of the deep, mellow Hagerstown silt loam, also brown in color. It is one of the most productive soils of the county. Besides small bits between the uplands, there are several areas where this soil has been deposited by water in times past, as in the Muscontecong Valley between Bloomsbury and Hampton, in the Clinton-Annandale region, and near Califon. A similar soil, nearly as good, is the Washington loam, derived from glacial drift. In some places it contains such considerable quantities of small stones as to be almost gravelly. The deposits of this soil are scattered through the north part of the county, the largest being adjacent to the Hagerstown silt loams in the areas mentioned above and in several places in the eastern part of Readington Township. In the areas outside of the Highlands where these soils are found, they have been superimposed on red shale and other Penn soils of the Piedmont. Fox gravelly loams, very similar in nature, are found near Little York, near Hampton, south of Norton, and at a few other isolated spots. Although they contain much wash, they are productive soils.

The most common soils of the Piedmont section of Hunterdon County are the red loams of the Penn series, which have resulted from the weathering of red shales, conglomerates, and sandstones. It is these soils which a German observer soon after the Revolution called "the red soil of Jersey, known generally by that name in America." [1] The Penn shale loam, locally known as "red shell land," is a silt loam containing fragments of shale. Both it and the clay subsoil underneath are of an Indian red color. In most places the parent red shale is within three feet of the surface. A soil less subject to erosion and more retentive of moisture is the darker Penn silt loam. These two soils together cover the greater part of the area between Sourland Mountain and the Hunterdon Plateau and extend northward to include most of Readington Township. A narrow belt between Cushetunk and the Hunterdon Plateau widens out between the Highlands and the Plateau, especially in the Frenchtown-Milford neighborhood.

Another soil of this series, the Penn gravelly loam, also occurs in the region between the Highlands and the Plateau. The red rock underneath is conglomerate, and the soil itself is filled with gravel and cobblestones. A large belt of this soil extends from Little York to south of Clinton and from there in a gradually narrowing strip to just east of Cherryville. This soil occurs chiefly on hills and ridges which rise to heights of three hundred feet above the other Penn soils. The ridges are known locally as "the Barrens," and the cobblestones are called "boother stones." The gravel in this soil makes it porous and prevents great erosion. Although the stone fragments make tillage difficult, it is productive except in dry seasons and is especially adapted to fruit raising. Small deposits of a fourth Penn soil, the Penn loam, are found in small areas, for example, near Rosemont and near Mount Pleasant. It is a good soil, closely akin in texture to the Penn silt loam.

The various heights of the Piedmont have soil types determined by the nature of their rocks. The weathering of the trap rock of Sourland, Cushetunk, and Round Mountains and of ridges near Stockton, near Byram, near Sand Brook, south of Copper Hill, and just west of Flemington has produced soils of light brown or reddish-brown color. These are the Montalto soils, ranging from stony and gravelly loams to silt loams. Because of the large content of fine rock particles, these soils are

[1] J. D. Schoepf, *Travels in the Confederation,* 1783–84, I, 21.

locally characterized as "mountain grit" or "ironstone." The stony loams of the heights and steeper slopes are thin, with bedrock often protruding through, and are fit only for pasturage. The gravelly and silt loams occur farther down the slopes and on the hilly land at their bases. They are porous and resistant to erosion and raise large crops of grain, hay, fruit, tomatoes, and potatoes.

On the Hunterdon Plateau the soils are of two main types, the Croton and the Lansdale. The Croton soils are grayish silt loams formed from the decomposition of argillite, and under them lie subsoils of compact clay. Since the bedrock in most places is only three or four feet under the surface, stream channels are shallow and drainage imperfect. An indeterminate section is still known locally as "the Swamp." The better areas are used for small grains, while others are used for pasture, sedges and rushes being the predominant natural grasses. Some portions of this soil belt are so unproductive that they might well have been left in permanent forest. The various Croton soils cover a large block atop the tableland, beginning west of Flemington and extending north nearly to Quakertown and southwestward nearly to the Delaware. Small detached areas of this soil type are in some places entirely surrounded by regions of Penn or Lansdale soils.

The Lansdale soils are brownish in color and derived from darker argillite, gray shales, and sandstone. A wide belt encircles the main section of Croton soils, and a narrower one surrounds Sourland Mountain. The two regions are connected by a strip of Lansdale soil which extends from near Lambertville to near Sergeantsville. There are also a dozen or so detached areas, some of them quite isolated. The best of the Lansdale soils are loams. A sizable section occurs between Stockton and Sand Brook, including the slight elevation locally known as Sandy Ridge. In other places, as in the Baptisttown and Locktown areas, appear the Lansdale shale loams, somewhat like the Penn shale loams in quality. Lansdale stony loams occur in a few rough areas. The Lansdale belt was once used extensively for the growing of peaches.

At the bottom of the deep valleys cut by the Delaware and its tributary creeks are low banks and islands of various materials usually classed as riverwash. Terraces now seldom flooded are the remainders of flood plains of earlier times. The soils of the terraces of the Delaware River are of Chenango fine sand and fine sandy loams. These brown soils, mostly alluvium from glacial deposits, are very fertile. In the Piedmont the soils of the first bottoms, subject to periodic overflow and still in process of change, vary with the areas from which the soil is brought. In the valleys of the South Branch and its tributaries just above Clinton occur Codorus silt loams, mostly wash from Chester and Montalto soils. A more common first bottom soil of the Piedmont is the Bermudian silt loam, wash from Lansdale and Penn soils. The flood plain of the South Branch below Clinton is largely of this soil. These and other bottom soils more limited in extent are usually fertile, though flooding causes serious farming problems. The second bottoms, terraces which only occasionally overflow, are of various types of silt loams. The Roanoke silt loam, a gray soil of gneissic origin, is found chiefly in the valley of the South Branch near High Bridge and in the lower courses of Mulhockaway Creek. It is not well drained and is used more for pasturage than for crops. Birdsboro silt loam, composed of alluvial deposits from Lansdale and Penn soils, is a good general-purpose soil. It is common in narrow strips along other Piedmont streams, and from Clinton downward on the South Branch it forms a considerable second bottom, greatest in extent near Flemington Junction.

This brief and incomplete survey of the soils of the county has left out a number of other soil types less extensive in area. For instance, west of Clinton there is a block of Berks shale loam, a brown silt loam containing fragments of a dark shale. It is an excellent soil for field crops and fruit, especially when lime and humus are added. Also, at a number of places and particularly on the bluffs of the Delaware, there are areas of rough and usually stony ground of various kinds, none of them of much value, even for forests. It is readily seen that the soils of the county vary considerably. Some are rather shallow, and those of a few areas are not well drained. But most of them are good, strong soils which need only care to keep them permanently in production as agricultural lands. For the most part they are well adapted to diversified farming. The keeping of livestock is especially beneficial, as it withdraws poorer lands from crops and furnishes enriching agents for the fields which are in use.

The whites who first came to what is now Hunterdon County found an almost unbroken forest except in the stoniest places and in the poorly drained areas of the Great Swamp. Small cleared fields, especially near the streams, were the result of the simple agricultural operations of the Indians, and there is a tradition that the site and neighborhood of present-day Quakertown was a treeless plain which they burned off each year in game roundups. There were also small openings here and there filled with blackberry vines, hazel and elder bushes, jewelweed, dogwood, honeysuckle, sumac, poison ivy, wild grapes, black haws, wild plums, sedges, and rushes, the kind of growth varying according to the nature of the soil and the amount of moisture.

A letter written by Thomas Capner in 1787, after the pioneers and their descendants had been hacking away at the forest for nearly a century, shows much about the thickness of the forest growth and its resistance to attack:

> When we had got within about 5 or 6 miles [of Flemington], our road was chiefly through woods. . . . I have been upon a hill at a small distance where there is a good prospect and there seems to be 4 times as much wood as cleared land. These woods are much like the English woods, only without underbrush. . . . I have been in a swamp which is about three miles distance. It was no other than a thick impenetrable wood without underbrush, but my brother says that he has been in a wood where there is not a tree higher than himself. In the general wood there are great quantities of trees fell down with age and are turning to earth. The different kind of trees about here are oak, ash, hickory, poplar, maple, chestnut, walnut, beech, gum, Button elm. . . . American ash looks like English maple and American maple like English Elm, but the leaves like ash. There are many kinds of oak White oak, red oak, black oak, pin oak and Spanish oak, etc. White oak is the best and the most common and pin oak the worst, which is used only to burn. There are many species of hickory. One kind very much resembles English walnut. They all bear a nut which resembles a walnut and eat as well. Poplar grows much finer and larger than in England. There are four kinds of maple. A sweet maple, rock maple, and sugar maple. These three if tapped will run a sweet liquor and make sugar if evaporated. Gum wood is not reckoned of any else [use] but to burn, as it will neither split nor bend and button wood is very little better. Gum elm a[nd] hickory button are never used except for burning.[2]

The varieties of trees growing in the original forest were practically the same as those growing in woodlands today. One type of tree once very common in the Highlands and on the Montalto and Lansdale soils was the chestnut, killed by the chestnut blight of a quarter of a century ago. Other trees common in the Highlands are the white oak, chestnut oak, maple, basswood, and ash. Areas with Hagerstown, Washington, and Penn series loams have in addition walnut, hickory, elm, and sassafras, with a high proportion of oaks on the Barrens. On the Lansdale soils the usual trees are chestnut oak, red oak, hickory, basswood, dogwood, and maple, with elm and beech at the foot of the slopes. Black oak, ash, birch, maple, hickory, ironwood, and post oak thrive on the Croton soils. The

[2] Thomas Capner to John, Samuel, and Roland Coltman, Nov. 17, 1787, Capner Papers.

y to see why traps, guns, powder, and shot were such standard
commerce in early days.

observers marveled at the numbers and variety of birds and wl. Turkeys, woodcocks, native pheasants, heath hens, quail, snipe, and other game birds were found in abundance at all nd ducks and geese in great numbers tarried for a while in spring . Passenger pigeons darkened the sky at several times during the d were shot or trapped in great numbers. In February, 1793, the irds brought seven pence per dozen on the New Brunswick mar- l were thought dear at that price. Redbirds, bluebirds, swallows, nd other song birds were far more numerous than today. The ing bird was regarded as something of a marvel by newcomers. s were provided for the purple martin, a very popular bird, be- its "anxious notes" were supposed to warn poultry of the prox- of hawks and crows. The chimney swallow was also one of man's ssociates. A number of birds were looked upon as of doubtful Woodpeckers were regarded as the enemies of fruit trees. Crows, birds, and redwings, which became more numerous with the ng of the forests, were such despoilers of sprouting and earing corn ounties were paid for killing them. Hawks, eagles, and other birds ey and the carrion-eating turkey buzzard were unpleasantly com- Feathered creatures seemed so plentiful that man did not consider ossibility of their depletion. It is a far cry today from the time an Amwell Township miller wrote in his memorandum book, ch the 27th day past middle of the afternoon the Skye looked black pidgeons."[4]

e streams of the new land were literally teeming with fish awaiting ngler. An early writer in West Jersey reported:

> vise in Delaware River are great plenty of Sturgion. . . . There are also Quantities of a sort of Fish like Herrings; with plenty of Fish called s, but not like Shads in England, but of another kind, being a much r sort of Fish. . . . And in the said Rivers and Cricks are many other of good Fish, . . . some of which are Cat-fish, Trout, Eales, Perch, &c.[5]

er common fish were rockfish, bass, shad, pike, sunfish, and flounders. e of these, particularly shad, were salted for winter use and had an ortant place in the economy of the people. Since mill dams inter- d with the running of the shad, clashes between fishing and milling rests were common.

'John Lambert's Almanack or Account Book, 1774."
Thomas Budd, *Good Order Established in New Jersey*, pp. 32–37.

post oak has a spreading root well adapt
drained places there are in addition willo
haw. Along the streams of the county the
more, elm, ash, willow, poplar, basswood, p

The wild life of the wilderness was no
the settlement of the land. The malaria-be
dangerous representative of the animal ki
chief predatory animals were wolves and p
were kept down by the policy of paying bo
lynxes and wildcats proved easy to eradica
and skunks, commonly called pole cats, wer
ing conditions and long remained a hazard t
other hand, many animals were of value in
furnished venison and skins, squirrels and ra
larders, and otters, minks, opossums, and racc
foxes, provided pelts for sale or home use.
making of hats, had a market at home an
colonial period. The muskrat furnished both
to use as a moth preventative.

Of much less value were the numerous grou
flying squirrels. Of the numerous snakes only
heads were poisonous. Hunters sometimes use
tect themselves when searching for them. Large
and eggs, but were otherwise harmless. Thei
worn about the body, as they were supposed t
interesting, though unscientific, observations
found in a letter which a Flemington child wrot
land in 1793:

> James killed a viper snake with 24 yongans in it—
> —as they lay higgs and when hacht to screen ther
> keep them with her she Opens her mouth and they ru
> yongans were from 8 to 10 Inches long a piece—these
> as is Except rattle Snakes an pilot Snakes as we have
> of Snakes.[3]

Everything considered, the animal life of the re
small and temporary effect on pioneer life. Thoug
squirrels, raccoons, deer, and even tree toads, none
was added to man's collection of domestic animals.

[3] Robert Choyce to Mr. and Mrs. Richard Choyce, Sept. 5, 1793

it is eas
items o
Early
wild fo
plover,
times,
and fal
year ar
dead b
ket an
larks,
humm
House
cause
imity
close
value.
black
cleari
that
of pr
mon.
the
when
"Ma
with
T
the

Like
grea
Sha
bett
sort

Oth
Son
im
fer
int

4
5

The climate of this region is in general moderate. Temperatures below zero are exceptional and seldom of long duration. Precipitation is fairly evenly distributed through the months of the year, averaging about four inches per month. Extended droughts are rare. Winters are usually mild and without great amounts of snow, but alternate freezing and thawing sometimes cause "heaving" and winter killing of crops. The average growing season is about 187 days at Lambertville and a number of days less in the northern part of the county. In some years there is considerable variation, among regions of different types of air drainage, as to the occurrence of killing frosts in spring and fall. The climate here, as everywhere under the sun, causes complaints and grumbling, but on the whole it is equable and favorable to agriculture. Everything considered, the first newcomers to this region found it a good land. There were forests to supply timber and fuel and household furnishings, virgin soils to produce abundant food and fiber crops, streams to drive the mills, and a wealth of game and fish to provide other food supplies. By arduous labor a good living could be wrested from the land.

When the first white men penetrated into present Hunterdon County slightly before 1700, they found various Lenni-Lenape or Delaware Indian groups in possession of their future homeland. These Indians lived in small, semipermanent villages here and there in the forest. Having been badly defeated by the Iroquois in times past, they were not very warlike and proved no obstacle to settlement. A partially nomadic, Stone Age race, part hunter and part farmer, they derived a living from forest and stream and from their small truck patches of corn and beans and squash. Each group probably had its traditional hunting ground, within which were one or more villages and numerous camp sites used on occasion. An archaeological survey of the county made in 1914 and 1915 has thrown much light on the whereabouts of Indian habitations.[6] A number of factors entered into the location of the more permanent villages. Nearness to water, shelter from winter winds, ease of access to their primitive thoroughfares, and good soils for tilled fields were of first importance. Since stream valleys best provided the desired topographical features, as well as good fishing and hunting, it is not surprising that nearly all villages were situated on rivers and creeks. Though a whole village could be moved in a few hours, good sites were used again and again, until a new choice was dictated by con-

[6] Max Schrabisch, *Archaeology of Warren and Hunterdon Counties,* pp. 9–34, 47–48.

venience or necessity. In general the regions most thickly populated by the Indians have since become those of densest population for the white man, although all of the factors involved have not been the same.

The valley of the Delaware had many decided advantages for the red man, and there are evidences of settlements of some permanency at or near present-day Lambertville, Stockton, Raven Rock, and Byram, and again in the Frenchtown-Milford area. Another region of comparative concentration of population was in the valley of the South Branch at or near High Bridge, and still another at Oldwick. There were apparently other scattered villages just south of the Rockaway Creek, at Quakertown, near Kingwood, and at two or three spots in the Musconetcong Valley. But by far the most important district of Indian habitation in Hunterdon County was an area of some twenty square miles in the neighborhood of what is now Flemington, extending north to Klinesville, east to Three Bridges, and south to Copper Hill. It was a region of good soils and many streams, it lay on main trails, and was protected from the winds by the Hunterdon Plateau. The chief attraction, however, was the large deposit of argillite found nearby. This stone was to the Indians of this and surrounding regions what flint, jasper, and quartz were to the Indians of most of North America before white traders ruined the native crafts by bringing in tools and weapons of iron and steel. Quarrying and manufacturing made this area a bustling hive of primitive industry. Argillite chippings and rejects show that Indian artisans worked on the banks of Mine Brook, the South Branch, and the Upper Neshanic. The large springs just north of Flemington determined the location of another group. Because of the comparative ease with which argillite could be made into weapons and tools, Indians came long distances to secure and work the stone. Argillite from this area has been found as far away as Port Jervis, New York.

The traditions of the countryside do not always agree with the findings of the archaeological survey, though the discrepancies are not serious. According to Dr. C. W. Larison in his "History of Ringos [*sic*]," old stories, supported by the wording of early deeds, placed an Indian settlement, Wishalmensey, at or near present Rocktown. The Indian trail from this place to the village of Wichcheoke, near present Stockton, was said to have become the Rocktown-Sergeantsville-Stockton road. Another tradition placed an Indian community a mile west of Centerville, where as late as 1870 a large bed of live-forever, said to have been used by both Indians and pioneers for making a healing salve, was thought to mark the spot.[7] A plant known as "Indian fig bush" was in like manner believed to locate a village at the site of what is now Holland. The "Indian Patch," a stony field on Round Mountain, was traditionally thought to mark either a village or a burial ground, and Round Valley and Mount Lebanon were said to have been Indian strongholds. Two large basins for crushing corn along with numerous arrowheads, bowls, and other implements were once found at a "house rock" on a ridge between Mountainville and Fairmount, giving credence to the local belief that there had been an Indian settlement at that place.[8] There are several reasons why the versions of tradition and science do not fully agree. The archaeological survey was made somewhat hastily and at a time after many Indian relics had been removed. On the other hand, traditions have a way of growing and becoming distorted. The records of earliest days, for their part, are sadly inexact. For example, John Reading, a surveyor in 1715, mentioned "Essakauquenamenshehikkon, an Indian plantation back of the Great Swamp." [9] An early deed, quoted in "Traditions of Our Ancestors," refers to "an Indian path which comes from Lukkazunning to Allamuncha," but gives no slightest indication as to the whereabouts of either place named. Allamuncha was probably Allamuchy in what is now Warren County.

The various Indian settlements were connected by trails, which were merely foot paths. The approximate location of some of the more prominent ones is fairly well established, but the whereabouts of the dozens of secondary ones is largely conjectural. In general Indian roads followed routes of least resistance, usually staying near watercourses and circling natural obstructions. Many of them are now followed in a general way by railroads and highways. One important path led up the Delaware Valley, connecting the line of villages and camp sites along that stream. The river itself was used for canoe travel and gave direct contact with settlements on the opposite shore. Naturally there was considerable interchange of population, and certain places, such as present-day Lambertville, were regular points of crossing at times of seasonal migration. On the Hunterdon side many lesser routes followed the various streams. An important one led up the Musconetcong, with trails branching off at either side. Paths to the south probably diverged near where the Riegel paper mills are now located and again at Bloomsbury, both providing short cuts overland to the Delaware.

7 "Traditions of Our Ancestors," *Hunt. Rep.*, Jan. 13, 1870.

8 *Hunt. Co. Dem.*, Jan. 14, 1920, Sept. 23, 1925.

9 "Journal of John Reading," *Proc. N.J.H.S.*, 3d ser., X (1915), 45.

Another apparently led through the gap north of West Portal, later forking so that one branch went to the High Bridge area by way of the Mulhockaway and the other to the Flemington settlements by a route which forsook stream valleys but which was so logical that it is still followed by highways. Still another path branched from the Musconetcong trail at Hampton, crossed the ridges to Spruce Run, followed down that stream to High Bridge, then led up the South Branch toward Long Valley. It was this route which was later followed by the New Jersey Central Railroad.

The South Branch and its valley naturally provided the Indians with important water and land routes. Because of the numerous villages between High Bridge and Three Bridges, there were probably trails on both sides of the river most of the way. The connections with the Musconetcong path have already been mentioned. Other trails branched off at numerous points. Camp sites indicate that one or more roads led from near Califon across country to the Lamington River Valley. A path up Prescott Brook to its headwaters gave access to Round Valley. Another led up Capoolong Creek to connect with those leading down to the Delaware. The Oldwick and Rockaway villages were joined to the South Branch settlements by a trail which skirted Cushetunk and Round Mountains. Another trail up the North Branch of the Rockaway apparently formed a junction with the Califon-Pottersville route.

A very famous trail, the importance of which may have been overemphasized by tradition and local history, was the one leading from the South Branch villages to the crossing at Lambertville. The exact location of this route is problematical. Archaelogical findings and the rules of probability would have such a thoroughfare follow the line of Indian village sites from Klinesville down Mine Brook, then up a tributary of the Neshanic, across a divide, and down the Alexauken to Lambertville. However, according to an early and persistent tradition, there was a trail across the county from Three Bridges to Lambertville by way of Reaville, Ringoes, and Mount Airy, along the line of the York Road. A path from the Flemington section is said to have made a junction with this route at Ringoes. It is possible that these were alternate routes. Undoubtedly there were other crisscrossing trails of which we have no knowledge. At any rate, it is clear that the Indians had a network of primitive roads, and this was a matter of some importance to the early settlers, who made good use of these paths through the forests. In many cases they simply widened and sometimes straightened them, and thus created highways for vehicles. Modern engineering and labor-saving machinery have made possible shorter routes in some cases, and convenience has caused changes in others, but many of the roads of today still follow the old Indian trails.

Before white men brought them the curses of firewater and disease, the Indians were a strong and active people, hardened by vigorous outdoor lives. Their complexions, naturally dusky, were tanned by sun and wind to a deep bronze. Their eyes were black, their teeth very white, their hair dark and coarse and kept well greased. The squaw wore her hair in a knot on the back of her neck, sometimes enclosing it with a sack. The brave cut off his hair except for the scalp lock and kept his face bare by the painful process of plucking. The fundamental articles of clothing were a garment of deerskin hung from the shoulder, a leather "lap" for men and a sort of skirt for women, moccasins, and sometimes leggings. The children wore few or no clothes except in cold weather. The parts of the body not covered were decorated with the juices of bloodroot, elderberries, etc., and with iron ore, clay, and other mineral pigments. Skins and furs, and to some extent woven fibers, were used for clothing. After the coming of white traders, these articles were replaced by gaily colored trade cloths.

The economic life of the Indian of this region was well adjusted to his environment. Since he was a wanderer, his possessions must be easily renewable or else movable by squaw power. His house was a simple hut or wigwam, conical in shape, made by covering a framework of poles with bark or thatch and perhaps lining it with mats. A simple vent hole at the top served as a chimney. Sometimes for winter protection the Indian dug a shallow hole, pitched the hut within, and packed dirt around the base. The floor was covered with leaves, skins, and mats and served the purposes of table, chairs, and bed. Handmade pottery and soapstone bowls were used for cooking, and wooden dishes and woven baskets for supplementary purposes. Heavy stone implements, such as pestles and mortars, were not carried on long treks, and tools which would not be needed may also have been hidden near the home village, to be redeemed later when the tribe returned to care for the crops. On hunting trips the braves took only strict necessities with them. For shelter they constructed tree huts or used rock shelters or caves. While they were away, the home village was left in the care of the women, the young boys, and the old men.

In making his tools and implements the Indian used many materials furnished by animal and plant life about him. Only the more durable have survived, and present-day collections of artifacts are for the most part of stone, including arrowheads, spear points, axes, hatchets, knives, chisels, hide scrapers, net sinkers, club heads, pipes, mortars, and pestles. The principal stone worked in Hunterdon County was argillite, though occasional relics have been found of jasper, flint, chert, slate, and copper. Bones were made into fishhooks, drills, needles, and crude agricultural tools. Sinews, especially of the deer, were essential for bowstrings, for fastening parts of tools together, and for use as heavy fishlines. Various kinds of wood were used in particular ways, hickory for bows, dogwood for arrow shafts, sassafras and other soft woods for cups, plates, and bowls, and harder woods for pestles and mortars to supplement those made of stone. Tree trunks were utilized for dugouts and birch bark for canoes. Native hemp, nettles, other fibers, and the stems of rushes were woven into mats, quilts, ropes, twine, bags, baskets, hats, and lining materials. Pottery and pipes were fashioned by hand from clay and pounded shells, ornamented with symmetrical designs, and sometimes painted. Wampum and shell ornaments were made on trips to the seashore or gained by trade with other groups. Indian workers, men and women, produced from the raw materials at hand the tools, weapons, and utensils which were indispensable to their lives. Their processes seem slow and laborious, but their whole tempo of life was much slower than ours. Time was the one ingredient of their existence of which they had a great sufficiency.

Indian agriculture, though requiring great toil, was not very productive. Fields were small, methods of culture primitive, and the depredations of animals sometimes serious. The principal crops were Indian corn, beans, pumpkins, and squash. A sort of wild tobacco, used for ceremonial smoking, may have been cultivated to some extent also. Only the crudest implements were used to break the ground and care for the growing crops. Worn-out fields were abandoned and others created out of the wilderness simply by girdling trees and allowing them to die. Corn was preserved for winter use and for the next year's seed by burying it in underground pits. It was eaten in a number of ways. Sometimes it was simply parched or roasted in the ashes. More often it was crushed in a mortar, then boiled to a kind of mush which might be made into cakes. Whole corn was made into hominy by soaking it in a solution of wood ashes and water, then cooked by boiling. Beans were usually boiled, and pumpkins and squash either boiled or roasted. For winter use beans were kept dry in small containers, while pumpkins and squash were sliced and dried and suspended on strings. Wild fruits and berries could be preserved for the winter by drying. Other wild plants, later largely destroyed by hogs and cattle, yielded edible roots and seeds. The Indian's only domestic animals were small native dogs of vicious temper, useful as watchdogs and as a reserve food supply. Turkeys and other wild fowl were sometimes tamed and kept about the wigwams, but never on a large scale. Reserve supplies of fish were kept in spring wells. The New Jersey Indians as agriculturists were not very far advanced and were greatly handicapped by lack of a variety of field crops and of domesticated animals. Nevertheless, with crude tools and simple methods they made their small plots effectively supplement the food supplies gained from forest and stream.

Historical records contain little data concerning the relationships between the Indians and the whites of Hunterdon County. John Reading, a surveyor and land speculator, states in his "Journal" that he used Indian guides and often stayed at night in "whigwams" on Indian "plantations." According to family tradition the relationships between Philip Kase, a German immigrant who came to the Flemington area in the 1730's, and Tuccamirgan, the local chief, were those of genuine friendship. At his death the chief was interred in the Kase burying ground following a wild Indian ceremony. The stones which the Indians piled on the grave were eventually removed, but were replaced in 1925 by a marble shaft. Despite individual friendships with white men, those Indians who did not move to other areas after the coming of the whites were subject to some suspicion in times of Indian wars. Since many of New Jersey's Indians moved across the Delaware, this region was safe from serious attack at a fairly early date, though traditions tell of Indian alarms at both Ringoes and High Bridge.

Fears throughout the colonies during the French and Indian War were reflected in New Jersey by the creation of the reservation at Brotherton in 1758. To this place went most of New Jersey's remaining three hundred Indians. Their descendants moved to New York in 1801 and on to Wisconsin in 1822, by which time there were only forty left.[10] The few Indians who chose to live among the whites rather than with their fellows seem to have been innocuous individuals, who made a precarious living by hunting and trapping and solaced their souls by periodic visits to the rum jug. One tradition tells of a small group at the junction of the First and Second Neshanic rivers, the chief event of whose lives was the weekly trip to the Flemington stores for tobacco and

10 Nelson, *Indians of New Jersey*, pp. 115–19.

rum.[11] There is no evidence of any intermarriage between Indians and whites in this county.

The white man benefited far more than the Indian from the contact of the two races. A few merchants and traders profited from trade with the natives, and Indian hunters were sometimes employed to kill game and predatory animals. The seed for the first crops of corn usually came from Indian stores. By purchase or seizure Indian corn fields came into pioneer hands and Indian hunting grounds became the permanent homes of a new people with a different economy. The Indian, on the other hand, gained nothing. The white man's rum was too tempting, and the new diseases which he introduced, worst of which was smallpox, were not curable by the herb remedies, the sweating, and the cold baths which the Indians used for all afflictions. Rapid settlement of the country pushed the red man into a subordinate position almost immediately. His whole economic setup was changed very suddenly, his game killed off, and even his native crafts made useless by the appearance of trade goods. It is little wonder that the Indians felt lost in a rapidly changing world to which they were unable to make adjustments, or that they succumbed rapidly to newly acquired vices and to new diseases. The Lenni-Lenape, despite their long existence in the region, dropped out of its history almost without consequence. They affected the economic life of the whites very little, their social life almost not at all. Only the many Indian place names, often changed drastically through alien usage, now remind us of the race which once called this region their homeland.

11 Theodore Bellis, "Recollections," *Hunt. Co. Dem.*, Aug. 28, 1924.

CHAPTER III

The People of Hunterdon

ONE of the most striking characteristics of the Middle Atlantic seaboard in Colonial days was the diversity of origin of the settlers. Peter Kalm, a Swedish observer at the middle of the eighteenth century, wrote of the "great mixture of people in these colonies." [1] Just before the Revolution St. John de Crévecœur, a Frenchman who had lived in America for several years, advanced an opinion not unlike the modern "melting pot" idea. He said, "Here individuals of all nations are melted into a new race of men, whose labours and posterity will one day cause great changes in the world." [2] In Hunterdon County as elsewhere, many strains were blended together to make what is often called the "native American stock." Amalgamation was necessarily slow at first because of ties of blood relationship, language, religion, and similarity of folkways which held together certain groups. However, a few generations of intermingling and intermarriage broke down the barriers which separated the various elements of the population, and in time these elements became thoroughly merged.

The adventurers who everywhere preceded the line of true pioneer settlements left in this region, as elsewhere in America, little record of their brief stay. In Hunterdon County there is little to remind us of them except a few fragmentary traditions. One very persistent story has it that Alexauken Creek is named after Alex (or Ellis or Elias) Hawkins (or Sauken or Hocking), a legendary hunter and trapper who is said to have ranged along the course of that stream in presettlement days. Such adventurers paved the way for the surveyors and land agents who followed close on their heels and who in turn were followed by the true pioneer farmers. The latter were adventurers too, but in an entirely different sense. They risked their labor, their strength, their health, and their very lives in the hope of creating homes and carving farms from the wilderness. Because they were willing to work hard, to accept hardships, and to adjust to new conditions their aims were

1 Peter Kalm, *The America of 1750; Travels in North America,* I, 224.

2 J. H. St. John de Crévecœur, *Sketches of Eighteenth Century America,* p. 43.

eventually gained. They did not always foresee that the end result of their labor would be the formation of a settled society with strong social and economic ties, but that was to be the final accomplishment of their combined efforts.

The first penetration of what is now Hunterdon County occurred soon after the year 1700. One of the first settlers was John Reading, a large landowner, who settled near present-day Stockton in or soon after 1703 and called his plantation Mount Amwell after his home village of Amwell in Hertfordshire, England. The name Amwell was adopted by the new township which was formed in the southern part of what is now Hunterdon County in 1708, and when Hunterdon County was set off from Burlington in 1713 Amwell Township became a township of the new county. Thus the name of Amwell, still used by East Amwell and West Amwell townships, was perpetuated.

John Reading was only one of the many land-hungry settlers who came into this newly opened area. One stream of early pioneers trickled across the low ridges of the Sourlands from the settled regions farther south in West Jersey, and another came from East Jersey by way of the Raritan Valley. The two currents, one largely Anglo-Saxon, the other largely Dutch, converged in the central part of the county. The southern and eastern portions were thus dotted with settlements at a time when the rest of the county was a comparative wilderness. There was also an infiltration of English Quakers and later of Germans from across the Delaware. The large German groups which settled portions of the county at a later period usually came from Europe by way of the ports of New York and Philadelphia.

Hopewell Township, just to the south of Amwell, and itself part of Hunterdon County until 1838, was a sort of steppingstone for settlers who later came into present-day Hunterdon from the south. The records of the Hopewell Baptist Church show that several members gave Amwell Township as their place of habitation in 1715, and not long after that other members were registered as living in Kingwood. The increase of population soon led to the establishment of ferries at present-day Stockton and Lambertville, and by 1714 there were enough settlers in Amwell Township to form a company of militia. Early deeds show that many of the landowners were actual farmers in the Field Tract, at or near present-day Ringoes, in 1720. In 1726 a census reported 3236 whites and 141 Negroes in Hunterdon County,[3] but probably only a small fraction of that number lived in present Hunterdon. An election return in

3 "Census of the Province of New Jersey, Anno 1726."

1738 named 165 voters in Amwell Township.[4] Kingwood Township had been created in 1728 and Readington in 1730 by a division of Amwell Township. The three townships at that time contained all of the territory now in Hunterdon County.

Most of the settlers coming in from Monmouth and Burlington counties and other districts to the south were of English, Scotch, and Scotch-Irish extraction. The Amwell and Kingwood members of the Hopewell Baptist Church bore such names as Stout, Curtis, Green, Hunt, Fowler, Woolverton, Drake, and Warford. There were English Baptist groups on Spruce Run in 1738, at Baptisttown in 1741, and at Flemington by about 1768. About 1730 a colony of English Quakers from Burlington County settled in the region of what came to be called Quakertown. Their records, which have been printed in pamphlet form by M. C. Vail, show such names as Stevenson, Rockhill, Emley, Doughty, Large, Willson, Webster, Robeson, Kester, Lundy, Atkinson, Willets, Heath, Williams, and Schooley. Occasional Quakers made their home in other places. For instance, John Holcombe, whose descendants are legion, came from Devonshire, England, to the Lambertville area at a fairly early date. At Smithfield (Oldwick) the original settlers before the German inundation bore such names as Smith, Johnson, Thompson, Plat, Ireland, and Carlistle. Scotch and English names also predominated among the sixty-nine members of the Barker Tract on the Musconetcong during the years from 1765 to 1775. Followers of the Church of England were to be found in the Ringoes area and in the region where the boundaries of Alexandria, Kingwood, and Franklin townships now intersect. Two English and Scotch Presbyterian groups lived near present-day Ringoes and Mt. Airy in Amwell Township at a very early date. Others resided at what is now High Bridge before 1730, in Kingwood and Bethlehem townships by about 1735, and at Mt. Pleasant by about 1750. The English and Scotch were thus strong elements in the population of Hunterdon County. John Rutherford in 1786 stated that many of the English came from Yorkshire and other parts of northern England. An example of the retention of old-country jealousies was the use of the term "Yorkshire trick" in a derogatory sense, much to the amusement of the Capners who had just arrived from Leicestershire.

Another basic element in the early population was that of the Holland Dutch. In some cases Dutch families intermingled with the English before coming to this county. John, Joshua, and Benjamin Opdycke were members of the Hopewell Baptist Church, in a community largely

[4] J. P. Snell, comp., *History of Hunterdon and Somerset Counties, New Jersey*, pp. 192–93.

English. These brothers later became pioneers in present Hunterdon County, at Headquarters, at Baptisttown, and in Bethlehem Township respectively. Hendrick Van Dolah, who settled at Sandy Ridge in 1725, probably came in with English acquaintances. At any rate, he took root far from the main body of his countrymen. The same is true of another Dutch settler in Amwell Township, Adrian Hogeland, and doubtless there were other examples. Occasional "Dutch books" appear in early inventories, but the term may refer to German books in some cases.

Most of the Dutch, however, came in by the Raritan route and settled within commuting distance of their churches and their friends. The settlers from New York and Long Island had pushed slowly up the river, and were in some force in the Somerville area by the last quarter of the seventeenth century. From there they gradually worked up to the junction of the North and South Branches, then followed up those streams and their tributaries. Their progress was not rapid or spectacular, but continuous because of increasing population. The building of a Dutch Reformed Church near the confluence of the North and South Branches of the Raritan about 1720 indicates a substantial number of settlers in that area. In 1737 the site was changed to the village of Readington a few miles away in Hunterdon County.

During the following years Readington Township became an outpost of the large Dutch settlement of Somerset County. In 1786 John Rutherford wrote, "All the most valuable Lands in Somerset are possest by the low Dutch, also the large township of Reading in Hunterdon." Among the Dutch settlers there were doubtless descendants of Walloon, French, and German émigrés who at one time or another had taken refuge in Holland. By the time they reached Hunterdon County, however, association and intermarriage had changed these descendants of outlanders into good Dutchmen. Among the settlers in Readington Township were families bearing the following surnames: Ten Eyck, Van Horn, Van Fleet, Van Etta, Van Doren, Van Sicklen, Voorhees, Stull, Lott, Hall, Fisher, Louw, Wyckoff, Suydam, Latourette, Covenhoven (Conover), Devore, Montfert, Lequear, Dubois, DeWitt, Hoagland, Jansen, Johnson, Kline, Messler, Smock, Brokaw, Nevius, Farlee, Quick, Melick, Sutphin, and Schenck. Most of these names were Dutch, but some were undoubtedly of other origin.

The Dutch-American stock is still one of the basic components of the "native American" population of the county. Solid citizens, literally as well as figuratively, the Dutch passed on from generation to generation their property and their characteristics. As Anburey, an English observer, reported in 1778, the Dutch gave to their children their "principles, industry, frugality, and an assiduous perserverance in the means of thriving." A very fair estimate of the Dutch was made by John Rutherford in 1786: "The low Dutch are a quiet, frugal people, possess considerable Property, are afraid to run in Debt, without being fond of Law or Offices or Government."

The third large component of Hunterdon County's early population was German. The same combination of persecution and economic depression which brought Rhineland Germans to other colonies brought them also to New Jersey, particularly to Hunterdon, Somerset, Morris, Warren, and Sussex counties. Many of them came directly from overseas, or had at most a short conditioning period before coming to these parts. A beautiful tradition for which no evidence can be found is that a shipload of German refugees embarked from Holland for New York in 1707, but were driven south by storms so that they were put off at Philadelphia. The stubborn Germans are said to have decided to go overland to join other German groups in New York, but to have changed their minds when they came to the beautiful, rolling country which is now Tewksbury Township. It is a good story, and the date nearly corresponds with the arrival of a small group of Palatines at New York. However, there are no records of German Lutheran services in New Jersey prior to 1714, and no New Jersey German groups were strong enough to support church organizations until the 1730's. Smithfield did not become New Germantown until after 1750. The story is probably apocryphal. At any rate, if Germans came into this section so early they must have been few in numbers. John Rutherford wrote of the Germans that they came "late and poor."

Traditions which say that the early settlers of the Ringoes area were German are given some support by a survey of 1721 which mentioned "the palatin's land." [5] But it is probable that most of the Germans came in later than the English and Scotch settlers. However, by 1747 there were thirty members of the German Reformed faith in that locality. The settlement was evidently scattered, as Johann Adam Boellisfeldt (John Bellis), one of the pillars of the church, lived at Copper Hill, where he had settled about 1730. A list of the twenty-two landholders located on the Field Tract in 1754 contains many German names, though there were a few English, one Dutch, one French, and two or three of doubtful origin. Slightly to the west of the Ringoes community was

[5] T. F. Chambers, *The Early Germans of New Jersey*, p. 31.

a second German group, scattered through the region about Sand Brook. William Rittinghuysen (Rittenhouse), one of the group, is said to have arrived as early as 1719. It is supposed that this German Brethren group was an offshoot from the large "Dunker" settlement in Pennsylvania. Two well-known figures in American finance, John Wanamaker and John D. Rockefeller, were descendants of members of this congregation. Besides the groups mentioned, there were other isolated German families in the southern part of the county. The names of a few of them have come down to us. Pieter Prall, founder of Prallsville (Stockton), is said to have been a second generation German from Long Island. Peter Fisher, ancestor of the numerous Fisher tribe, arrived in what is now West Amwell Township from Germany in 1730. Johann Philip Kase (John Case), supposedly the first pioneer at Flemington, settled there in the 1730's. German names and signatures in German script are not unusual in probate records of Amwell Township, especially after 1740.

German settlers in the north part of the county came at about the same time as those of the south. German Lutheran services were held by a traveling minister in the Rockaway (Potterstown) region in 1714, and by the 1730's there were large enough groups at that place, at Leslyland (Whitehouse), and at Fuchsenberg (Fox Hill) to warrant the founding of churches. Still another group made their homes in the valley of the Spruce Run in present Lebanon Township. Germans of the Reformed faith were even more scattered than the Lutherans. Churches at present-day Lebanon and Fairmount were in existence in the 1740's, and another was built not long after by an isolated group in the Mt. Pleasant-Everittstown region. All in all, it would seem that a substantial portion of the early population of the northern part of the county was German. The lessees on West Jersey Society lands as early as 1735 were largely German. One writer, A. B. Faust, has estimated that one third of the population of Hunterdon, Morris, and Somerset counties was German in 1790.

Among the Germans there was occasionally a man of means. Balthazer or Baltus Pickel owned a thousand acres of land between Cushetunk and Whitehouse, and his wealth made him a pillar of the New Germantown church for nearly half a century. Most of his countrymen, however, were poor when they arrived, though not lacking in ambition. Rutherford in 1786 spoke of the "laborious, ingenious and parsimonious Germans, who . . . are daily acquiring Estates in the large counties of Hunterdon and Sussex." An observer in Pennsylvania just after the Revolution praised the "industry and economy" of the Germans there, but was worried about their lack of education, culture, and political awareness.[6] The same criticisms would probably have applied to the poor German peasants who settled in Hunterdon County. They were the least cultured and most poverty-stricken of the groups which settled there, but eventually they raised themselves, first economically and then culturally. Like the Dutch they were in the end absorbed into the amalgam of "native Americans." They lost their identity as a group, Anglicized their names, and dropped their native tongue. Among their descendants have been many agricultural, political, professional, and business leaders of the county.

Of less importance but perhaps acting as a leaven were the immigrants of French blood. Walloons from modern Belgium had come to the Dutch settlements from an early day, and French Huguenots, whom persecution had driven to all parts of Europe, came into this region along with the Dutch, the English, and the Germans. The French groups readily intermarried with other nationalities. For example, Emmanuel Coryell, the owner of Coryell's Ferry, was a French Huguenot, but his wife was at least partly of Dutch extraction. The French immigration to what in time became Frenchtown was a result of the French Revolution. The moving spirit of this settlement was Paul Henri Mallet Prevost, a moderate republican who had fled to avoid the guillotine. About 1795 he purchased a large tract in Alexandria Township and proceeded to settle his family, his retainers, and other refugees. The group was never large, but the Prevosts in particular were for a long time very influential in that part of the county. Other refugees from France and the French West Indies, mentioned in letters of the period, may have taken permanent root in this region also.

Some of the motives which brought a more or less continuous stream of immigration to America are obvious. Hope of economic improvement, desire to escape from constraint and oppression, and longing for adventure all played their part. There were, of course, any number of individual reasons for making the move. The decision, one of vital consequence in the life of the individual, was probably in every case the result of a combination of causes. The immigrants who came to Hunterdon were part and parcel of a huge exodus, and their reasons for coming were the same as for others in the American colonies. The strongest urge was doubtless economic. Gabriel Thomas in 1698, writing in the interest of the landowners of West Jersey, declared that even the poor who came over would "have a fair prospect of getting considerable

6 Schoepf, *op. cit.*, pp. 103–10.

Estates, at least of living very Plentifully and Happily." Even in his special plea to the French Huguenots he placed much stress on the economic advantages involved. In the case of the Germans, religious persecution and political oppression were important causes of removal. However, when Peter Jung (Young) came to Amwell Township from Raubach in 1727, the recommendation which he brought from the Reformed minister of that place stated that he was "pious, honest, and upright," and mentioned that he was the son-in-law of the mayor of the town.[7] It rather looks as though Peter Jung emigrated to America for other than religious or political reasons, and the same may well have been true of others.

The well-kept correspondence of the Capners and other English families who came to Hunterdon County after the Revolution brings to light many individual reasons for coming to America. Hunterdon County was no longer a true frontier, of course, but crossing the ocean and settling in a new land was still a serious matter. Perhaps the reasoning of these Englishmen was not far different from that of other immigrants, even those from other nations and at other periods. Joseph Capner, a farmer of Leicestershire, was so carried away with reports of America and "on Tiptoe to be an American Farmer" that his neighbors declared he was "not right in the head." His uncle, John Hall, in an earlier letter to Philadelphia friends, had slyly mentioned one of Joseph's grievances: "Mind ye the parson at Higham takes Tythe in kind and milks their Cows ever [*sic*] 7th meal." At any rate, Joseph crossed the ocean in 1786, and after looking about a bit, bought a farm just west of Flemington, acquired cattle, and started the production of cheese. His young brother Thomas, cheated by an employer, and his sister Ann, jilted by her lover, found themselves at loose ends and soon followed, bringing along their mother. Brief training in England qualified Thomas to be a millwright in America, and Ann easily found a husband. Their uncle John Hall, a retired miller and something of an inventor, came to Flemington not long after in order to be with his relatives. His friend Isaac Passand, a retired business man, listed six good reasons why he would not come to America: business ventures in England, bad health, the hardships of the long voyage, "the fear of bugs and lice," the difficulty of finding a good housekeeper in America, and the fact that American women were "called ignorant and idle." However, love of adventure soon caused him to join his friends. His health improved in the new land, and he soon found outlets for his energies and

[7] J. Wilhelm Horde, certificate.

fields of investment for his money. Whether his opinion of American womanhood changed is not known. At any rate, he remained a bachelor.

Other letters of this group mention poverty and unemployment, disgust with the course of the war in Europe, the opportunities in America for artisans and tradesmen, the possibilities of profits in dairying, and the advantages of apprenticing children as causes for emigration. One account tells of the man in England who became more worried with each addition to his family. After the birth of the seventh child, he brought his family to America where children were an advantage rather than a handicap. A letter of John Hill in 1793 contained a plea which probably sounded interesting to liberty-loving friends in England: "Now we have a great deal to bragg off [*sic*] in this country, that we have no Poor people when compared with yours, that we have all of us plenty, that we are infested with no ravenous Beasts cal'd kings, that we have none of those Reptiles cal'd Lords, and that we have none of those Infernal imps cal'd Bishops and Tythe Gathers [*sic*], nor any of the Hellish Tribe." During the economic slump after the Napoleonic wars there was another wave of English immigration. Letters to Hunterdon County residents stress the many bankruptcies in England, "the misery and poverty and want of employment," low wages, high taxes, and low prices for farm produce. A number of the new immigrants visited or tarried in Hunterdon County for a time, but most of them eventually went on to the frontier in Pennsylvania and western New York.

The case of Hugh Exton, a brother-in-law of Joseph and Thomas Capner, who came to Hunterdon County in 1795 after withstanding the arguments of friends and relatives and even his wife for several years, illustrates how a combination of circumstances might cause even a man with roots deep in his native soil to move to a strange land. His letters list his grievances: tithes, poor rates, rising taxes, constant "altercations" with his landlord, and interference with free speech and other liberties by the government. The lease to his farm ran out in 1795, and his landlord decided to raise his rent, thus taking advantages of improvements made by Exton. Both landlord and tenant were hot-headed, so no compromise was possible. At the same time Exton's political activities and his flouting of laws regulating the press and public gatherings had put him in danger of arrest. The letters which he had been getting from America suddenly took on new meaning. Their glowing praise of that land eventually won him over. In America he could become a landlord rather than a tenant, and in addition he could have that liberty which he desired. He sold his cattle and other property in

England and brought with him to this country his family, his servants, and a group of relatives and friends. After a period of looking about, he settled near present-day High Bridge, where he leased for several years and later bought the historically famous "Union Farm." A dairyman and sheep raiser, he became one of the most prominent farmers of Hunterdon County.

To pick up and leave for a strange land took considerable daring on the part of the immigrant and a great deal of careful planning. Possessions which could not profitably be taken along had to be ruthlessly disposed of. Those which were kept, augmented by purchases, had to be carefully crated, checked, and guarded. The contents of the large chests which the Capner family brought in 1787 totaled several hundred items. The immigrant had to provide for a voyage of from eight to twelve weeks in a day before artificial refrigeration. A not unusual arrangement was for the captain of the ship to furnish water, biscuit, beef, pork, peas, flour, potatoes, oatmeal, cooking utensils, and perhaps "pudding twice a week." Or the passengers might arrange to take everything which would be needed. Lists of stores taken by emigrating groups included everything from condiments and liquors to livestock and poultry to be butchered during the voyage. The passengers had little to do on shipboard, and often became bored with the monotony and tired of the cramped conditions. Joseph Capner in 1786 spent his spare time gazing at the sea and recording every strange sight in his journal. His sister, when she came the following year, flirted with the ship's officers and read *Tristam Shandy, Perigrine Pickle,* and *Roderick Random* on the way across. Talking and quarreling with other passengers, taking care of children, and nursing seasick and other ill persons probably disposed of much time. Mary Choyce, a relative of the Capners and a delightfully frank young lady, wrote to her parents in 1795, "When I was sick, there was nothing I wished for so much as a bason [*sic*] of water from our pump in the kitchen."

After having made several voyages across, the ever-observant John Hall felt qualified to act as adviser to others. His detailed instructions throw much light on the problems faced by individual immigrants. Two long extracts from his letters are quoted:

I will now put down a Cattalogue of the articles Nessessary for a Single Sterrage passenger either man or Woman. [I]t must be always reconned on 8 weeks passage here. If 3 or 4 Joyn in Stores it will make the Expence less as [the] same Cooking utensils will do for the whole—I shou'd wish them to take a receipt of the money pd. to the Captain for passage and I shoud.

specify in said receipt that the Captain to find—Bread or Bisquet—Water —Beef, Irish is the best—pork, American is the best—peas for pudding and soup how maney times a week—Flower as maney pints a week as you can get—pottatoes without limmit—Salt and Oatmeal if you can or you must provide it—and the privilidge of his Baking utensils as often as they Can make it Conveient [*sic*]. Then you are to provid—A bed or Matrass and two Blankets (people going to housekeeping will take as maney domestic utensils as they please in Chests, pay no freight nor duty if they have been used), ... pewter plate—knife and fork—pewter spoon—A saucepan—a Tin Teapot—a pewter Bason to make a dumpling and pudding in—a Tin pint and half pint—a pewter Chamber pot—Hand Towels—A pudding bag —a pound or less of Candles—Bit of soap—a Stone or two of Best Flower— two or 3 pound of Currants one or two—Rice 6 or 10 pound—Coffee ground in a tin Cannister, one pound—Tea a pound—Suger 4 or 5 pound or more —Eggs washd. in cold Water clean and a pippin with a flooring of Butter at Bottom . . . (then put on a flooring of Eggs, then strew over a Quantity of salt almost to fill up the Chasms or about half full, then another floor, will keep good—The Butter to use must be washd. from its salt &c.)—Chees —Bacon, 5 or 6 pound—Oinyans if in the fall—Herrings a Quarter of a Hundred—Gingerbread—Cyder and porter 2 doz: each—a Quart or two of Spirrits—Vinegar a quart—pepper wood in Box—mustard—If further things are thought necessary—Apples for Dumpling or eating—Almons & raisings—Jar of picles (Cabbage as cheap and simple as any)—preservd. Gooseberries and Damasins—a Ham—Small Jars of potted Beef well Coverd. with suet or it moulds. A person or a Company uniting in a mess of the Greater part of the above articles may take a passage as Comfortable nearly as in the Cabbin the Lodging excepted. . . .[8]

And a[s] maney of your domestic utensils that are good you may pack up as Pewter, Iron Candlesticks, Knives, forks but no fid fad articles of Brass. [C]andlesticks and such aristocratic ware wd. make an indifferent contrast if wee put you in a Log house—[W]ee think your brass ware Kittles might be proper but not bell mettal nor cast iron ware and a Chees pan wd. be two [*sic*] Cumbersome as our Cedar Tubs are very good ones . . . or you may buy brass pans here as we have one that holds more than 20 Cows milk &c—[Y]our small mazareen kettles bring tho I said no bell metal, but bring no patchd. old ones, sooner sell and buy new ones and have them builled as used. . . . A scale beam you might each of you bring. . . . Each of you a hand saw and a guage and 2 or 3 Chisels (Our Screw Augers are better than yours.) and each of you a goss hook—Spades the lightest and neatest made . . . every one even to Nancy and Tom boy, for women do most of the gardening here—Scythes half a dozen each. . . . I have sometimes thought of each of you bringing a plow but I think they wd. be two Cumber-

[8] John Hall to Mr. and Mrs. Richard Choyce, Apr. 30, 1793, *ibid.*

some Carriage but however bring such dimensions as you may hereafter direct the making them for I think you will both dislike them used here. Likewise the dimensions of Cowsheds as you may direct thier [*sic*] building here—I have had some thought of each of you bringing a set of light Harrow tines. [I]n that do as you please. . . .[9]

The newcomer's first task in the new land was to wash the shipboard linen. That done, he was faced with the problems of a new environment. The first reaction, at least of immigrants to Hunterdon County, was often unfavorable. Stumps and rails, waste land, poor cattle, hot summers, and mosquitoes were grounds for complaint. John Hall reported that the differences in farming methods, ways of doing things, and especially manners "at first disgust all Emigrants." Mary Choyce wrote to her parents, who were thinking of following her across the ocean, that they could not be "gentlefolks here," and John Hall suggested to his brother-in-law that he had better learn to shave himself and to drink brown sugar in his tea. The English at Flemington were amused as each newcomer in turn became "land sick" and then gradually recovered, but they were exceedingly scornful of those who went back to England without giving America a thorough trial. John Hall once said, "Wee know that time reconciles most emigrants," and Isaac Passand expressed himself by saying that the immigrant had "to be born again." [10]

Europeans newly arrived in this country sometimes had to cope with a breakdown of health, but more often with homesickness. Mary Choyce pleaded with her parents for long and regular letters giving "all the news." Old ties of friendship were dropped, however, all too soon. Thomas Capner, for instance, neglected to write to his best friend for eight years. Mrs. Capner told of a German woman near Flemington who received letters from home just often enough to be made to feel very sad each time by news of the death of more old friends and relatives. Aliens in a strange land had serious problems of a social nature also. Mrs. Capner wrote at one time, "We eat a cheerful breakfast laughing at some of the customs of the country." [11] Her sense of humor doubtless helped her to make adjustments more easily than most people. Her daughter quoted a neighbor as saying, "People in years can't learn the ways of this country but their children will." [12] The first generation

[9] John Hall to Hugh Exton, Nov. 2, 1795, *ibid.*

[10] John Hall to Mr. and Mrs. Hugh Exton, July 5, 1795, Mary Choyce to Mr. and Mrs. James Choyce, Apr. 26, 1795, *ibid.*

[11] Mrs. Mary Capner to Mrs. Hugh Exton, Feb. 7, 1791, *ibid.*

[12] Ann Capner to Mrs. Hugh Exton, Apr. 6, 1794, *ibid.*

immigrant, unless he settled with a group of acquaintances, was likely to feel that he belonged neither to the old nor to the new.

The inhabitants of present-day Hunterdon County for over a century could trace their origins chiefly to Great Britain, Germany, Holland, and France. Newcomers during the period were from the same general parts of Europe as the earlier settlers, and interregional migrations brought in no large alien groups. Immigrants doubtless thought it a strange land, but in time they or their children came to feel and act like their new-found neighbors. All became members of the "new race of men" of which Crévecœur spoke. But despite the rather large influx of other nationalities, it was the English characteristics which were to prove dominant. English law came in as part of English rule. Spoken English established itself early because of the need of a common language. English institutions and customs were slower to take hold in some of the more segregated groups, and won sometimes only as the result of compromise. The success of the Dutch Reformed Church might be looked upon as an exception to the general triumph of English institutions. The Dutch churches, however, in time adopted the English language, and their membership soon embraced many persons not of Dutch ancestry. Since the creed and ritual of the Reformed Church were much like those of the Presbyterian, no compromise was necessary on that score.

The Dutch settled in fairly compact groups and were separated from other nationalities by religious and social ties. Their connections with the large Dutch settlements of Somerset County were close. Nevertheless, amalgamation was inevitable. Dutch children associating with English children or attending English schools picked up the new tongue, and adults often felt the need of learning the common language. The next step was that of dropping the old for the new. In 1784 the Dutch Reformed Church at Readington tried an interesting experiment. A subscription list was circulated, and each doner was to make his choice as to the language which he preferred. The number of services in each language was to be prorated according to totals subscribed. As a result those favoring English secured eight more services in that language during the coming year than were to be given in Dutch. At the same time the following resolution was adopted: "Resolved: also, that hereafter the minutes of this Consistory shall be recorded in English, in order that the succeeding generation may have them in a language which (as it now seems) will be better known to them." [13] The new pastor in

[13] H. P. Thompson, *History of the Reformed Church at Readington, N.J., 1719–1881*, p. 71.

1789 did not speak Dutch fluently and soon dropped the use of that language. Some of the older people angrily locked him out of the church, but his decision was accepted by the consistory. The graybeards were conciliated by having occasional Dutch sermons by visiting ministers. In the end the English language entirely displaced the Dutch. When a translation of the old records was wanted in 1873, an itinerant teacher from Pella, Iowa, did the task. The change in language was symbolic of a wider change brought about by intermarriage and constant association with Americans of English stock.

Despite a certain amount of grouping, the Germans of this area were on the whole rather scattered, or, as John Rutherford put it, "dispersed." This probably accounts for the fact that they were so thoroughly absorbed into the population, while many of their kinsmen, the "Pennsylvania Dutch," remain to this day a people apart. Close contact with the English and a need to know the tongue in most common usage caused the German language after two or three generations to "wear fast out." Already in the 1750's Dr. Henry Muhlenberg, the spiritual leader of the Lutherans of the region, urged that it was desirable for pastors to know English. The German Reformed churches began having services in English soon after the Revolution. An amusing story is told concerning Rev. Caspar Wack of the church at Lebanon. A stranger who spoke only English visited the church one Sunday and was well pleased that he could understand a word now and then. He did not know until later that the sermon had been in English. The church at Larison's Corner near Ringoes, slower to change than the others, had German services only until 1798. Eventually, however, German went out of general use throughout the county. A great many German names were simplified, sometimes so much as nearly to hide their origin. A factor of some importance in the complete amalgamation of the Germans was the comparative poverty of the German Reformed congregations. This caused these groups to cooperate and eventually to combine with English and Dutch religious groups.

During the latter part of the eighteenth and the early part of the nineteenth centuries, when European immigration was small, the county's population had ample opportunity to become homogeneous. The first alien group of any size to disturb the racial composition was the Irish, not to be confused with the Scotch-Irish of an earlier day. The number of Irish coming to America became sizable in the early 1830's and increased greatly during the potato famine of the 1840's. Their coming coincided with the period of canal digging and railroad building in this country, and many became day laborers at such tasks. Some of them came to Hunterdon County in time to help dig the feeder canal of the Delaware and Raritan, and Irish workers did most of the work in constructing the New Jersey Central, the Belvidere Delaware, and the Lehigh Valley Railroads. The roistering Irishmen of these labor gangs often caused distress to the peaceful citizenry of the county. The Irish and English miners brought in during the copper mining craze at Flemington in the 1850's and early 1860's were a disorderly crew who caused several riots. Racial antipathies between Irish and Negro laborers working on the Lehigh Valley tunnel through the Musconetcong Mountain in 1872 culminated in the "Pattenburg riots," in which several men were killed. It is probable that the troubles caused by these labor gangs colored to some extent the attitude of the people toward the more peaceful Irish families who settled down in their midst.

Though most of them came from peasant families, the Irish did not usually choose to become farmers on this side of the ocean. Lack of capital was probably the chief cause, but another was dislike or distrust of the occupation which had given them a bare subsistence in Ireland. The schedules of the local census takers show that unattached young men and young women often served as hired help on farms and in homes, however. James H. Blackwell, a farmer near Flemington Junction, irately recorded in his diary on December 7, 1854: "They have neither honnour, principal or gratitude, but are lazey, wastefull and careless. No wonder that Ireland is a poor starved nation." In the villages the Irish served as laborers in factories, livery stables, taverns, and stores. There were also a few Irish artisans, usually unmarried journeymen. Occasional first-generation Irish became successful in business, in the professions, or in politics. For example, John Corcoran, a poor immigrant boy who served as a farm hand for fifteen years, later became in turn railway employee, sheriff of Hunterdon County, and owner of a large stone quarry near Stockton.

The greatest concentration of the Irish was at Lambertville, where there were 230 natives of Ireland in 1850.[14] Somewhat segregated in a section of the village known locally as Dublin, they brought a touch of new color with them. A local paper thus describes a Catholic picnic of 1859: "They had a merry time, the lassies bedizened in their finest trimming, and the laddies with their new corduroys, going through the double quick motions of the Dannybrook [*sic*] jig." [15] Starting poor, the

[14] Hunt. Co. local schedules, *U. S. Census, 1850,* Schedule 1, Lambertville.
[15] *Lambertville Press,* Sept. 1, 1859.

Irish of the first generation seldom acquired real estate or any large quantity of worldly goods. Usually illiterate themselves, they seldom showed much interest in educating their children. A by-product of the Irish immigration was the bringing of Catholic churches and missions into Protestant communities. According to the United States Census, the number of natives of Ireland in the county increased to 1233 in 1870, but decreased thereafter. By 1890 the number had fallen to 891, and in 1910 there were only 382 Irish immigrants and 718 persons who claimed Irish parentage. The Irish who did not move to other areas have been gradually absorbed.

The second immigration of the Germans to American shores paralleled that of the Irish, but continued much longer. Hunterdon County was not entirely passed over by this group, and by 1850 there were a few German hired men on the farms, seemingly well-liked by their employers, and small German colonies in some of the towns. Census figures show 262 natives of Germany in the county in 1870, and a small but steady flow has continued since. By 1930 there were 675 native Germans and 1145 other persons of German parentage. During the past century there has been a small trickle of newcomers from other countries of northern and western Europe, but never large enough to affect greatly the population of the county as a whole.

The coming of the "new immigrants" from southern and eastern Europe, however, has brought new problems of assimilation and Americanization, as well as an oversupply of cheap labor. In 1930 there were in Hunterdon County 3942 persons born in foreign countries and 6411 of foreign parentage. Except for the Germans mentioned above, nearly all of these newcomers were from southern and eastern Europe. One person in four in Hunterdon County was thus from a foreign home, a sufficient ratio to upset the normal social development of the county. The problem is further complicated by the fact that most of the new groups have tended to become segregated, either in villages and towns or in closely knit rural settlements. However, there are already indications that within a few generations they will probably be absorbed into the population as completely as have been the earlier immigrants. Special efforts of churches and other groups and the influence of public schools, newspapers, radio, and ordinary social intercourse are rapidly Americanizing the newcomers. The second generation usually makes a great effort to conform to the American pattern.

The Italians, like the earlier Irish, first came to the county as itinerant laborers. A few were employed on the Pattenburg Tunnel in 1873, and a small colony worked the Prallsville stone quarries in the 1880's. The number increased slowly, until by 1910 there were 711 Italians in the county. In 1930 the native Italians numbered 585 and persons of Italian parentage 1020. Exploited as cheap labor, the Italians constructed the sewers and pipelines of the county, repaired its railroads, and worked in its factories. Like the Irish they seldom became farmers. Though the native Americans attended Italian picnics to see the folk dancing and the fireworks, they did not at once accept the strangers as full equals. Differences in customs sometimes caused serious difficulties, and racial and religious antipathies prevented much intermarriage. The history of the Italians in the county bears a striking similarity to that of the Irish at an earlier period, and probably in time they will become as completely absorbed as have the Irish.

Magyar or Hungarian laborers appeared in the county in the 1880's, but few came there for permanent residence before the turn of the century. In 1903 the people of Flemington saw their first Hungarian wedding, and in 1907 they witnessed a flare-up of Old World hatred when fighting broke out between young Hungarians and Poles of the "Slav colony." [16] Though a little alarmed at the lack of restraint of a few of the newcomers, the people of the town made an effort to bridge the gap by encouraging a Hungarian orchestra and by providing church services and Sunday school classes in the Magyar language. By 1910 there were 239 natives of Hungary in the county, and immigration continued until the outbreak of the World War, when it stopped except for those who had tarried for a while in New York City. The 1930 census showed 340 natives of Hungary and 400 American-born children of Hungarians. The Hungarians, less segregated and fewer in number than the Italians, are becoming rapidly Americanized, and already there is a tendency toward simplification of names. They have a real love for the land and usually prefer farming to other occupations.

The greater part of the Slavic immigration came after the first World War. In 1930 there were in the county 563 natives of Poland and 821 children of Polish parents. Some of them have settled in the factory towns, others on small lots of land, particularly between Whitehouse and Baptisttown. Both the rural and urban groups of the first generation have been somewhat clannish, but in recent times the tendencies toward segregation have been weakening. The same is true for the small Czech and Slovak groups. There are also several hundred persons who name Russia as their homeland. Many of this group are Jewish, as are

16 *Hunt Co. Dem.*, June 30, 1903, Feb. 5, 1907.

part of those who gave Poland as the land of their birth, and during the last few years there have been enough Jews in Flemington to support a synagogue. Religious and "racial" barriers, though noticeable, are not so marked as in many places. Various Baltic and Balkan countries have small colonies in Hunterdon's factory towns, and other nations have a few representatives. The local census taker for West Amwell Township in 1930 reported people from seventeen European countries. That the beginning stages of Americanization were already under way was shown by the fact that only two families were unable to speak English.[17]

One component of the county's population, the Negroes, has as yet received no notice. Since this group will be given considerable attention in a later chapter, only a hasty survey will now be made. In 1790 one-fifteenth of the population of the county, as then bounded, was Negro. As the Negroes were freed, many migrated from the farms to the villages, where they usually became laborers. The number of Negroes in the county has been steadily decreasing for a century. In 1850 there were 817 Negroes in the county; in 1880 there were 552; in 1920, only 359. During the period of slavery and the early days of freedom in Hunterdon County, the Negro, though in an inferior position, had a definite and not too unhappy place in society. Animosities were created by the political campaigns of the 1850's, by the Civil War, and particularly by the Fifteenth Amendment. Since that period, however, matters have improved greatly, and racial antipathies have long since faded. The Negroes of Hunterdon County consider it their home, as well they might.

Thus far we have considered those who came into Hunterdon County from the viewpoint of nationality or race. A heterogeneous group of newcomers has come in recent times from the large cities not far distant. The movement was of enough importance in the years just prior to the World War and again for a time during the postwar slump to cause considerable comment. It slackened during the 1920's, but resumed again during the depression of the 1930's. The *Hunterdon County Democrat* ran a series of articles for the "back-to-the-landers," emphasizing especially things not to do, and the State Department of Agriculture also furnished much good advice. As was natural, many of those without supplemental income failed and drifted back to the city. On the other hand, some of those who had a little capital and plenty of persistence have made themselves over into true dirt farmers. City modes

[17] *Ibid.*, May 8, 1930.

of speech sound a bit odd in the countryside, but the newcomers have usually found only welcome in the attitude of farm groups.

The purchase of Hunterdon County farms by wealthy outsiders has become a common occurrence in recent years. In many cases the newcomers have restored fine old houses and, by buying marginal farm land, have helped to solve the problem of abandoned farms. But differences of wealth, social position, and culture divide them from the older inhabitants and from one another. The majority live in Hunterdon County only a portion of the time, and many of them after a few years sell their land to other outsiders. Some have made valiant efforts to become true local citizens, but the greater number are and probably will remain "city folks" to the natives of the county.

Thus far in this chapter we have been entirely concerned with the coming of people into Hunterdon County. The other side of the picture, that of emigration from the county, is nearly as interesting. Almost from the beginning of settlement adventurous persons pushed on beyond the county's borders. In the early days they went into the wilderness of northwest New Jersey. For example, Quaker families from Quakertown began emigrating to Great Meadows in Warren County about 1740, and the records of the Kingwood and Mt. Pleasant Presbyterian churches show that they, too, were weakened in the 1750's by "removal of many members to distant parts." Soon after the Revolution a great exodus to the Lake Country of western New York began. Abijah Gilbert, a friend of the Capners, became a pioneer near Lake Otsego, and wrote letters to his friends in Hunterdon County stressing the richness of the soil of the new country, the beauty of the scenery, and the rapidity of settlement. He minimized the "fever and ague," the insects, the lack of comforts, and the shortage of labor. The same kind of coaxing which had brought immigrants to Hunterdon County was now used to draw its people to the frontier. Theophile Cazenove, a French traveler who went through Easton in 1792, in his journal commented on the fact that there was "much emigration . . . to the Genessee country."

During the same period the Pennsylvania frontier was being extended westward. Quakers from Quakertown and Presbyterians from Amwell were dismissed to church organizations in many parts of that state and to what is now West Virginia. People in Flemington in the 1790's were much interested in a place called Butternuts in Pennsylvania, a land development by one John Priestly. Several journeyed to the place to look it over, and a few settled there. After an arduous trip and a careful inspection of the land for sale, Joseph Capner decided that he preferred to remain on his farm near Flemington.

Hunterdon County like all New Jersey contributed to the settlement of Kentucky. In 1786 John Rutherford ruefully wrote that "annual swarms" emigrated to the southwest in order to escape rents, taxes, and debts. Beginning in 1785 the Hopewell Baptist Church dismissed many members "to western frontiers" and specifically to Kentucky. Since several persons often left the church at the same time, it is probable that neighboring families emigrated together. Hunterdon County place and family names are still to be found in many places in Kentucky and Tennessee. Sometimes related family groups left *en masse.* There have been no members of the Ringo family in the county for over a century, but the name is very common in Kentucky. Individual reactions to frontier life differed. A homesick woman from Readington Township wrote from Kentucky:

If you wish ever to move to this place you must Come and Se it fust and your one [own] eyes will Satisfy you. [I] do not like the plase. I wish to be in Jersey a Gane for the Chicnes [chiggers], flees & ticks keeps us in full employ. . . . [I] wod never never advise any person to Come to this plase. . . . [We] live in loog Caben.[18]

Quite a different story was that of a successful Kentucky farmer who had come from near Ringoes:

We are all well at present blessed be God. . . . Would be glad to see you all again but I Do not expect to See you in this wourld. . . . I now live in a new Country. . . . I like the Country much. I think it Will Make a fine Country in time. It in proves [improves] very fast. [T]he tax is low and I have 605 akers of land. [M]y Children is all married but tue and has land of thare one [own].[19]

It is little wonder that relatives and friends followed him to the land of promise.

There are many indications that the Old Northwest drew a great many people from Hunterdon for over half a century. A typical casual reference to such a migration is the following note in one of the day books of the Case Tannery near Flemington: "Joseph Case started from Philip Case's Aug. 16 to the miammi [Miami] cuntry to see his uncle John Case In the year of our lord 1812." A worn scrap of paper tersely finishes the story: "Arived Safe in the Indianna Territory August 30, 1812." It

[18] Sarah Huff to Sary Morehead, *circa* 1794, Hunt. Co. Hist. Soc., miscellaneous correspondence and papers.

[19] David and E. Palmer to Burtis Ringo, Sept. 28, 1799, *ibid.*

is said that there was a large migration of Kingwood and Delaware Township people to the Symmes Purchase in Ohio. Many others left the region about Ringoes, especially during the hard times after 1815. In various accounts, letters, advertisements, and newspaper stories, mention is made of Hunterdon County people or their children in Warren, Preble, Richland, Darke, and Wayne counties of Ohio, in Franklin and Shelby counties of Indiana, and in Grundy, Jersey, and Clarke counties of Illinois. The same records also mention specific towns, some of them in the counties just named, where natives of Hunterdon were to be found at one time or another: Uniontown, Urbana, Milan, Sandusky, Wooster, Shiloh, Plymouth, and Lima in Ohio; New Trenton, Marion, Remington, Corienna, and Greenwich in Indiana; Jerseyville, Morris, Quincy, New Rutland, Dewitt, Oswego, Joliet, Galesburg, Fairview, Springfield, Edwardsville, Genesco, Pana, Cordovo, Carlyle, and Fairfield in Illinois; and Hudson, Waterloo, Jackson, Mason, Eaton Rapids, Farwell, and Pontiac in Michigan.

A number of these places were veritable colonies of Hunterdon County people. When Dr. George Rex visited Morris, Illinois, in 1856, he found members of the Hoagland and Reading families and other old friends very much in evidence. The editor of the *Grundy County Herald* was Henry C. Buffington, former editor of the *Hunterdon County Democrat,* who employed the same printer that he had had at Flemington. Emigrants sometimes kept up the ties with their homeland for two or three generations. The people left behind were always much interested in the success of their wandering friends. When Charles and Nancy Burkitt returned from a visit to a Hunterdon County colony in Ohio in 1837, many relatives and friends of those who had gone West came to inquire about them. Such contacts naturally encouraged still others to emigrate. The possibility of settling among friends robbed the adventure of going west of most of its terrors.

Changes in transportation methods gradually made much simpler and easier the task of moving families and possessions. One record says, "It took our friends the Mouts in 1811 . . . 6 weeks hard travelling to get to Cincinnati O." [20] In 1839 a group who went by wagon train to Ohio made much better time. The log of the journey was kept by one of the men. On the last day of May he wrote, "Took the Glade road, crossed Dry Ridge and stopped all night on top of Allehany Mountains. Made 21 miles the 31st." A few days later he recorded, "Passed Triadelphia and landed in Wheeling at noon. At two oclock crossed the Ohio

[20] Copy of letter to Oliver H. Blackwell, Apr. 10, 1860, Blackwell diary.

River, passed through St. Clairsville, County Town of Belmont County. Made 25 miles the 5th. June." [21] It took the group just over three weeks to get to Montgomery County, Ohio. There they rested with friends before pressing on to their destination in Darke County. The coming of the railroad brought even greater changes than had the building of good roads. In 1860 it took Oliver H. Blackwell just thirty-eight hours to travel from New York to his destination at Lafayette, Indiana.

Before the Middle West had acquired more than a small fraction of its ultimate population, pioneers were pushing on beyond the Mississippi. Already in the 1830's Hunterdon County people were to be found there. One of the Blackwells then lived in Louisiana, and one of the Rittenhouses had taken his family to Dubuque, Iowa. Opdyckes, Fishers, and others had emigrated to Iowa, Missouri, and regions beyond. James W. Marshall, a restless Lambertville boy who had tried Indiana, Illinois, and Missouri in turn, was the discoverer of gold in California in 1848. The number of men from the county who took part in the gold rush was apparently fairly large. On December 29, 1850, Blackwell recorded in his diary the visit of an old friend "who has lately returned from California—with the dust." Some of the letters sent back from California are very entertaining. Samuel L. Bonnell, one of a group from Clinton, who went quite early by the Panama route, wrote as follows in 1849:

> The gals here are not of the best. . . . We have almost every necessary of life with the exception of Buckwheat Cakes and Sausage. This country will do to make money . . . but not a cuntry to settle down as dirty as the mischief and now the Rainy season has set in and . . . as mudy as the duce. . . . [A]ll that makes anything pleasant in my estimation is the dimes. . . . I believe all the Clinton boys can rays money enough to Get back but they are satisfied to stay and get a little more of the stuff.[22]

Bonnell and some of the "Clinton boys" returned after two or three years with many interesting stories but no great riches. Rings made from California gold are still owned by Bonnell's descendants.

Before the gold rush was entirely over, the call was "Ho! for Kansas." Emigration societies, desirous of populating that territory with antislavery people, offered free passage to those who wished to go. During the decades after 1850 many Hunterdon County people went not only to Kansas but also to Nebraska, Arkansas, Colorado, Utah, Texas, Oregon, Minnesota, and the Dakotas. Several young men from the county participated in the gold rush to the Klondike. Later emigrations from

[21] David Lair, "Log of a Journey to Ohio by Philip Gordon Family, 1839."
[22] Samuel L. Bonnell to F. Honness, Dec. 14, 1849, E. E. Bonnell Papers.

FIG. I. Indoor bake oven with utensils.

FIG. II. One type of tin kitchen. Number scale on this and succeeding illustrations represents inches.

FIG. III. TOP: Typical trivets made by local blacksmiths. MIDDLE: Skillets used for various cooking purposes. BOTTOM: Two types of Dutch ovens. The one on the left could be set in coals or suspended from a crane.

the county have been so scattered as to be difficult to trace. In 1889 J. G. Williamson, minister of the Bethlehem Presbyterian Church, wrote, "Our former members are widely dispersed among the States and Territories of the Union." It is safe to say that today there are Hunterdon County people or their descendants in every state of the Union.

The question might be raised as to the results of this constant drain on the county's human resources. A rather sad part of the story has been the constant breaking of family ties. In 1855 a woman wrote, "James is full of going West again, but I trust it will end threatening as it heretofore so often has." [23] She was probably wrong, for once a boy got "western fever" he was hard to dissuade, though he might well come back if disappointed in his venture. Another result has been the sudden disruption of communities. Farmers going west sold their personal goods and their farms; and hired men quit work, asked for their pay, and headed for the frontier. Lawyers, doctors, ministers, teachers, and other professional men also felt the lure of the West. Church organizations were disrupted and church finances hurt. The constant agitation undoubtedly had its bad effects, but a high birth rate and immigration gradually repaired the population loss. Perhaps the story of one family may help to show something about final results. The eldest of the four sons of Jeremiah L. Lott of Kingwood Township went to Ohio, the second to Illinois, and the third to New York, but the fourth remained at home and in time became the father of a large family of his own.

The newest frontiers have been the great cities of the nation. In 1872 the members of the Union Farmers' Club showed the general concern already felt in regard to the cityward migration by devoting a discussion period to the subject. From that day to this, editorials and letters in the county papers have discussed the problem, usually emphasizing the advantages of staying on the farm. Correspondents have pointed out the deserted churches, abandoned farms, and empty homes, and ministers have often preached against "the tendency to leave the old homestead for the feverish excitement of a great city." In the 1920's the Hunterdon County Board of Agriculture campaigned against the trend cityward. Now that the population of the county is again on the increase, the alarm has subsided. In 1940 a former Hunterdon County man living in New York City wrote approvingly about the "fresh looking, pretty, stylish . . . girls who come to New York from small towns and prefer working here to remaining in idle comfort back home." [24]

23 Unsigned letter, Mar. 11, 1855 (copy), Hunt. Co. Hist. Soc., miscellaneous correspondence and papers.

24 *Hunt. Co. Dem.*, Mar. 21, 1940.

There are no population statistics of the colonial period for Hunterdon County as now constituted. The figures for the county as then bounded were 3377 in 1726 and 5507 in 1737. By 1745 the number had increased to 9151, though the county had by that time been cut down to its present boundary on the north.[25] A computation of the population of present-day Hunterdon County by townships as recorded in the national census records for 1790 gives 14,849. The number had increased to 20,427 by 1820, and despite western migration rose slowly to 24,789 by 1840, to 33,654 by 1860, and to 38,570 by 1880. Then began a slump which lasted for nearly half a century. In 1890 the figure stood at 35,355, in 1900 at 34,507, in 1910 at 33,569, and in 1920 at 32,885. After that time the trend was reversed. By 1930 there were 34,728 people in the county, and by 1940 the number had increased to 36,766, only 1804 less than in 1880. The population once lost has been slowly regained during recent decades, largely as a result of immigration into the county.

25 "Census of New Jersey, 1726, 1737, 1745."

CHAPTER IV

Land Ownership

THE REGION later known as New Jersey was a part of New Netherland under the Dutch. The English had never recognized Dutch claims to the region, and in 1664 seized the colony in the name of James, Duke of York, to whom Charles II, his royal brother, had granted it as a proprietary. Before the year was out, the Duke transferred the New Jersey portion of the lands seized to two of his friends, Lord John Berkeley and Sir John Carteret. In 1674 Berkeley disposed of his interest to Edward Byllinge and John Fenwick, both prominent Quakers. A quarrel as to their respective shares soon ensued, and the situation was made worse by the financial difficulties of Byllinge. Quaker friends of the latter, hoping to make his part of New Jersey a refuge for people of their faith, rescued him; and three of them, of whom William Penn was most prominent, became his trustees. In 1676 an advantageous division of territory was made with Carteret, the Quakers accepting the western portion, henceforth known as West Jersey. In the meantime Fenwick had led a colony to that region and had founded Salem. The trustees secured his consent to a plan which would allow piecemeal sale of the lands of West Jersey and thus extricate Byllinge from his financial difficulties. The ownership of the whole territory was divided into one hundred shares or "proprietaries." Ten of these went to Fenwick and ninety to Byllinge. Most of the shares were then put on the market, a substantial portion being bought by Quakers. The purchasers, known as the Proprietors of West Jersey, were to receive land dividends with each successive division of territory.

The one hundred proprietaries were bought and sold in both England and America, and as a result ownership became widely dispersed. Shares and part shares often changed hands rapidly. The largest single owner was the West Jersey Society, a London corporation formed in 1692. In addition to some holdings in East Jersey, this company owned twenty of the proprietaries of West Jersey for a long period. In one of the periodic parcelings of land to the owners of proprietaries, this group received a large tract of land in Hunterdon County. Many proprietors actually occupied part of the lands assigned to them, and others kept possession of their portions for a considerable time, hoping for a rise in land values. But sooner or later, through sale, inheritance, or court process, their holdings passed into the hands of other persons. All land titles in West Jersey stem from deeds granted by the proprietors or their heirs. Contrary to popular belief lands in this area were not passed out to individual owners by royal patents, though titles were in final analysis derived from the crown. The fine parchment deeds which are still preserved by many families give the numerical year of the reign of the monarch then on the throne, but are merely transfers of title from one individual to another and in no sense royal grants.

The proprietors and those who purchased from them thus held their lands as successors to the crown. There was in this situation no recognition of any rights of the Indians to the soil. In practice, however, the proprietors saw fit to mollify the red men by purchasing the land from them. Thus they made doubly sure of their ownership, except in instances where others illegally negotiated with the Indians before they did. In such cases and in those in which the same land was transferred two or more times by the Indians, grounds for future controversy were provided, though the individual proprietors or those to whom they sold had every legal right on their side. Naturally, the amounts paid to the natives had little relationship to land values. In 1703 trinkets and trade goods worth about seven hundred pounds were paid for a tract estimated at 150,000 acres. This purchase included much of what is now the southern part of Hunterdon County. John Reading, well acquainted with the Indians, was one of the negotiators. Sometimes a proprietor was granted authority to purchase directly from the Indians the lands already assigned to him. In 1709 George Willox was granted such a license in order to clear of Indian title his lands in Readington Township. Willox gave the Indians trade cloth, trinkets, weapons, and whiskey in return for a clear title. In similar manner all lands of West Jersey were eventually acquired by the whites.

The West Jersey venture constituted a huge and, on the whole, very successful real estate speculation. By 1765 the holders of the one hundred proprietaries had participated in five land dividends, totaling some 2,625,000 acres. Absentee owners, among them the West Jersey Society, in many cases lost money through the actions of purchasers, lessees, and squatters, but individual proprietors on the spot, such as John Reading, probably made a substantial return on their investments. Proprietaries, fractional proprietaries, lands, and rights to lands changed hands rapidly.

Landowners subdivided and re-subdivided their lands for sale or lease to those who could utilize them. In some cases land was acquired by actual tillers of the soil at an early date, but in others such acquisition was long delayed, sometimes until the early nineteenth century.

The Council of Proprietors kept well ahead of the wave of settlement, though not always ahead of the demands of speculators. Surveyors, lawyers, and conveyancers were in great demand by landowners. John Reading and his son of the same name, owners of several proprietaries, were among the surveyors and agents who ranged through lands now in Hunterdon and counties to the north. The "Journal" of John Reading the younger, now printed in the tenth volume of the *Proceedings of the New Jersey Historical Society,* gives a fascinating account of their adventures. Naturally, the Readings used their opportunities to select likely looking plots for themselves and for those whom they represented. Some of the early surveyors' lines remain nearly unchanged to this day. The locally famous Bull's Line, which extended from Sandy Ridge to what is now the Mercer County boundary, remained an unbroken dividing line between the farms on either side until about 1875. Most of the original lines, however, have long since disappeared. The surveyors who dragged their Jacob's ladders and measuring chains through the forest were often careless. A few rods one way or another did not seem very important then. Larger landowners when subdividing and selling their holdings sometimes hired surveyors to check old lines and mark off new ones, but there were many transactions in which acreage was merely paced off by the parties or their agents. As time went on, farmers often extended their boundaries at the expense of absentee owners. After a few decades surveyors had much trouble in finding original survey lines. In 1762 one landowner wrote to his surveyor and agent, "Though they were laid down they were never done with care," and cautioned vigilance in selecting chainbearers, especially stipulating that no relatives of purchasers might serve.[1] A century later the finding of old markers and the accurate retracing of early lines was a most difficult and sometimes an impossible task.

Close behind the surveyors came the speculators who flourished for a century by subdividing and selling land. The proprietors sometimes sold or leased their lands in small plots, but more often disposed of them in large tracts to others, who sold them to small purchasers. Many stories, some of them doubtless apocryphal, have come down to us concerning the land speculators of the county. Thomas Lowrey, a well-

1 William Coxe to John Emley, 1762, Emley Papers.

known land dealer before and after the Revolution, is said to have sold lands under a time arrangement by which the purchaser paid only one third down. Those who were unable to continue their periodic payments found him as merciless in foreclosing as he had been affable at the time of the sale. Lands often went through the hands of several speculators before getting into the possession of actual dirt farmers. Farmers themselves, for that matter, were not averse to land speculation on a small scale. The dividing line between investment and speculation depended largely on the point of view.

The first land dividends within the bounds of present-day Hunterdon were distributed almost as soon as they were clear of Indian title in 1703, and it is possible that a few grants were actually surveyed and even sold in part before that time. The location of all the various tracts as parceled out is a large-scale jigsaw puzzle which has thus far been only partially solved. The approximate location of many of the assignments is known, however. Dividends amounted in each case to several thousand acres per proprietary. A grant to a proprietor, especially if he owned a number of shares, might be in several pieces. John Reading located one of his many tracts on the Delaware River in the Stockton area, and what were thought to be the vestiges of a blazed "R" still showed on a tree near Barber's Station in the early nineteenth century. Original deeds are said to have shown that other holdings in the western part of old Amwell Township were owned by Thomas Stevenson, Gilbert Wheeler, John Callow, Thomas Wilson, and Richard Bull. Some assignments were evidently small, probably resulting from the division of land among the owners of fractional proprietaries. William Biddle, Thomas Olive, and Daniel Wells were together granted some five thousand acres in present Kingwood Township, and Francis Collins, Richard Mew, and John Bull received a like allowance. In each case the three men together owned a proprietary. Benjamin Field was apportioned about three thouand acres in the Ringoes area, and Robert Dimsdale twelve hundred acres just to the west.

At Flemington the Penn, Coxe, and Kirkbride holdings came together. The Penn Line is said to have run through present-day Flemington in an east and west direction at about the place where the Civil War monument now stands. South of this line William Penn held approximately five thousand acres, and north of it Daniel Coxe owned over four thousand acres, known as the Mt. Carmel tract. Dr. Daniel Coxe, physician for Queen Anne, was one of the greatest proprietors of West Jersey. Part of his shares were sold to the West Jersey Society, the rest passed to his son, Colonel Daniel Coxe, at his death. Colonel Daniel Coxe was a prominent figure in West Jersey politics for many years. To the east of Coxe's lands Joseph Kirkbride owned five thousand acres, including the south part of Readington Township; to the west, in the Quakertown area, were the Thomas Gardiner tract of 2225 acres and the Thomas Stevenson holdings. George Willox held a plot just north of Holland's Brook. In the Whitehouse-Oldwick-Pottersville area was the 3275-acre tract of James Logan, and other lands belonging to John Budd. The east part of Tewksbury Township was a proprietorial grant to James Parker, and nearby were the Livingstone lands. John Bowlby, Daniel Coxe, and Robert Johnston owned tracts in Lebanon and Bethlehem Townships, while the Hunterdon County holdings of Thomas Bryerly, something over 7300 acres, were in present Alexandria, Holland, and Union townships. Bryerly sold the entire area to William Allen and Joseph Turner of Philadelphia in 1749. Between this tract and the Musconetcong lay the lands of Sir Robert Barker.

The largest single proprietary holding of the county was that of the West Jersey Society, commonly known as "the Society's Great Tract." This large plot was nearly surrounded by the lands listed in the above paragraph. Stretching from the Delaware to the Lamington River, it comprised about one third of the present county. In the return of the survey of 1711, it was described as follows:

> Situate, and being in the county of Burlington, in the province of West Jersey, beginning at the north end of Coshanton [Cushetunk]; thence north to Lamington Falls, on the North Branch of Raritan [the Lamington], where Peapack path crosses said river; thence along said path; thence west to the South Branch of Raritan River; thence southwest to Delaware river; thence, as the river goes, to where the Sokaway comes into the Delaware, to the top of a hill on the south side of the upper branch of the Reshasakaway; thence to a fork of a run, being a branch of Capooaulong; thence down said run to another fork; thence up the same to a fork of the said run, to a hickory sapling by the old field of Sekapus, to a corner near his plantation; thence east to Raritan River [South Branch]; thence northeast to the end of a run at the foot of the said mountain Coshanton; thence southeast to Mr. Willock's corner.[2]

In 1752 the Society sold its lands to an American company of some half dozen men. A summary by John Emley, agent in charge in 1762, shows how the most of the Society lands were disposed of:

2 *N.J. Arch.*, 1st ser., VII, 325.

Calculation of the Society's Tract by Jonathan Hamton February 7, 1761, and found it to contain Strict measure 92,513 acres

Cut off by the Vew in October 1760	9,163
Allen and Turners Purchase	10,880
Coxe and Kirkbride Do.	11,337
Hoff and Bonnell Do.	500
Rosecrants	598
Surveyed into farms in 1762, 63 & 64	32,000
Alexander Morris White & Dunster	10,500
	74,978
Remains of the Society tract undivided by my Calculation Errors Except'd	17,535

John Emley [3]

The lands which William Allen and Joseph Turner purchased from the West Jersey Society formed an irregular plot extending from Califon to Pattenburg. On the west was the Bryerly tract, which Allen and Turner purchased in 1749. The owners were interested chiefly in minerals, but did not hesitate to turn an honest penny by the sale of lands. The plots which the Society sold to Coxe and Kirkbride lay south of Clinton and adjoined large holdings already owned by these men, as mentioned above. The Hoff-Bonnell tract was at present-day Pittstown, known for several decades as Hoff's Mills. James Alexander was the father of the American Revolutionary general known as Lord Stirling. He and his partners held much of what is now Alexandria Township. The 32,000 acres which Emley mentions as being "surveyed into farms" brought large profits to the new company. Plats among the Emley papers show that there were 150 of these farms. Emley was agent not only for this company but also for Daniel Coxe, Sir Robert Barker, and others.

Proprietors in the southern part of the county, because of the fairly rapid rate of settlement there, had a ready market for their best holdings. Sale of farms of a workable size began early. Nevertheless, it took some time to parcel out the lands, and impatient owners often sold large blocks to other speculators. The Benjamin Field tract at Ringoes was being apportioned out to purchasers prior to 1720. A release in 1758 mentioned twenty-two purchasers, who together held nearly two thirds of the original plot of something over three thousand acres. The unsold portions had changed hands at least twice prior to that date. Original deeds show that the Penn lands south of Flemington were being subdivided as early as the 1730's. Probably because of trouble with squatters,

[3] John Emley, undated memorandum, Emley Papers.

the Penn family eventually sold what remained of the tract to Thomas Lowrey and others, who divided it into farms for sale. Various owners of the Olive tract in Kingwood Township sold individual farms from time to time for a century. Similar sales were evidently made from most of the large tracts of the county over an extended period of time.

A great deal of investigation would have to be done in order to determine to what extent land was held under various forms of tenure in this area. A sampling of old deeds and of deed records makes it appear that most sales gave unrestricted ownership except as limited by nominal quitrents. A quitrent was a periodic payment made by the owner of land to the proprietor from whom it had been originally purchased or to his heirs. Land held subject to quitrents was held under socage tenure, which was the freest tenure recognized under English law.[4] The matter of quitrents in West Jersey was left up to the individual proprietors, who in most cases did not bother to collect the nominal sums agreed upon. In effect socage tenure often gave as complete ownership as did title in fee simple, a form of ownership recognized by law after the American Revolution and the usual form today. There were some proprietors, however, who tried to insure themselves and their heirs a permanent income by charging and collecting quitrents. Still others wished to retain title until lands had increased in value, and preferred, therefore, to lease their holdings rather than to sell them. Settlers without capital to purchase land elsewhere were often willing to become tenants, at least for a time.

One proprietor who sold his lands subject to quitrents was James Logan. Another, in Lebanon Township, used a system of "perpetual lease," which to all intents and purposes gave socage tenure. Ralph Smith of Smithfield (Oldwick) gave contracts by which the purchasers paid quitrents over an extended period of time. The Lutheran church at that place agreed to pay annually the rather exorbitant sum of nine shillings three pence per acre for 103 years. In 1768 the organization purchased relief by a lump payment. In another case Smith sold a farm of 183 acres subject to an annual quitrent for ninety-seven years. The West Jersey Society preferred to sell its holdings in large blocks, but on occasion both sold and leased farms to actual settlers. The successors of the West Jersey Society, however, preferred small purchasers and soon disposed of their lands by sale of farm-size lots. The Allen and Turner domain was eventually subdivided also, though more slowly.

[4] Socage, by feudal law, was the tenure of lands by any fixed or determinate service. In practice this service might be commuted into a money payment or quitrent by which the "socman" was freed or quit of other services.

The last plot of the old Bryerly tract, for instance, was not sold until 1810.

The landowners who attempted to lease land and the proprietors who tried to collect quitrents had many difficulties. The aims of farmer settlers and land speculators were diametrically opposed. A recent writer speaks of the "lawless and ignorant squatters of Hunterdon and Morris Counties, who talked of Indian titles." [5] The indictment seems somewhat severe. Indian titles were often a mere subterfuge, it is true, but squatters might well raise the question of why the proprietors had a better moral right than themselves to the land which they had improved. At any rate absentee owners found themselves unable to prevent land-hungry men from seizing and using idle lands. When such squatters were faced with eviction, they naturally bound themselves together for protection, and by uniting were often able to force compromises. Settling on lands prior to purchase was not in the least unusual. It is probable that owners did not frown on a practice which increased the value of their holdings, and arrangements for sale were not difficult to make later in most cases. On the other hand, those proprietors who held out for high prices or who wished to impose quitrents often found themselves opposed by force.

The field notes of a survey of the Bryerly tract in 1710 mention "the land whereon William Titfoot the Blacksmith is settled, who is no proprietor." [6] In 1735 Lewis Morris, Jr., acting for the West Jersey Society, made a tour of its lands and reported that there were ninety-eight families there, each of them claiming a farm. Their total holdings were 12,535 acres. These people had no legal title, but the Society did not evict them. Instead, Morris persuaded most of them to sign short-term leases. The men who accepted leases were later offered advantageous terms of purchase. Morris's report also mentioned that there were other squatters on the tracts which had been sold to Coxe and Kirkbride. In the same year squatters with blackened faces drove off Coxe's agents and even made threats on the life of the proprietor. The West Jersey Society, which had not done very well financially, became discouraged and tried to dispose of all of its holdings, but did not succeed at this time. The Council of Proprietors attempted to ease the situation by declaring another land dividend in 1737. For a time the large landholders took heart from favorable court decisions. This may explain the large purchases of Society lands by James Alexander and others in 1744. The

[5] D. L. Kemmerer, *Path to Freedom,* p. 187.
[6] Snell, *op. cit.,* p. 415.

lands which they acquired contained many squatters who were later to cause them a great deal of trouble.

Probably it was a more determined stand on the part of the proprietors which precipitated worse troubles. Various groups of squatters, lessees, and those in arrears on their quitrents cooperated, and by pressure on legislators virtually controlled the colonial assembly for several years. Though the worst land riots of the colony occurred in its northern counties, there was also considerable agitation in parts of Hunterdon County. Squatters who had long been on the Allen and Turner tract claimed the wood lots from which the proprietors were cutting fuel for their iron furnaces. In 1749 they precipitated a fight characterized by bickering, broken heads, arrests, and rioting. The climax was the destruction of the Spruce Run furnace by the incensed pioneers. Five years later Allen and Turner were still bringing action of trespass against squatters, though it is doubted that they gained anything thereby. Even when they won in the courts, they were unable to have the decisions enforced. In the meantime the West Jersey Society was having serious troubles. In 1749 the Society claimed that squatters "by open force withheld the Estates from the true and Lawfull Owners, cutting down the Trees, committing all manner of Waste and spoil thereon." [7] Evidently the pioneers were clearing their farms for cultivation. A little later Governor Belcher reported to the English government that disgruntled squatters and lessees were holding meetings and had joined a "Club" which had a large membership in Essex, Bergen, and Somerset counties. Mounted representatives kept up contacts among the various scattered groups. Since the Allen and Turner rioters were called "Club men," it is probable that they too belonged to this land association. In 1750 the proprietors asked the British government to send ships, troops, and artillery to help them, but received no response.

It is little wonder that the West Jersey Society gladly sold its remaining lands in Hunterdon County in 1752. The ability of the new owners to divide the land into farms and to sell them at a profit shows that the squatters were willing enough to make reasonable settlements in order to get complete and unrestricted title. In practice they were allowed to elect which lands they wished to receive, and naturally each man wanted his own farm. Apparently other owners also moderated their terms, for there were no more land riots in Hunterdon County. Sir Robert Barker wrote to his agent, John Emley, in 1764 as follows: "My desire is that you may use your best endeavour, by gentle and per-

[7] *N.J. Arch.*, 1st ser., VII, 316–17.

suasive means, to get the livers on the estate to come under lease for one year, or, at most, three."[8] Landowners were now willing to lease on reasonable terms and to sell where possible. The progress of farmers in becoming absolute owners varied greatly in different parts of the county, though the ideal of unrestricted title was common from the start. By the latter part of the eighteenth century that ideal had been realized for most farmers of the county.

Americans generally have not had that love for the land which is so characteristic of many of the peasants of Europe. The farmers of Hunterdon County have been no different from others in that respect. There have been, it is true, numerous cases in which farms have remained in the hands of the same family for many generations. The Cyrus Van Dolah farm at Sandy Ridge, sold to Paul Whiteman in 1933, had been in continuous Van Dolah ownership for the previous 195 years. The county agricultural agent's office at Flemington has a considerable list of farms which have been in the hands of the same families for a century or more. However, these are exceptional cases. Since earliest days a farm has been to most owners merely a piece of property. If offered a good price or an advantageous trade, the owner has seldom allowed sentiment to prevent acceptance. There always has been in Hunterdon County, except during brief periods of stagnation, considerable transfer of ownership in land. Probably this has been particularly true during boom times when land could be sold at a profit and during depression periods when mortgages caught up with those operating without sufficient capital. The great number of recorded deeds—and, of course, all transfers were not recorded—is some indication of the rapidity with which land has changed hands.

The process by which proprietaries eventually became the property of farmer-owners naturally swelled greatly the number of land transfers in early days. Even after making allowances for the disparity between general price levels of that day and this, land prices of the colonial period do not seem high to us, though they were often looked upon as outrageous by squatters and tenants. Various records mention amounts of from one half to slightly over one pound sterling per acre for outright sales. Large blocks of land sold at wholesale brought much less per acre, of course. Naturally the price of actual farms varied with the fertility of the soil and especially with the extent to which improvements had been made. Moreover, as time went on and the country filled up, the general level of land values advanced somewhat despite the com-

[8] Snell, *op. cit.*, p. 415.

petition of the frontier. In 1786 John Rutherford reported that good lands were worth from two pounds ten shillings to five pounds per acre in paper money, and poorer lands from twenty to thirty shillings. The fluctuations in currency values during this period naturally make comparisons difficult. In 1792 a good farm near Flemington brought over eight pounds per acre, and in a transaction of the following year $20 per acre is mentioned. Cazenove in 1794 found that Warren County farms, partly in cultivation, were worth from three to four pounds per acre. In the period of general exuberance and prosperity of the 1790's it is not surprising that land valuation increased rapidly.

In the period following, the lure of cheap lands on various frontiers was a continuing factor in the situation. At times of greatest migration many farms were thrown on the market, doubtless causing an oversupply of land and a depression of prices. On the whole, however, land values tended to rise. The coming of new immigrants and the production of large families filled up the gaps left by those who went west, so that farms seldom remained without operators for long. The Extons, who purchased Union Farm in 1811, stated in 1826 that they had benefited by the general rise in land values. A lot in Kingwood Township brought about $25 per acre in 1829. The slump of the late thirties and early forties, during and after the Panic of 1837, was marked by many mortgage foreclosures and by falling prices, but was eventually outlived. A Fox Hill farm brought nearly $35 per acre in 1848, and one at Sand Brook almost $40 in 1853.

During the prosperous years of the 1850's land valuations increased in Hunterdon County as well as elsewhere in the United States. A number of farms in various parts of the county sold at from about $40 to over $100 per acre in 1856. When the holdings of Dr. John A. Taylor in Union and Bethlehem Townships were sold in 1858, the improved lands brought from $60 to $80 per acre, though woodlands were worth far less. Prices reported in the *Hunterdon Republican* for twenty-five farms sold during 1859 and 1860 make it appear that most lands were worth from $50 to $90 per acre, while desirable farms were sometimes bringing over $100 per acre. During the Civil War land prices in this area did not keep pace with the rapid rise in general price levels. By the end of 1867, however, they had increased materially, with good farms bringing as much as $135 and even $170 per acre in greenbacks, corresponding approximately with $95 and $120 in gold. In November, 1869, William S. Riley, for over a half century one of Hunterdon's best-known auctioneers, reported to the *Hunterdon County Democrat* that nine farms, totaling 1038 acres, had sold at an average price of $108.50 per acre, or about $78 in gold.

In the early 1870's came a slump in values which was to continue for over thirty years. By the latter part of the decade Riley's lists show many farms selling at less than $50 per acre. During the more prosperous 1880's this movement was temporarily checked, and Riley's average was $58.40 in 1886 and $58.74 in 1887. However, the downward trend was soon resumed at an accelerated pace. In 1890 he reported an average price of $49.50 per acre, and stated that many farms could not be sold for sufficient amounts to cover their mortgage encumbrance. By 1898 his figures had fallen to $31.65 per acre, and good improved farms were selling at $20 per acre at forced sales at the turn of the century. This long depression of land prices cannot be explained alone on the basis of the general fall in price levels. The great expansion of American farm lands in the Middle West, the development of cheap transportation, and the earlier adoption of labor-saving agricultural machinery in other places all played their part. Accompanying characteristics were the decrease in rural population, the scarcity of farm labor, the deterioration of soil fertility, and the almost total cessation of voluntary farm sales.

Soon after the beginning of the twentieth century signs of improvement began to appear. By 1908 farms were bringing up to $40 per acre at auction. In 1909 the *Hunterdon County Democrat* stated that farm values had doubled during the previous ten years. The upward trend was largely the result of a general improvement in the farm situation, though purchases by "back-to-the-landers" and "city parties" made some difference. One aspect of importance was the dividing of ordinary-sized farms into plots of a few acres each. This was done on a large scale by the Kline Realty and Improvement Company, which bought up over a score of farms in the Whitehouse area and subdivided them. The plots were sold for the most part to Polish and other foreign families, at greatly advanced prices, if popular report can be trusted. However, real estate men and farmers agree that land values in Hunterdon County did not increase generally during the World War period because of difficulties being experienced by agriculture.

The prosperity of the 1920's slowed up the back-to-the land movement, and at the same time increased the problems of the farmer. The scarcity of labor, the lure of the cities, and the tendency of those with some education to go into occupations other than farming were among the causes for the great number of sales at this time. So far as can be ascertained, the prices of farm lands did not advance to keep pace with general prices. The depression of the 1930's brought back-to-the-landers again in new force, and modern roads have brought people of means seeking country homes. Because of the secrecy which marks most transactions, the situation in regard to recent farm price trends is not clear. One of the most important factors in any given sale is the type of purchaser. Real estate men say that two levels of prices are maintained, one for farmers and the other for city customers. They do not think that until recently there was much change in prices where lands were bought primarily for farming purposes. For the most part this section has been spared in recent times the economic upsets caused in some parts of the country by the rising and falling of land prices. The rapid rise in prices during the second World War may change that situation, however.

Whether land prices were high or low, there have always been many cultivators of the soil unable to purchase farms. For such farmers renting was one rung in the "tenure ladder," *i. e.,* a step in the direction of eventual ownership. In colonial days renting was a common arrangement. However, ambitious farmers always wished to become landowners, and there was little tendency toward the formation of a permanent tenant class. Rutherford in 1786 stated that only the more thriftless farmers remained renters and that they were unable to pay rentals of more than three or four per cent on the owner's investment at a time when money was worth six. And if Doctor Logan of Philadelphia, who owned land near Flemington in the 1780's, was a fair example, ignorance of farm methods and farm problems on the part of landlords did not help their own position. The fact that there were always cheap lands on the frontier was the chief cause of the scarcity of good tenants.

Though renting as such was never very profitable, speculators could not allow their possessions to lie unused. A farmer with idle capital often bought more land than he could profitably tend in order that he might be able to leave each of his children a farm, or, as Cazenove put it, from mere "pride of being considered a large landowner." In any case he was usually forced to rent out part of his holdings. Other farmers, wishing to retire in their old age, either from sentiment or because land prices were depressed at the moment, chose to rent rather than sell. Still others were themselves changed into renters when mortgage foreclosures took their farms from them. A common type of renter for small plots of land was the artisan who was a part-time farmer. But most renters, at least until recent times, were men with insufficient capital to buy land for themselves. Renting has never been entirely satisfactory to either party, but there always has been a considerable amount of it.

Unfortunately, there are no statistics as to the actual amount of renting until a comparatively recent date. The following table shows the number of renters as given by census enumerations since 1880, and gives for purposes of comparison the total number of farms in the county at each time:

Census Year	*No. of Renters*	*No. of Farms*
1880	878	3155
1890	867	3109
1900	1034	2930
1910	806	2907
1920	752	2795
1925	391	2605
1930	363	2465
1935	346	2509
1940	321	2431

Over one fourth of the farms were occupied by renters in 1880 and in 1890. How many were on their way up the tenure ladder and how many were to remain tenants, it is impossible to guess. At any rate, by 1900 progress up that ladder has been arrested. The number of renters had increased and the number of farms decreased, until one farmer out of three was a renter. With better times after 1900 the number of renters decreased. In 1910 and again in 1920 the ratio of renters to the total number of farms was not much different from what it had been in 1880 and 1890. After 1920 the number of renting farmers apparently declined rapidly. Since an increasing number of plots did not qualify as "farms" under the census definition, a part of the reduction was more apparent than real. Nevertheless, there was an actual reduction in the amount of renting, in part due to the fact that the back-to-the-landers and the immigrant groups preferred ownership to renting.

Not a great deal is known concerning the renting agreements of early days, though a few leases are to be found among recorded deeds. A contract between the West Jersey Society and Robert Breeden, a former squatter, in 1740 throws some light on the situation, and illustrates well the temporary nature of renting arrangements. The tenant was to lay out a tract of one hundred acres which would include his own improved land but not that of his neighbors. He must improve and fence the whole tract and plant an apple orchard. He was not to waste or sell timber, though it is difficult to see how he could clear his farm and at the same time fulfill this requirement. The lease was for five years, and the renter was to pay annually "four Shillings Proclamation Money as a Rent." The Society was to pay all costs of lawsuits in defense of his possession of the land. Nothing was said about what might happen at the end of the period. Insecurity of title and attempts to collect rents and quitrents were the chief causes of the land riots, as we have seen. The renter without doubt considered the farm which he had improved his own, and probably awaited only the chance to acquire title by payment of what he considered a reasonable amount of money.

A renting arrangement for a farm near Ringoes in 1787 provided for a cash rental of thirty-five pounds in specie per year plus all taxes and levies. In this case renting was a step toward ownership, and the renter bought the farm in 1792. An indenture of 1811 between Rebeckah Compton and William Birch illustrates the share tenancy which was probably most common at the time. In this case the tenant of a farm in Kingwood Township was to give the owner one third of all grain, flax, and potatoes, one half of all hay and fruit, and all the firewood which she might need. He was to pay the taxes, but would have no other cash outlays. Similar contracts for renting farms in Readington Township, one in 1814 and the other in 1818, are copied into the docket book of Ezekial Cole, Justice of the Peace. In all cases provisions were included for the care of buildings and fences. Renters were allowed to cut firewood for themselves, but not for sale. Neither hay nor manure was to be hauled from the farm. Some of the contracts were very specific as to what crops were to be grown on particular fields.

The main characteristics of share leasing, drawn from Old World practices, probably evolved very early. According to a Readington tradition Peter Schomp, a farmer of the middle of the eighteenth century, received so much corn from his tenants that he was nicknamed "Corn Pete," and passed both his lands and his nickname on to his son. Cazenove in 1794 reported cases in Warren and Morris counties in which landlord and tenant halved expenses and yield, and others in which the tenant furnished the stock and gave the landlord half of the produce of the land. The features of share tenancy remained largely unchanged throughout the nineteenth century. A rental agreement of 1898 is very similar to those of a century earlier, except for a provision as to the costs of fertilizer. Renting on shares, to use the expression common in advertisements in the county papers for over a century, remained the most usual method of renting in Hunterdon County until about 1920. In 1880 there were 654 share and 224 cash tenants in the county, and the proportion was not greatly different forty years later in 1920. During the next decade the total number of renters decreased greatly, as already noticed, but most of the loss fell to the share tenants. In 1930 there were 198 cash tenants remaining, but only 165 of all other types. The 1940 census listed 220 cash, 57 share, 6 "share-cash," and 38 other tenants, the last-named probably renting under some of the new leasing arrangements which have been recently worked out. The number of farm managers in the county increased from 25 in 1900 to 69 in 1920, but fell to 43 by 1940. The relative and absolute decline in share tenancy in the county and in rural New Jersey generally is apparently one of the concomitants of the increasing commercialization of agriculture.

The same economic factors which prevented early landowners from retaining their large holdings militated at a later time against the building up and retention of large landed estates. During Colonial days, and particularly in the period just before the Revolution, a class of landed gentry existed for a time. Judge Samuel Johnson of Sidney and his son-in-law, Charles Stewart of Landsdown, were members of this local aristocracy. The names of their plantations are preserved to this day by the two villages Sidney and Landsdown. Another would-be aristocrat was Thomas Lowrey, land speculator and merchant, who lived in turn at Flemington and Milford before moving on to Philadelphia. To this list should be added Moore Furman, the man who developed Pittstown, and perhaps some members of the Reading family. Undoubtedly there were others of higher or lower degree. Large mansions, fine clothes, fancy coaches, and numerous servants set this group off from the common people, but economic forces were to prevent that continuance of ownership which is needed to create a class of true aristocrats.

Great landowners were sooner or later handicapped by their inability to find dependable laborers and tenants to till their lands. Large estates of early days, therefore, had much undeveloped land which was hard to protect from encroachment and was of no great potential use to the owners. Land-poor men often yielded to the constant pressure put on them to sell part of their possessions. Division of estates among heirs, except as interfered with by primogeniture and entail in Colonial days, also cut down the size of holdings. From the days of the proprietors to that of the attempt of the Pidcocks to build up a large estate in Readington Township in the latter half of the nineteenth century, large landowning seemed doomed to failure. Energetic men like the Frenchman Paul H. M. Prevost and the Englishman Hugh Exton worked large estates, but in the end their holdings were divided through sale or inheritance. In the case of the Pidcocks other sources of income bolstered up the effort, and for a long time they were able through renting and the hiring of managers to keep in full operation over a score of dairy and orchard farms. The Pidcock lands were ultimately acquired by the Kline Realty and Improvement Company and subdivided into small plots, as previously mentioned. Several large holdings have been built up within the county in recent years, and it will be interesting to see whether, by using modern farming methods and utilizing outside capital, they can long be maintained as units.

The average farm of early days contained considerable timberland and was probably somewhat larger than the unit of a later period. A tax list of 1722 for Hopewell Township, now in Mercer County, probably presents a picture not greatly different from that across the line in Amwell Township. Of the 102 farms, 46 were under one hundred acres, 38 between one hundred and two hundred acres, 12 between two hundred and three hundred and 6 over three hundred acres.[9] The ninety-eight squatter families reported on the West Jersey Society lands in 1735 held farms averaging about 128 acres each. In 1758 purchasers in the Field Tract held on the average 103 acres each, if plots of less than twenty acres are left out of the computation. A sampling of the probate records of the eighteenth century and of newspaper advertisements of the nineteenth shows a preponderance of small farms and only occasional ownership of large tracts. Of twenty-seven farms offered for sale in the *Hunterdon Gazette* between September 15 and December 6, 1826, only four were over two hundred acres, and several of the larger farms were to be subdivided for sale. As pointed out previously, an occasional farmer owned several farms, but seldom did he leave them all to a single heir when he died.

The local schedules for the 1850 census, a most fertile source of information on many subjects, list every farm of the county and record ownership, acreage, production, livestock, and other details for each. There were by that time a surprising number of farms under fifty acres, and the majority were under one hundred acres. Farms of over two hundred acres were exceptional, and usually contained considerable timber or other uncultivated land. The family-size holding was the general rule. Beginning with the federal census of 1860, comparable statistics are available for farm acreage, and the following table gives a summary of the number of farms of various sizes as of dates given:

[9] Snell, *op. cit.*, pp. 185–86.

ACREAGE OF FARMS

Census Year	*Under 3*	*3–9*	*10–19*	*20–49*	*50–99*		*100–499*		*500–999*	*Over 1000*
1860	—	60	151	454	1213		772		1	0
1870	—	292	248	494	1246		868		0	0
1880	2	191	205	508	1092		1153		2	0
1890	208		222	508	1066		1105		0	0
						100–174	*175–259*	*260–499*		
1900	11	126	165	464	1055	971	120	16	0	2
1910	9	189	195	434	990	925	140	23	2	0
1920	16	178	206	410	977	856	120	27	5	0
1925	20	220	218	447	848	726	100	25	1	0
1930	30	178	186	425	804	710	98	29	3	2
1935	13	217	205	399	827	700	113	28	5	2
1940	31	258	220	406	692	648	120	46	9	1

The figures do not include all waste land. The total number of farms after 1880 has been given in a previous table. The amount of farm land and the number of farms have both declined markedly in recent decades. It will be noticed, however, that it is the farms between fifty and 174 acres that have had the greatest decrease. The small farms, apparently becoming smaller, have about held their own numerically, while the number of large farms has increased noticeably. These developments correlate with the changing agricultural conditions of the region which are to receive attention in the following chapters. Statistics for the state show parallel trends in other parts of rural New Jersey.

CHAPTER V

Forests and Arable Land

A DENSE FOREST covered the land which the pioneer wished to utilize for crops and pasturage. His first task, therefore, after providing a shelter for his family, was to clear away the growth. Although the trees furnished him with material for buildings and fences and even for tools and utensils, he looked upon the forest growth as an encumbrance, a nuisance to be dispensed with as rapidly as possible. Johann Schoepf in 1783 told of the young American who, when he saw treeless western Ireland, exclaimed, "What a wonderful country! What a lucky people, with no woods to plague them!" His attitude was that of most Americans, and one which led to the reckless destruction of our virgin forests. Today we are inclined to judge harshly this action, but a word of extenuation must be said for the pioneer. His work with ax and fire, though very wasteful, was dictated by absolute necessity, for in no other way could he insure a livelihood for himself and family. From a long range point of view he did less real harm than his descendants, who slashed through the forest growth of nonfarming and marginal lands in a later period. The process of changing timberland into crop land was not an easy task, nor one soon completed. Undergrowth and small trees were cut off and burned. The same might be done with larger trees, but often they were merely girdled and left to die. The chief trouble with the latter method was that dead limbs were continually falling on the crop planted beneath. The stumps and the trunks of girdled trees were often left to deteriorate, but sometimes the process was speeded up by the use of fire. Newcomers from abroad were appalled at the appearance of the fields, with their protruding stumps and fire-scarred skeletons.

The tools with which the pioneers felled the forests were very simple. Among the most common items in the inventories of estates during the first century of the county's history were the felling ax, used for cutting down and trimming trees, and the broad ax, used for squaring timbers. Occasionally a man owned a saw, and nearly always wedges and a beetle or beetle rings. The beetle was a large, homemade wooden hammer or maul, the driving surface of which was protected on either side by an iron collar or ring made by a local smith. Most farmers also had mattocks, usually spelled "mathooks" in the records, for clearing out roots and stones. With these few tools the pioneers laid waste the forests and in due time left in their place open fields. By constant practice they became very proficient with these implements. Thomas Capner, himself skilled in the use of tools, wrote from Flemington on November 17, 1787:

I have sent Samuel two little wooden axes. They are all the kind that are used here. With the least they fell trees, cut wood, etc. If they want to square pieces of timber or trees, they cut them down or they line out the thickness they want. They then stand the tree out, notch two [*sic*] the line about a foot apart and split it off in small blocks. They then with the crooked ax stand on one side of it and hew it so true that you would think it was plained [*sic*]. They are likewise very dextrous with the small ax. Brother John says an American will cut down a tree whilst an Englishman is looking where to begin.

It was natural that the individual farmer should clear first the portions of his land which were best suited for crops, leaving the less suitable parts till a later time. Hilly, stony, or infertile lands were left for wood lots or used as undeveloped pasture lands. A will of 1773 called such land "backwood land." [1] As long as there was plenty of wood everywhere, woodlands were not considered as of much value. But as the forests dwindled, it became increasingly important to each farmer that he have a wood lot to serve as a source of rails, building timber, and fuel. Advertisements of farms for sale in the 1790's and thereafter emphasized the existence of timberland or wood lots if there were any. The accounts of Dr. John Bowne of present-day Bowne Station show that he hired men each winter to split rails and cut cordwood. In 1810 he sold to the local blacksmith 12¾ cords of chestnut wood at $1.50 per cord. A farmer with more rails than he needed could usually find a ready market for them. A by-product of the cutting of fuel and fencing material was bark of various kinds sold to tanning establishments, such as the Case tannery near Flemington. Some wood lot owners sold white oak for staves and headings and hickory timber for ax handles. Shellbark hickory nuts, walnuts, butternuts, and chestnuts were in constant demand, but the owners of wood lots seldom profited from the crop as much as the men and boys who tramped the woods when the nuts were ripe. In 1856 shellbarks were worth 75¢ per bushel; in 1880 they were worth $2.00; and by 1890 they were still higher in price. The rise re-

[1] *N.J. Arch.*, 1st ser., XXXIV, 571.

flected the destruction of the hickory trees of the area by tool makers and spoke factories.

As specific types of timber became more valuable, men with wood lots stressed in advertisements that their tracts contained valuable timber. Oak and chestnut, considered as best for rails, building materials, and firewood were common. An owner in 1828 stated in the *Hunterdon Gazette* that a stand of timber which he wished to sell would yield from two to three thousand rails per acre, and another in 1831 averred that his trees were large enough to make six or eight rails to the cut. Hickory timber, important in making tools and vehicles, and walnut, used by furniture makers, were always considered worthy of special mention. The largest wood lots were located on ridges of the Highlands, in the Barrens, in the Great Swamp, and in other areas not considered good farming land. By the late 1830's there was a marked tendency to subdivide these lands into plots of from one to twenty acres and to sell them separately. Sometimes only the timber was sold, sometimes both land and timber.

If a man did not have a wood lot on his farm, it was important that he have one as near to his home as possible. For example, according to an 1831 advertisement in the *Hunterdon Gazette,* a farmer of Alexandria Township had 22.19 acres of woodland only one mile from his homestead farm. Others, however, were not so fortunate. George O. Vanderbilt, speaking of the Barrens in the 1850's, in 1912 recalled that "many farmers living in the lower part of said Alexandria township and in Kingwood and Franklin townships owned chestnut lots in this 'barren' section of the county from which they cut and hauled yearly their posts and rails for fencing materials." [2] According to a tradition of the Frenchtown region, the road to the Barrens had been built by the cooperative efforts of these farmers. Its popular name, "Brandy Road," originated from the large amounts of applejack consumed during its construction. The wood lots when cleared were converted into farm land, thus decreasing the wood supply. In less arable areas wood lots which were cleared were allowed to grow up again, and during the period of growth were designated as "sprout lots." In due time the sprouts became trees, ready for another harvest.

In the 1840's Blackwell each winter recorded in his diary the hauling of white oak, black oak, walnut, and other logs to the local sawmill. When the ice on the South Branch was hard enough to bear the weight of crashing trees, he felled even those at the water's edge. At about this

[2] *Hunt. Co. Dem.,* Jan. 23, 1912.

time portable steam sawmills were introduced. This fact, plus an increasing demand for wood lot products, led to the total destruction of forest after forest. Large oaks were cut for timbers and lumber, while smaller ones were felled and hewn into railroad ties. Spoke factories and the makers of tool handles offered such high prices for good hickory that large trees of that species soon became rare. In 1868 an advertisement in the *Hunterdon County Democrat* of a wood lot for sale near Ringoes stated that the buyer by cutting the timber could net $250 per acre and have the land left.

It is not surprising that in Hunterdon County, as throughout the United States generally, poor forestry practices were already denuding regions which should have remained in woodland. Edgar T. Bush, in his reminiscences, describes one such tract in the Great Swamp. Neighbors said that it had once been a fine oak woods, but he remembered it in the 1870's as typical "cut over land," covered with big stumps, straggling brush, and thrifty blackberry briers. Its only usefulness was to serve during part of the year as an indifferent pasture for cattle. With intelligent care this land could have been kept profitably in permanent timber. Some of the tracts which were cleared proved to be good agricultural land. For example, what had formerly been a large woods in Readington Township was in 1870 producing bumper crops of corn and wheat. But many areas cleared after the Civil War proved of little use as farm land and might well have remained in permanent forest from which the mature trees could be periodically harvested.

A fate similar to that of the woodland of which Bush wrote awaited practically all the patches of forest which had been spared and preserved. The "long woods between Pleasant Run and White House"; the "leafy stretch" from Barley Sheaf to Centerville; the pleasant woodland between Flemington and the Fairgrounds where camp meetings were held; the Holcombe Grove near Lambertville, long used for large scale outdoor affairs; the grove near Mt. Airy which was used in similar fashion; the Voorhees Grove between Reaville and Three Bridges, well-known to several generations of Sunday School children; and many other woodlands long in use for community affairs fell one after another.

In the 1870's wood-using industries in Lambertville were cutting heavily into the nearby forests in West Amwell Township, and the steam sawmill of the Rev. C. S. Conkling at Stockton was destroying the big white oaks and other trees for miles around. One huge log, fifty feet long, was hauled twelve miles to this mill. Elmer Bonnell in his diary during the 1880's spoke of "Dilley's" and other sawmills near Clinton, and several times mentioned "new ground" planted to crops. During the next decades the remaining forests of the county were slashed down without mercy. Small and inaccessible tracts were spared longest, but eventually they, too, were razed. The Lamington Swamp near Whitehouse was cleared in 1898. Small patches of woods near Quakertown were cleared one after another in 1906. A wood lot in the Baptisttown area was not cut until 1913. By this time timber was getting scarce, even in regions formerly heavily forested. The stave factory at Bloomsbury, which employed twenty men, was forced to move from Hunterdon County in 1914 because of shortage of oak and chestnut. The few forest tracts which had survived that long succumbed for the most part during the first World War. The most notable example was that of the Carhart Woods, a tract of 133 acres west of Clinton, said to be the last large stand of virgin white oak in the state. It was cut in 1918 to make stems, keel shoes, and rudders for the wooden ships then being built by the federal government. The deforestation of Hunterdon County illustrates on a small scale what was happening throughout Northeastern United States and to a greater or lesser extent throughout the nation. The bad effects of our wastefulness are already apparent.

As if man's efforts were not enough, nature lent a hand in the destruction. Soon after the turn of the century the hickory bark beetle became a serious pest. The remaining hickory trees of Hunterdon County suffered from its inroads, but not to the point of extinction. Far more serious was the infection of the chestnut trees of the region with the blight, a fungous growth which appeared in widely separated areas from Virginia to New England about 1906. Its progress was not very rapid at first, and losses in Hunterdon County were slight for several years. Some parts of the county reported an unusually large chestnut crop as late as 1913. An erroneous story which went the rounds was that nuts from afflicted trees were very poisonous. However, by 1915 most of the trees of the county were dead or dying. A newspaper correspondent who loved flowery phrases wrote in June of that year:

From Pittstown up through "the Barrens" . . . the sad work of the chestnut blight is painfully evident. Looking across the valleys to the near and distant hills, which used to be beautiful in their fringed and beflowered robes of forest-green, we see only dead, unsightly poles pointing cheerlessly toward heaven. . . . A careful observation of a tract of many acres revealed only two live specimens. Since there is practically no other timber on much of this area, what was once a beautiful country . . . looks sadly bedraggled now.[3]

Old chestnut stumps and snags still hopefully send up sprouts, and these young trees sometimes bear chestnuts; but the blight invariably kills them sooner or later. The loss caused by the chestnut blight was a serious one, especially since it killed off the young trees as well as the older ones.

There still are left a few small tracts of timberland which serve as a reminder of the luxuriant growth that once covered the whole surface of the county. And many cutover lands and abandoned fields are regaining a natural growth which in time may become forests of some value. Even more important for the future, perhaps, is the increasing interest in reforestation of rough and infertile areas. Partly because of governmental encouragement, there have been some planting of small tracts and the application of good forestry practices to others. The coming generation has been made more conscious of the need of such practices through the work of 4-H Forestry Clubs. Through the efforts of the successive county 4-H club agents and of rural teachers who have acted as leaders, several thousand children during the past two decades have learned much concerning the planting and care of forests, along with other practical lessons in nature study. It will, of course, take a long time for the county to recover from the results of the wasteful practices of former days. The happy solution would be to have all unprofitable or badly eroding farmlands put into permanent forests. That may happen in the future if owners of such lands can be shown that reforestation is a paying proposition.

The virgin soils of the new land were used nearly as improvidently as the forests. The early American viewpoint on the maintenance of land fertility was one of indifference based on ignorance. New lands were cropped as if their riches were inexhaustible, and few men made any effort to replace the elements taken from the soil. Peter Kalm severely criticized land care in New Jersey as early as 1748. According to his description, a farmer first cleared a field, then proceeded to crop it to death. He kept it in grain until it was "worn out," then made it into a pasture. Manure, instead of being put on the land, was generally wasted. The only effort made to replace soil fertility was to give the land a rest. However, since no attempt at weed control on fallow and pasture land was made, the first crop after cultivation was resumed was not very productive. A more serious matter for the future was the fact that little

[3] *Dem.-Adv.*, June 17, 1915.

effort was made to prevent erosion. The white man had adopted, on a much larger scale, the wasteful land practices of his predecessor, the Indian.

As early as the middle of the eighteenth century there were a few Americans who through correspondence and reading kept themselves informed about the experiments of gentlemen farmers in England. Some of them adopted the new farming practices and carried out investigations of their own. Their work was of great importance for the future, but only slowly affected the procedures of the average farmer. In 1786 John Rutherford, himself a progressive farmer, gave a very sad picture of New Jersey agriculture in general. Lands were overcropped, little effort was made to restore depleted fertility, and careless practices led to washing and gullying.

English immigrants like the Capners and Extons brought with them better soil and crop practices, but their example was only slowly followed. Letters of these newcomers show that Rutherford's general description fitted Hunterdon County all too well. When Mrs. Mary Capner in 1788 asked a farmer why Americans "spoiled the land," she was told that the land simply "wore out" and that there was no remedy for it. Her informant also made it plain that this was the land of liberty and that a man might ruin his land if he so chose. Her son Thomas was at first surprised at the large amount of "bushy land" which had been allowed to grow up to sprouts and briers. John Hall complained in 1792 that Americans permitted their top soil to go "down the river," and that, though they overcropped their land, they often did not bother to haul the manure from their barns onto the fields. At another time he wrote, "The Land as I observe of this country to answer to improvements in more Burthen of Timothy Grass and Clover than [E]nglish Land of the same Quallity. But such a long series of Beggarly plowing and cropping has reduced below bearing wheat all about us in general and no dung is produced so how is it to be brought about?" [4] Cazenove in 1794 had a similar story to relate in regard to agriculture in Morris and Warren Counties, but reported the use of some manure on wheat lands.

It is possible that observers exaggerated in stating that so little use was made of what were obviously natural fertilizers. In earliest days manure may have been pitched into the streams or otherwise wasted, but as land became less fertile at least some farmers made use of it. At the very time of John Hall's grumbling, which was in the main justified, chicken manure was a not uncommon article of commerce in Hunterdon County,

[4] John Hall to Hugh Exton, June 17, 1795, Capner Papers.

and "drawing dung" was a regular farm job. John Hall's relatives set the native Americans a good example in the careful use of manure. As time went on, farmers were increasingly faced with the problem of soil depletion. At the same time the growth of an agricultural press in this country led to a wider knowledge of proven agricultural practices, among them the use of manure. Farmers throughout the East turned to the utilization of this farm by-product, especially in connection with the growing of corn. Blackwell, at the middle of the nineteenth century, and Bonnell, thirty years later, saw to it that all manure produced on their farms went onto the land; but both followed the old custom of hauling manure only twice a year, spring and fall. Through their county and agricultural papers both probably knew that exposure to weather caused considerable loss of fertilizing elements, but like others they were slow to adopt practices which would upset their regular routine. In recent decades the constant reminders by agricultural writers, and more recently by county agricultural agents, have brought results. Most farmers now haul their manure onto the land daily or every few days. In modern times the use of motor vehicles and motorized farm machinery has cut down materially on the amount of horse manure, but specialization in poultry raising and dairying has increased other kinds markedly.

Another home-produced material utilized to some extent as a fertilizer was wood ashes. In pioneer days ashes not used for making lye were probably wasted, but it seems likely that they were used to some extent on crop land at a fairly early date. At Flemington in 1798 they were a regular item of commerce at a shilling per bushel. Most buyers doubtless wanted them for making lye or soap, but they were probably also used on garden patches. The *Pennsylvania Farmer* in 1804 recommended wood ashes for use on clay soils. In his diary during the 1850's Blackwell tells of dropping ashes on hills of corn at planting time, a practice common in other parts of the East in this period. A dealer advertising in the *Hunterdon Gazette* in 1856 suggested that ashes were a good fertilizer to use as a top dressing on fallow ground. When locally produced ashes became scarce, they were imported from regions which were still destroying their forests. In 1882 Bonnell hauled from Clinton at corn-planting time ashes brought in by rail. The Flemington agent of a New York firm advertised "Canada unleached ashes" in the *Hunterdon County Democrat* as late as 1888.

There were on the farms other waste products which careful farmers spread on their land. Leases as far apart in time as 1787, 1814, and 1898 specify that the tenant must take no hay, straw, or other enriching agent from the farm rented. Blackwell carefully scattered on his fields even the chaff from his threshing floor and yards. Each year he made large compost heaps, using weeds, straw, swamp muck, and other waste vegetable products. The material was turned over occasionally and allowed to deteriorate. The resulting compost was used in the family garden and spread as a top dressing on wheat ground. In 1845 he tried an ingenious way of manufacturing fertilizer. His entry for July 17 and 18 reads, "Built and fired two stacks of Bogs." Four weeks later he wrote, "Overhauled the ashes and mixed with swamp mud. The heaps when opened were still on fire." His record for September 9, 10, and 11 was as follows: "Hawled 31 loads of the above on about 4 acres of wheat ground." His laborious practices would have been ridiculed by most farmers a half century earlier, but were observed with interest and even copied by his neighbors. Many of his ideas apparently came from the agricultural papers which he read. The making of compost was being urged strongly by all of them during this period.

The soils of the Piedmont are not naturally very rich in calcium, and even the Highland soils were depleted of this necessary element by continual overcropping. A sure sign that land was becoming sour was the appearance of the meadow weed known as horse sorrel. The solution of the problem was the application of lime to correct the soil reaction, as was well-known by those who kept up with agricultural developments in England. But farmers here were somewhat slow in turning to this method. John Rutherford wrote in 1786, "Limestone is good and plenty, and if not so dear calcining might be used to great advantage as a Manure." According to John Hall the average farmer in 1792 did not use lime except for small quantities at wheat-sowing time. High labor and interest costs were deterring factors, but habit and inertia also played a part. When Isaac Passand offered to buy lime for his farm in 1793, he found no answering enthusiasm on the part of his tenant, who apparently did not relish the job of spreading it on the land. Jedidiah Morse in his *American Universal Geography* in 1796 mentioned as "almost an insurmountable obstacle to agriculture" the way in which New Jersey farmers clung to old methods. Ignorance of the principles of soil chemistry naturally made many farmers reluctant to spend work and money on what seemed to them a mere fad. It was this same lack of knowledge which later caused lime to be regarded as a cure-all for every soil ailment.

Important sources of lime for Hunterdon County farmers from the earliest days of its use were the large deposits of limestone just across the river in Bucks County, Pennsylvania. The calcining of lime and its use on the land were common there much earlier than in New Jersey. The firing of limekilns consumed enormous quantities of wood, and the early ones in Bucks County were often run on a cooperative basis. Some of the limestone was probably brought to Hunterdon County for burning, as both the Kugler and Lequear families had traditions of burning Bucks County lime in kilns on this side of the river at an early period. In the 1830's the Delaware Division Canal brought anthracite coal to the Bucks County limestone section, and as a result the calcining business expanded greatly. Toll bridges across the Delaware at various places provided passage for lime wagons going to and returning from the kilns. The owners of the Centre Bridge at present-day Stockton granted them special tolls, and the owners of other bridges may have done likewise. To lessen the monotony, farmers often banded together when hauling lime, and sometimes "lime frolics" were organized on a neighborhood basis. Strings of lime wagons were a common sight in Hunterdon County for a long time.

The Pennsylvania lime kilns attempted to build up their New Jersey trade by constant advertising. In 1838 the price at the kilns, according to advertisements in the *Hunterdon Gazette,* was 10¢ per bushel for cash or 11¢ on credit. To facilitate trade the kiln owners accepted farm produce and cordwood in payment for lime. In 1843 one of them offered it at the ratio of two and one-half bushels for one of oats, four for one of corn, nine for one of wheat, and five for one of buckwheat. He was also willing to accept hay or pickled pork. In one issue of the *Hunterdon Gazette* five separate concerns in Solesbury Township of Bucks County advertised lime for sale. Some of them told their Hunterdon County customers to come by way of the Centre Bridge, others by way of Lambertville. Various Pennsylvania lime dealers had agents and storage facilities on the New Jersey side of the river. In 1840 a *Hunterdon Gazette* correspondent who called himself "An Eye Witness" commented on the great volume of the lime trade at Lambertville. Canal boats, loaded at the kilns, were brought down the Delaware Division Canal, pulled across the Delaware River by cable at a point just below New Hope and Lambertville, then shunted into the feeder canal of the Delaware and Raritan. Naturally, much of this lime was destined for areas beyond Hunterdon County. When the railroads were built, the problem of transportation was solved for farmers who lived far away from limestone deposits, though hauling by wagon was by no means stopped. The Pennsylvania kilns ceased to advertise in the Flemington papers, but much of the lime sold by the dealers of the county continued to come from across the Delaware. It was delivered by rail to such points as Ringoes, Copper Hill, and Flemington. In 1860 the *Hunterdon Republican* advertised calcined lime by the carload at 12½¢ per bushel delivered in Flemington.

The late Edgar T. Bush in his reminiscences wrote a very interesting description of lime hauling as practiced in the neighborhood of Croton during his boyhood. This was published in the *Hunterdon County Democrat* on April 21, 1932. Young Edgar made several trips with his father to the limekilns at Uhlerstown, a mile west of Frenchtown, then was allowed one day to make the trip alone. Determined that he would be among the first to arrive, he started from home at two o'clock in the morning. When he neared Frenchtown, where he was to cross the river by the covered bridge, he was disappointed to see dozens of wagons there ahead of him. Farmers of two different neighborhoods had organized lime frolics and had started on their journey at an early hour. At Uhlerstown he found it necessary to wait in line for several hours, as only one kiln was open. When his turn came at last, he put on his modest load of twenty-five bushels and started on the return trip. The fatigued boy and jaded horses did not reach home until evening. It took two such trips as this to transport sufficient lime to cover one acre of ground. Neighbors of the Bush family hauled limestone direct from the quarries and had it calcined at a kiln then operating near Croton.

The plentiful limestone deposits of the Highlands in New Jersey were not utilized to any great extent until after those of Pennsylvania had been in use for some time. When the Allen family advertised the Union Farm near High Bridge for sale in the *Federalist and New Jersey Gazette* in 1799, they emphasized the supply of limestone, "deemed valuable for manure," but said nothing about kilns for its utilization. However, the extensive deposits of that region were gradually opened up as farmers learned of the benefits of lime. In 1831 the owner of a farm west of Quakertown advertised in a broadside that most of his 204-acre farm had been limed within recent years, and that there was on the farm a limekiln in good condition. His lime supply may well have come from the pits near Clinton. The census enumerator for Clinton Township in 1850 made the following notation on his schedule: "There is a great quantity of Lime Burnt in the Township. . . . Lime is used abundantly . . . which makes the township very productive." George H. Cook in his *Geology of New Jersey* in 1868 reported that there were many quarries and large kilns in the Clinton area and in the Musconetcong Valley. By this time some lime from the large deposits in Warren County was being brought in by rail.

Farmers had accepted lime somewhat hesitatingly, but during the heyday of calcining they used it in enormous quantities. The accounts kept by Dr. John Bowne well illustrate the change. For some years immediately after 1800 he bought comparatively small amounts each year, probably to spread on land sown to wheat. By the 1820's his yearly purchases totaled many hundred bushels, though he was not a large-scale farmer. In 1839 the *Hunterdon Gazette* spoke of the universal use of lime by Hunterdon County farmers. Many of them apparently considered lime a specific for any kind of soil exhaustion. Jesse Buel in his "Notes on New Jersey Farming" in the *Cultivator* in 1839 observed that New Jersey farmers still overcropped their land and did not get maximum benefits from manure produced on their farms. As for the use of lime he had this to say, "Lime is applied at the rate of 50 to 100 bushels per acre, the poorer land receiving the smaller, and the richer the greater dressing. It is generally mixed with arable lands by the harrow. Its benefits are palpable; and the increase of the first crop often pays the outlay." [5]

Lime burning was an important industry for nearly a century. So far as can be ascertained, there were no great changes in its use during that period. Informed persons even in early days knew from English practice that lime should be slaked with water or by long exposure to the air before being applied. Otherwise the caustic effects were such that elements of the soil were torn down and the land in the end made poorer than before. The father of Edgar T. Bush always carefully slaked his lime with water as soon as it was unloaded and before piling it in the field to await the day of spreading by hand. Elmer Bonnell in the 1880's followed exactly the same procedure. But many farmers, observing that large crops were usually obtained in the year following an application, applied lime which had been only partially slaked, perhaps merely by a brief exposure to the air. Because the deleterious effects of large applications were so marked, the practice of using smaller amounts at shorter intervals became common. A great deal of land in the county was probably seriously harmed by the use of poorly slaked lime.

During the first half of the nineteenth century many farmers calcined their own lime. As time went on, however, a tendency toward concentration of the industry was observable, and large kilns near the quarries served considerable areas of the county. In other places individual farm-

[5] Quoted in *Hunt. Gaz.*, Sept. 17, 1839.

Fig. IV. Lard lamps, candle molds, and homemade lanterns indicate methods of lighting common as late as the Civil War.

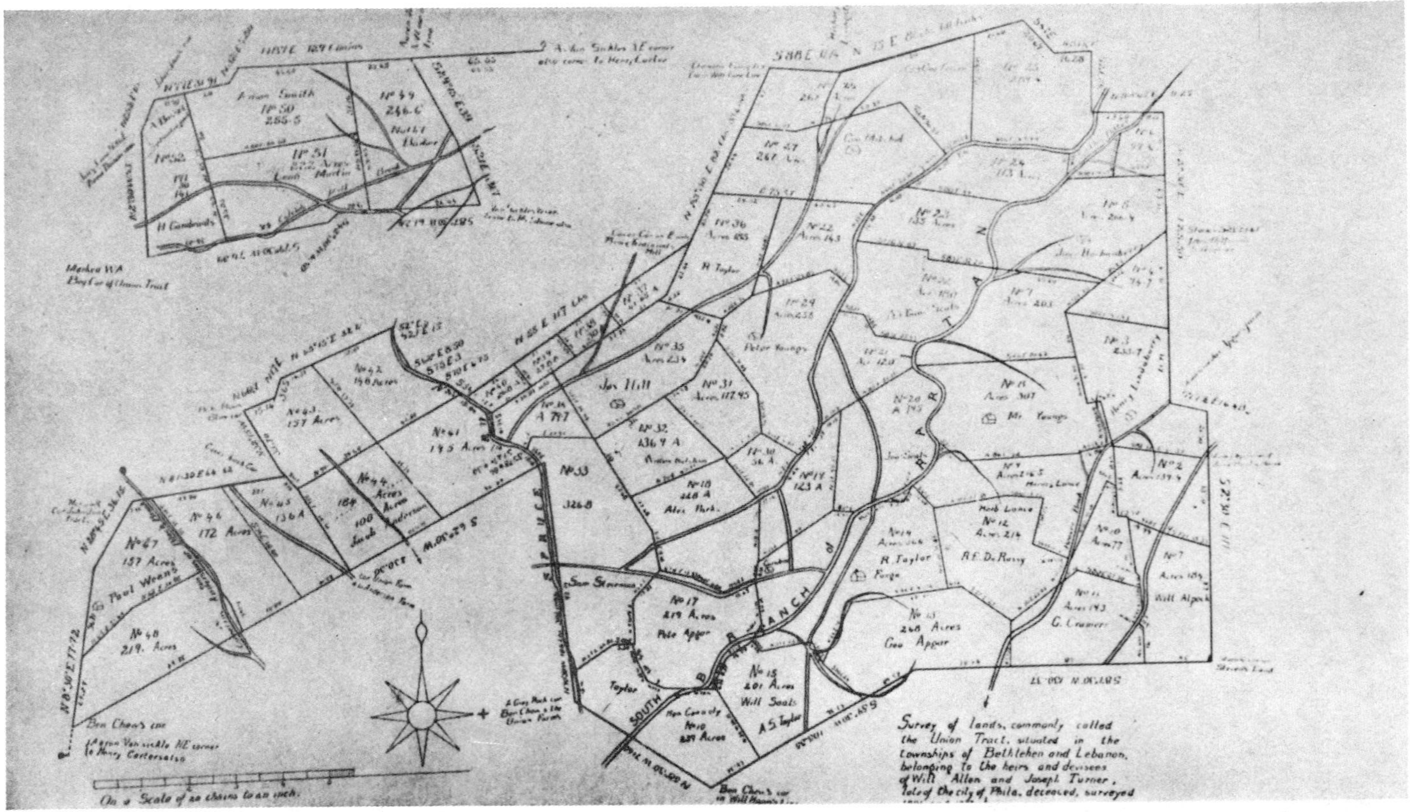

FIG. V. A resurvey by David Frazer in 1801 showing the division of an estate into farms.

ers took over the calcining of lime for their neighbors on a custom basis. A Holland Township farmer in 1877 advertised as follows: "Four good Lime Kilns stand on the line of the two farms above described, where the manufacture and sale of lime for the surrounding country has been an established business for the last thirty years." [6] In the latter part of the century the use of calcined lime declined, and only a few of the local kilns survived into the twentieth century. Better knowledge concerning the part played by lime in the chemistry of soils, learned through tests made at the New Jersey and other experiment stations, led to an increasing use of pulverized limestone in place of calcined lime. Machinery for its production soon replaced limekilns to a considerable extent in the preparation of agricultural lime. Since the limestone deposits of Hunterdon County are comparatively small, local quarries have nearly all fallen into disuse. Today farmers may buy either pulverized limestone or hydrated (water slaked) lime in neat paper bags, on the outside of which is printed an analysis of the contents. The application of limestone or lime is now an almost universal practice. In recent years the use of pulverized limestone has been stimulated by the program of the Agricultural Adjustment Administration.

Another fertilizer used at a fairly early date was pulverized gypsum or sulphate of lime, at first known as "plaister of Paris," later by the shorter name of "land plaster." The use of gypsum as a fertilizer was first practiced in this country about 1770, and it is thought that the idea came from Germany. Gypsum was imported from Nova Scotia or brought from quarries near the Hudson River and was pulverized in plaster "breakers" at local mills. The first positive record which we have of the use of land plaster in Hunterdon County was in 1793, when Isaac Passand purchased some for his renter to use when sowing clover. In 1798 the Cases and other farmers near Flemington hauled plaster from New Brunswick by wagon. A farmers' handbook, The *Pennsylvania Farmer,* spoke thus of land plaster in 1804: "Plaster of Paris is at present in very general use, the beneficial effects of which have contributed greatly to the improvement of agriculture." The author especially recommended plaster for use when sowing clover, and suggested from one to four bushels per acre. To support his thesis, he quoted many testimonials which vouched for the benefits obtained from its use. Hunterdon County farmers had evidently accepted land plaster by this time. Dr. John Bowne in 1802 paid 10*s.* for two bushels. By 1810 he had apparently satisfied himself as to its value, for in that year he bought

[6] *Hunt. Co. Dem.,* Nov. 6, 1877.

22½ bushels. The account books of the Cases show that they continued their use of plaster, probably because they grew so much clover. A typical notation by one of the family, written on the flyleaf of a ledger reads, "Day 13 of June plastered my Stubel field 1809."

The day book of a store at New Germantown (Oldwick) in 1816 shows that it sold its customers two grades of plaster, one at 4*s*. per bushel, the other at 6*s*. 6*d*. In 1824 the Holcombe store at Mt. Airy was charging only 50¢ per bushel. The cost of transportation seems to have been the chief factor in setting the price, as the cost of grinding was small. Bergun Berkaw, a Readington Township miller, crushed gypsum at his plaster breaker for $1.50 per ton in 1825. As the roads were improved and the number of local breakers increased, competition brought down the price of gypsum. Blackwell paid only 35¢ per hundred pounds in 1848. He used it for a top dressing for wheat ground and also put it in hills of corn at planting time. Bonnell put plaster on his timothy and clover fields in the spring, though a more usual practice was to do it in September. The advertisements of millers and merchants who had plaster for sale usually appeared in the *Hunterdon Gazette* during the seasons for sowing wheat and for planting corn. Plaster was also sometimes used for other crops. The winner of a prize in a potato yield contest in 1875 stated in the *Hunterdon County Democrat* that he had used land plaster freely around his potato plants. During the 1870's and 1880's importers of Nova Scotia plaster played up their product as being superior to the American variety. The use of concentrated commercial fertilizers caused land plaster to fall gradually into disuse soon after this time, but gypsum is today a constituent of some of our modern fertilizers.

A soil-building element long used farther south in New Jersey was greensand marl. It needed no preparation and was spread on fields near the source of supply, sometimes in as great quantities as one hundred wagon loads per acre. After the coming of railroads, commercial companies shipped it to regions outside of the marl belt. On April 19, 1855, Blackwell recorded in his diary, "Forenoon spread a load of Squancum Marl on our wheat—25 bu. on 1700 square yards which we are told is too thick." In the 1860's and early 1870's the Squankum and Freehold Marl Company advertised their product in the Hunterdon County newspapers. In 1870 marl sold for $3.10 per ton in Lambertville and for $3.40 in Flemington. It is probable that the large quantity of marl needed per acre deterred Hunterdon County farmers from using it to any great extent. At any rate it was seldom mentioned in the lists of fertilizers advertised in the county papers after 1875.

A fertilizer which was used extensively for a time about the middle of the century was guano, the accumulated excrement of birds found in large deposits in Peru and other Pacific coast countries of South America. In 1860 a Flemington firm advertised that it had on hand fifty tons of "No. 1 Peruvian Guano." [7] Guano was a very rich fertilizer, but so costly that its use called for a considerable capital outlay. Since fertilizer advertisements after the Civil War seldom mention it by name, it is probable that its use, except for special purposes, was discontinued. Another enriching agent used during the same period was "Poudrette," manufactured from the night soil of New York and other large cities. Similar products were "Pablette" and "chemical compost," which were advertised in the *Hunterdon County Democrat* at $35 and $40 per ton respectively in 1865.

Chemical fertilizers began to appear just before the middle of the century. Their popularity all over the United States was a result of the printing of an American edition of Liebig's *Chemistry in Its Application to Agriculture and Physiology* in 1841. Liebig's mineral theory was accepted by agricultural writers and passed on to their public through farm papers, which were being widely circulated by this time. A very interesting figure in the development of the chemical fertilizer business in New Jersey was "Professor" James J. Mapes of Newark. A "consulting agricultural chemist" and experimental farmer, he patented a "superphosphate of lime" and placed it on the market in 1852. This fertilizer was composed of bone dust dissolved in sulphuric acid and mixed with sulphate of ammonia. The cost of production of the new fertilizer was sufficiently low so that it could compete with guano, the price being $40 per ton in 1853. Superphosphate of lime was recommended for all grain crops, for gardens, and for orchards, and was spread on the land at the rate of 250 pounds per acre. In 1859 Allen and Needles, a competing firm, began selling a cheaper fertilizer under the same name, despite Mapes' patent, and in 1865 a commercial preparation known as "phosphate of lime" was being marketed at $60 per ton. In the 1880's, or perhaps earlier, Lister's "standard superphosphate" and "special wheat fertilizer" began a long period of popularity. Other phosphate fertilizers appeared on the market from time to time under a variety of names.

With the popularity of phosphate so great, it is not surprising that its manufacture was tried locally. The most successful venture was that of an enterprising miller, William Hill. At Kuhl's Mills, east of Copper Hill, he began in the 1870's the manufacture of "bone phosphate." For

[7] *Hunt. Rep.*, Mar. 20, 1860.

many years he provided farmers of the countryside a market for bones, and Hill's Fertilizer became locally famous. Other concerns within the county tried similiar manufacturing with varying success. Andrew Bartles of Flemington used his farm north of the fairgrounds as a proving ground and demonstration center for bone products manufactured by his brother in Williamsport, Pennsylvania. Ground bone, bone dust, and other forms of processed bone were constantly on the market after the middle of the century, and fertilizers from packing-house residue became common in the 1880's.

The popularization of chemical fertilizers was due not only to the advertising of manufacturers but also to the more trustworthy work of agricultural extension workers. In 1864 Rutgers College, as a land-grant college, was required to give annually a series of lectures in each county of the state. Dr. George H. Cook was selected as the first lecturer in 1866. His talks in Hunterdon County and those of his successors, given at Flemington, Locktown, and other points, were often devoted to the subject of fertilizers. According to the *Hunterdon County Democrat,* Dr. Cook's discussion of "Commercial Fertilizers," given in October, 1879, at the courthouse in Flemington, drew a large audience, and many individual farmers had questions to ask. The experiments conducted on the farm which Rutgers College acquired in 1864 also did a great deal to change farm fertilizer practices. This was especially true after the founding of the Experiment Station in 1880. Actual field tests by cooperating farmers, a number of whom lived in Hunterdon County, began in 1881. Another problem attacked was that of fertilizer inspection. At the urging of Dr. Cook the state required in 1874 that all fertilizers offered for sale be accompanied by printed analyses. A chemist of the State Board of Agriculture was to test all fertilizers from time to time. In 1880 the work of fertilizer inspection was turned over to the newly created Experiment Station. Governmental agencies thus took over experimental, regulatory, and educational functions as regards fertilizers. As a result of their work the farmer has learned how to increase the fertility of his land without waste effort. The expanding program of the Experiment Station and the increasing dissemination of knowledge by the Agricultural Extension Service have eliminated guesswork and rule-of-thumb methods.

There is much evidence to show that farmers turned rather readily to commercial fertilizers as a solution to their problems. Testimonials of crop increases for various fertilizers were easy to secure and were often printed in the *Hunterdon County Democrat.* A man near Flemington in 1883 raised forty-three barrels of potatoes on half an acre by using Mapes' "Potato Manure." A Ringoes farmer in the same year produced thirty bushels of wheat per acre by the use of phosphate. "Corn phosphate" gradually took the place of land plaster as a fertilizer to put in hills of corn, and "peach tree phosphate" was used in peach orchards. In 1893 "AA" and "No. 3" phosphate were recommended especially for oats and rye. Elmer Bonnell in 1883 used a mixture of phosphate and hen manure as a fertilizer for corn, an excellent practice probably learned through reading farm publications. A Pittstown correspondent of the *Newark Courier* in 1875 spoke of the "improved methods of cultivation consequent upon a larger knowledge of the chemical properties of soils and the use of fertilizers." [8]

Although progressive farmers before 1800 knew that green manure was beneficial to the soil, the plowing under of growing crops was probably not a general practice until fairly recent times. The *Pennsylvania Farmer* in 1804 advised its readers to plow under turnips, vetches, growing weeds, and especially buckwheat or clover. Farm magazines preached the benefits of green manure during most of the nineteenth century, and Hunterdon County newspapers often printed extracts from these journals. Although regarded as a mere fad by many, the plowing under of growing crops became a regular practice of better farmers. In the fall of 1812 one of the Case family near Flemington recorded in his record book that he turned under a growth of clover which had sprung up since mowing time. In 1873 the members of the Union Farmers' Club discussed the question, "Should clover when plowed under for wheat, be crossed up?" Various clovers, rye, and growing weeds were long the most common of green manure crops. Sudan grass, because of its luxuriant growth, has been used to some extent, as have cow peas and sweet clover. The most recent addition is the soybean. The program of the Agricultural Adjustment Administration has increased the practice greatly. Experiment Station research on the problem began in 1891, when crimson clover seed was sent to a number of farmers, including Dr. Henry Race of Hunterdon County. Since that time the contribution of the Experiment Station to knowledge of various green manures has been very great. Two other farm practices bearing a relation to soil fertility are the rotation of crops and the growing of legumes. Both will receive attention in a later chapter.

The problem of erosion, a serious one in many parts of the county, was one which did not for a long time receive the attention which it

8 Reprinted in *Hunt. Co. Dem.*, Oct. 12, 1875.

deserved. Many farmers considered erosion as unavoidable or as something of no great concern. Fields from which the soil washed off very badly were abandoned or turned into pasture. Such a thing as contour plowing was, of course, unheard of. Nevertheless, some farmers tried to solve the problem, as when Blackwell built stone breakwaters to protect his lowlands from floods in the 1850's. A farm procedure of ancient origin was that of leaving occasional balks, or strips of grass, where erosion was serious. To some extent fence rows also provided protection. With the coming of power machinery it has proven difficult to preserve balks, and fence rows have often been removed. This may account to some extent for the serious erosion of recent years.

With growing concern over the loss of soil through erosion, more farmers have turned to the growing of cover crops, especially in corn fields where winter washing is often bad. In 1881 Bonnell sowed clover seed in his corn at the last cultivation. This practice had the added advantage of providing a green crop for plowing under in the spring. The most common cover crops until recently were rye and sweet clover. To these in recent years has been added rye grass. Extension Service tests show that this crop, because of its extensive root system, is the best possible cover crop. It is usually sowed at the time of the last cultivation of the corn. By 1944 D. M. Babbitt, the County Agricultural Agent, reported that approximately 1500 acres were being sown to rye grass annually in Hunterdon County.

The most hopeful sign regarding erosion is the changing attitude of farmers. Most farmers today readily admit its bad effects, and many are willing to do something about it. In 1935 twenty-one men met at Ringoes and signed up with the Soil Erosion Service of the Department of the Interior of the United States. A labor force was supplied, at first by the Works Progress Administration and later by the Civilian Conservation Corps camp established at Clinton Point. The first work was done on the farm of Fred Totten near Ringoes in May of that year. In November the County Board of Agriculture showed the growing interest by appointing a committee to inspect and report on work done. During the next two years 150 farmers, owning 10,896 acres of land in the watershed of the Neshanic River, cooperated in establishing erosion control and water-conserving measures. At about the same time 119 farmers in the Clinton area signed up for the same program; and by the time of the abandonment of the Clinton Civilian Conservation Corps camp in 1940, 12,682 acres there had been improved.

The work done in both areas varied from farm to farm and included the laying out of contours, the construction of terraces, gully structures, etc., and the planting of shrubs and trees. Cooperating farmers put in such soil-conserving practices as strip cropping, controlled grazing, and reduction of plowland. Those who allowed their farms to be thus improved were well-pleased with the results. In June, 1941, the County Board of Agriculture sponsored a move to have Hunterdon County included in the soil conservation district which already included Morris and Warren Counties and which had its headquarters at Hackettstown. The action was eventually successful; and in the following year a representative of the Soil Conservation Service was assigned to Hunterdon County to work with the County Agricultural Agent on soil conservation problems, and to make plans for those farmers who signed up with the Soil Conservation Service.

Generally speaking, it can be said that in regard to soil care Hunterdon County agriculture has shown marked improvement since the days of John Rutherford and John Hall. Farmers have for many years recognized the need of keeping up soil fertility. As a result of the work of the Experiment Station and the dissemination of information through the Agricultural Extension Service, farm magazines, and pamphlets of the United States Department of Agriculture, farmers of today are much better informed as to soil needs than were those of a few decades ago. The average of land fertility has probably been increased considerably. Increasing numbers of cattle and poultry now provide much greater quantities of manure. Marginal land has been converted into woodland or pasture, and some attempts have been made to solve the serious erosion problem. The interest shown by 4-H Clubs, by farmers who enroll in Extension Service soil fertility courses, and by the young men who belong to the Hunterdon County Soil Fertility Club points to even more careful consideration of soil conservation in the future.

CHAPTER VI

Buildings, Equipment, and Methods

THE FOREST GROWTH hindered the early settlers from exploiting immediately the wealth of the soil, but on the other hand it provided them with materials for buildings and much of their working equipment. Hewn timbers from trees near at hand were easily workable and made sturdy houses, barns, and sheds. The log cabin, which became the most common dwelling on successive American frontiers until the Great Plains were reached, was apparently evolved from the log house of Sweden, brought over by the Scandinavian settlers on the Delaware. The pioneers who settled in Hunterdon County were well versed in its construction, and log houses remained common for a century in some parts of the county. Many were still standing in the early decades of the nineteenth century, and one near present-day Stockton was tenanted as late as 1865. Nevertheless, it is probable that the log cabin was not so universal as once thought. In Hunterdon County, at any rate, houses of other materials were not slow in appearing, and often the cabin dweller was eager for the day to come when he could move into a better home and use the old cabin for a stable or corncrib.

The tools used for the shaping of timber and the construction of log buildings were largely of American development. The pitching or felling ax had a narrow blade and deep haft. It was used for felling trees and for rougher work generally. The broad ax, which was used for hewing down the timber, had a long blade with a tapering edge. It was possible to build a cabin by using only these two axes; but other tools, particularly the adze, the splitting ax, and the mortising ax, were often employed to speed up certain processes. A special tool, the frow, was used in splitting oak shingles. A homemade maul or beetle, a saw, some wedges and gouges, a hand augur, a drawing knife, and occasionally some finer tools completed the working equipment of the pioneer. The log to be hewn was cut to length, then laid on trestles for scoring and dressing down. Since green timbers warped, they were often laid to one side to cure, and later the "wind was taken out" by further dressing. If necessity caused the use of uncured timbers, the cracks which later appeared between them were cared for by extra chinking.

Sometimes for a year or two the pioneer would use a hastily constructed hut of unsquared logs, with earth floor and roof of poles and bark. But his more permanent cabin, built as soon as he could get to it, was of squared and carefully notched oak logs of a size to make a wall about ten inches thick. The roof was made of hewn rafters covered with riven lath and shingles, and the gable ends were covered with weather-boards. The floor was of split logs dressed down. In some cases the inside walls were covered with lath and plastered with clay and straw. One side of the cabin was dominated by the fireplace, made of stones and overtopped by an oak mantel. A lug pole crossed the room in front of the fireplace at a height to suspend pothooks and trammels. The walls were often pierced for one or two small windows. Partitions were usually of cloth hanging from ropes, but thin wooden interior walls were sometimes built.

Reminiscences concerning the last log houses torn down in the vicinity of Stockton give some idea as to size and appearance. A very old cabin demolished in 1837 was described as "a log house about 20 feet square one story high . . . covered with shingles. It had a door in front with a window front and back." Three other cabins in the same vicinity were of similar construction. A larger log house a mile north of the old ferry was thirty-five feet by twenty, one story high, with windows on three sides. Still another had a partial second story. "The logs were notched at the ends and laid up like a crib. . . . Other logs were split into small pieces and pined [*sic*] into the crevices, then filled in with clay morter [*sic*] to even up the inside and plastered over again with clay." [1] The timbers stuck out six or eight inches beyond the notched corners. Each cabin had one or more fireplaces made of stone with clay mortar. The doors were of planed split timber, dressed down, and were hung on wooden hinges. The original Case dwelling near Flemington, which stood for a long time as a stable, is said to have been thirty feet long and eighteen feet wide. A house of hickory logs, evidently built fairly late, was still standing near Barbertown in 1881.

Frame houses began to appear in the vicinity of sawmills within a decade or two after the first settlements. The simple "up-and-down" saws of that day were slow by modern standards, but they reduced greatly the labor involved in building. Frame dwellings of one or two stories eventually outnumbered all others in this area. Construction of such a

1 Dr. G. H. Larison, "Last Primitive Houses near Howell's Ferry."

house was simple. Heavy squared timbers provided the skeleton, and boards laid across the beams made a floor for the attic or second story. The outside walls were of thick sawed lumber, and the inside ones were of lath and plaster. Roofs were usually of oak shingles made near at hand. As transportation improved some building materials were imported. The building accounts show that when the first courthouse was built at Flemington in 1791, cedar or pine shingles were carted from New Brunswick; the hardwood lumber was produced at two local sawmills; and pine boards were brought from the bank of the Delaware, to which point either the logs or the lumber had been transported by raft. Frame houses were usually painted white or red. When Case the tanner built a new house near Flemington in 1793, he painted it red with white shutters, and Isaac Passand suggested that it looked "like a palace."

The stone houses constructed at an early date naturally have withstood the ravages of time better than those of wood. A number are still standing, while others have been torn down within comparatively recent times. Therefore, we have better descriptions of early stone houses than we do of those of frame construction, despite the fact that the latter were the more common. A stone house at Quakertown, built about 1725, was thus described by Scott Allen, an aged man who spent his youth there:

> It had kitchen and cellar on the first floor, the fireplace was very large, we could put in a back log eight feet long. The joists were large enough for girders in a barn. There was a long flight of steps on the outside to reach the second story, which made it look very odd. The second floor had three rooms, one large one and two bedrooms. The garret was one long room and was used to store grain in, and it was no light task to carry it up those two long stairways. The roof was quite steep or would be in our day.[2]

Another house near Quakertown, still standing though considerably altered, bears the date 1735. Originally it was a solid, gloomy structure with thick walls, cut only by two windows, and with massive chimneys. The skeleton of huge beams showed in all the rooms until ceilings were added at a later date. The living room was paneled to the ceiling, and all partitions were of wood. A winding interior stair led to the second floor.

Dr. George H. Larison describes early stone houses in the Stockton area as having been "laid up with clay morter [*sic*] and painted with lime and sand in broad point style and plastered inside with clay morter

[2] M. C. Vail, *History of Land Titles in the Vicinity of Quakertown, New Jersey*, p. 11.

and whitewashed with lime." Joists were of hewn timber and floors of heavy sawed planks. Stone houses were usually a story and a half, sometimes two stories, in height. A piazza covered by a projecting roof was a not unusual feature of both frame and stone houses. Building stone usually came from close at hand, but sometimes was transported several miles, as in the case of the courthouse already mentioned and that of the Presbyterian Church built at Flemington in 1793.

Houses of brick were not usual in early days. Brick clay was often near at hand, but other building materials were even easier or cheaper to procure. Only occasionally would a man set up the equipment and proceed to manufacture the bricks for his dwelling on the spot. Transportation costs precluded the importation of bricks from distant places. The fine brick residence built by John Reading the younger in 1760 is still standing near Flemington Junction, and a brick house in Flemington built at about the same time stood until 1940. Brick was also sometimes used for sidewalks, cellar floors, and garden walls.

Architectural developments in Hunterdon County during its first century and a half were, as in West Jersey generally, slow in coming. Carpenters and masons learned their trades through apprenticeship rather than from books, and were not much influenced by changes outside the general region. Within the area there was a tendency toward uniformity, as English, German, and Dutch influences faded or were merged. The rectangular frame, stone, or brick house of one and a half or two stories, with two or three rooms on each floor, became more or less standard. When more room was needed, an ell was often added. A cellar under part of the house became a usual feature, sometimes being used for cooking but more often for storage. Outside kitchens adjoining the house were common until late in the nineteenth century, and it was not unusual to have a separate "Dutch cellar" for storage of fruit and vegetables, leaving the cellar under the house for cider, dairy products, and meat. Often attached to the house was a shop for weaving, mending tools, and doing odd jobs.

As more and more operations were moved to outbuildings and as houses became larger, greater importance was attached to roominess. Persons advertising houses for sale in the local newspapers stressed the number of rooms and made great use of the adjectives "comfortable," "spacious," "substantial," and "commodious." The frames of wooden houses continued to be of heavy timbers, usually sawed but sometimes still hewn out with a broadax, until after the middle of the nineteenth century; and the use of lighter materials came in slowly even then. The

Rev. J. L. Stillwell, in his little volume of reminiscences entitled *Old Readington,* wrote as follows in 1935: "When the Herman Hageman house was erected the modern method was adopted, what was then known as the balloon frame was given a trial. Many prophesied it would not stand but it has stood for over sixty years." In more recent times, as rural living has become increasingly like urban living, newly built farmhouses have copied many of the features of village and city houses. As a result the homes of the countryside have become increasingly diverse architecturally.

Foreign observers in early days were struck by the fact that farmers of this area attached more importance to having good barns than substantial dwellings. That attitude has never been entirely lost. The barns of the first pioneers were usually of logs and built somewhat like their cabins. Logs were satisfactory for small buildings, but not adapted to large structures. Log barns, increasingly relegated to storage, lasted well into the nineteenth century, but those made of frame began to appear at an early date. The skeleton of the frame barn was constructed of hewn white oak timbers, which were covered over with rived or sawed weatherboarding. Buildings took a number of shapes but were necessarily commodious. A common type from an early day was the "Dutch barn," which Kalm described as follows:

> The main building was very large . . . , the roof was high, covered with wooden shingles, sloping on both sides, but not steep. The walls which supported it were not much higher than a full grown man; but on the other hand the breadth of the building was all the greater. In the middle was the threshing floor and above it, or in the loft or garret, they put the unthreshed grain, the straw, or anything else, according to the season. On the one side were stables for the horses, and on the other for the cows. The young stock had also their particular stable or stables or stalls, and in both ends of the building were large doors, so that one could drive in with a cart and horses through one of them and go out at the other. Here under one roof were the threshing floor, the barn, the stables, the hay loft, the coach house, etc. This building is used chiefly by the Dutch and the Germans.[3]

In Hunterdon County Dutch barns were found in Readington Township, in Round Valley, in Kingwood Township, and to a lesser degree throughout the county generally.

Building materials other than wood seem to have been little used, and the large frame barn of one type or another became standard. As was the case with houses, barns increased in size, and many of their functions

[3] Kalm, *op. cit.*, I, 118.

were given over to auxiliary buildings. At the Union Farm in 1799, in addition to the two dwellings and the large barn, there were a smokehouse, extra stables, and several barracks for hay. A farm of only one hundred acres near Stockton in 1827 had a large barn, extra cow stables, a hay barn, and a number of small outbuildings. An Amwell Township farm of two hundred acres had two barns, a cow stable seventy feet long, corncribs, a smokehouse, a carpenter shop, and a cider house and distillery. Wagon houses, ice houses, woodhouses, hogpens, and corncribs were common. A farm on the Old York Road in 1890 had wagon houses, poultry houses, granaries, a hay barn, a "cow house," and a barn eighty-six feet by forty with an annex forty by thirty-six. The barn's heavy frame of hewed timber, said to have taken one hundred white oak trees, supported a slate roof of thirty tons weight.

A very useful building was the barrack. A good description by Mrs. Mary Capner, written Nov. 13, 1787, for the edification of relatives in England, is quoted:

> Barracks are a building I have not described to you, tho I noticed them at first coming into the country. Tommy has made one for his Bro. [It has] four poles fixed in the ground at the distance of fifteen feet in a square. The poles are squared fifteen feet or more at top and five feet at bottom unsquared. This is all above ground. In the square part of the poles there are holes bored thro at the distance of twelve Inches big enough for a strong iron pin to be put thro to suport [*sic*] four wall plates which are tennanted at the ends, then some light spars are put upon the wall plates and thatch upon them. When it was only five feet from the ground, the room can be raised at pleasure 21 feet or any distance from the ground between that and five feet. These are to put hay or any kind of grain under and the roof is always ready to shelter it from hasty rains which is common hear [*sic*] in summer. Those that have only two cows have the bottom part boarded at the sides and a floor laid over and the hay at top and the cow stable under.

Barracks remained common features of the average barnyard for nearly two centuries, until changing methods decreased their usefulness, and a few are still in use today.

The large array of buildings so characteristic of the nineteenth century is still a feature of most of the farms of the county, and particularly of general farms. Naturally there have been changes. Wagon houses have become shelters for motor vehicles and other farming equipment of a machine age. Barracks have slowly disappeared with the increasing use of the hay baler, but the equally picturesque silo has appeared in its place on most dairy farms. The farm plant has been subjected to radical changes in many cases because of increasing specialization in dairying or poultry raising. Old buildings have been demolished or altered and new buildings, better fitted for specific purposes, erected. With increasing specialization the average farmyard may be expected to continue its evolution.

A farming problem from pioneer days to the present has been that of confining livestock, or at least so restraining them as to prevent them from eating and trampling crops. It is probable that the first farmers of this section used "stump fences," or those made from felled trees, as was done on other frontiers. Thomas Capner wrote in 1787 that some Hunterdon County farmers at that late date still used stumps to support their rails, but that most of them employed worm fences. The worm or snake fence, made of rails split from timber close at hand, was a natural answer to the enclosure problem. Another answer was the setting out of hedges, but the half-wild hogs and cattle were able too often to break through such barriers. On the other hand, worm fences could be made substantial enough and high enough to stop any stock. By a law of 1713, renewed in 1730, fences protecting crops must be four feet six inches in height. The post and rail fence, common by the end of the eighteenth century, required less timber than the worm fence, but its construction required much labor and time. The posts had to be squared by hand or at a sawmill, and the ends of the rails dressed with a broadax. Chiseling or boring out the holes in the post was a time-consuming task. This type of fence required more labor for maintenance than the worm fence, for a regular job each spring was that of straightening the posts and putting rails in place. Since the post and rail fence was neater than the worm fence, the rare farmer who was concerned with tidiness might place such a fence along the public roads and the less showy but less troublesome worm fence on his other boundaries. Another kind of wooden enclosure, used for gardens and other small plots, was that made of palings.

The amount of rail fence used on an average farm was large. In order that they might use any field for grazing purposes on occasion, farmers attempted to enclose all fields. A farm of seventy-six acres near Flemington, advertised in the *Hunterdon Gazette* in 1830, was "conveniently divided into 7 fields," mostly with chestnut fence. Eight years later a Kingwood Township farmer, "going to the West," advertised that his farm was "all fenced in small fields, with a lane convenient to water." Most of the rail fences were worm fences, and as long as there was sufficient timber the upkeep of such fences was only a matter of diligence. In wooded parts of the county and in parts accessible to wood lots many worm fences were kept up until after 1900, and a few are still used to this day. On the other hand, the Capners wrote that some parts of the county were feeling the shortage of rail timber before 1800. The fact that post and rail fences required less timber than worm fences was probably one cause for their increased use. As the cost of chestnut and oak rails increased, it is little wonder that areas not blessed with timber looked for substitutes. Some farmers built stone fences, but the labor involved was too great for them to become common. Joseph Capner at Flemington and Hugh Exton at Union Farm used hedges of English hawthorn or "quick," and Isaac Passand tried a combination of hedge with post and rail fence, using two rails instead of four between the posts. The *Pennsylvania Farmer* in 1804 advised farmers in areas short of rail timber to plant either the English or the American white hawthorn. A more successful hedge plant was the Osage orange, which became popular soon after the middle of the century. During the following decades farmers set out thousands of hedgerows, which were to be the despair of their descendants. Nurseries often planted and cared for hedges at a given price per rod.

The real successor to the rail fence was barbed wire. Blackwell, disgusted with his expensive and troublesome post and rail fences, tried some kind of wire fence in 1851, but true barbed wire did not appear until after 1880. Because of numerous cases of serious injuries to animals, particularly to horses, there was at first considerable opposition to its use. On March 2, 1886, the *Hunterdon County Democrat,* citing court decisions disapproving of the use of barbed wire, said, "The murderous contrivance ought to be abolished altogether." But farmers were not to be deterred, and many turned to the new device even in parts of the county where there was still timber. Elmer Bonnell, after laboriously splitting rails in his wood lot each winter for years, gave barbed wire a trial in 1887. In January, 1890, he put up a fence of six strands and recorded that it cost him "not quite 3¢ per foot counting posts and wire." "Knox patent" and other barbed wire soon became an accepted part of the rural scene. In rental arrangements a common solution of the fencing problem was for the landlord to furnish the barbed wire and the renter to cut the posts and put up the fence. During the past few years electric fences have been rapidly replacing barbed wire, despite objections to their use, and economy of construction and maintenance are very much in their favor. Woven wire fences have been used in the county for a long time, and in recent decades have become very important, chiefly because of specialization in poultry.

The farming equipment of the pioneers of this area was not elaborate, but less primitive than on many frontiers. The gum made from the cross section of a hollow tree, the dipper made from a gourd, and the lard lamp were common even in older settlements. Makeshift brush harrows, grapevine hay drags, and forks made from the branches of trees probably had their day, but the nearness of developed communities made it possible for most farmers to have better tools. Probate records of early Amwell Township men list iron-toothed harrows, harrow teeth, plows, plow tackling, scythes, grindstones, hoes, spades, and forks, in addition to the woodworking tools previously mentioned. The plows of the period had points or shares of iron and a moldboard made mostly of wood. The triangular harrow of that day had a heavy wooden frame, and though not adjustable was very sturdy. Cradles for harvesting grain were used by some farmers in this area before 1740. Everyone owned saddles and bridles, of course, and not a few had wagons, sleds, and other conveyances. The earliest windmill of which a record has been found was one belonging to Garret Vansickle of Readington Township in 1748.

For a century there was no really fundamental change in the working equipment of the Hunterdon County farmer. The long letters sent by the Capners to their relatives in England at the end of the eighteenth century mention tools and apparatus little different from that found at Mount Amwell, John Reading's homestead, in 1717. The most important change was that the "goods and gear" of farms had become increasingly elaborate as each farmer and his descendants purchased and made specialized tools and equipment. In that day when every farmer was a Jack-of-all-trades, each family acquired a formidable array of tools, ranging in size and function from the shoemaker's awl to the long hooks used in handling ice. Forks, rakes, axes, hoes, sickles, and scythes were made in various sizes and shapes. Harrows and plows were made for specific types of work, and wagons were made light or heavy according to use. Invention and adaptation of tools and implements to fit specific needs were the first steps, though not long ones, in the development of labor-saving devices.

The first farmers were hard put to it to get sufficient land under cultivation to support their families and livestock. With ax and fire they let in the light of day and with mattock and hoe laboriously planted their seeds. Heavy plows and harrows pulled by plodding oxen might be employed as soon as a clearing became sufficiently large. Later, when roots, stumps, stones, and other obstacles to cultivation had been removed, lighter plows and harrows were used; horses were substituted for oxen for most operations; and the amount of hoe work was greatly lessened. Whether "crossing," that is, plowing the land both ways, was practiced where feasible in early days is not known. At any rate, it was a common practice in Hunterdon County in the nineteenth century.

The methods of culture for small grains and grasses brought from Europe were largely borrowed from Old World practices. Grain was broadcast by hand or by means of some homemade throwing device, and then covered by use of the harrow. Supposedly very ancient machines for sowing grain, made by drilling holes in hollowed logs, were long preserved in the county, but it is doubtful whether their use was very extensive. Where timothy and other "artificial grasses" were grown, they were broadcast after the final harrowing. The planting of Indian corn or maize was naturally an entirely different matter. In the first clearings corn was planted here and there indiscriminately, Indian fashion. Later, in order to permit cultivation with plow and harrow, a more ordered system was worked out. In 1748 Kalm recorded that corn in West Jersey was commonly grown in checked hills five and one-half feet apart in either direction. Rye was often sowed among the hills in the fall, to remain after the corn was taken off the field. About 1800 a common method of cultivation was that of having rows five and one-half feet apart, with the individual stalks eighteen inches apart in the row. Narrow balks of grass were sometimes allowed to grow between the rows.

Gradually there came into use methods of corn culture which were to hold sway for nearly a century. With a light one-horse plow ridges were thrown up, first one way across the field, then the other; and hills of corn were planted where the furrows intersected. An alternate process, one increasingly used as the nineteenth century progressed, was that by which a "furrowing sled" was drawn across the field each way to "furrow it out." The crossing points of the marks left by the sled located the hills. The actual planting was done by hand. Young boys and girls dropped the corn and sometimes put fertilizer in each hill. The grain was usually covered by men with hoes, though some farmers used a device which could be pulled by a horse. A description of one field planted in this way reads as follows: "The rows were as straight as a line,—from end to end,—every hill contained exactly four plants as near as possible—and the regularity of the interstices were as true as the spaces on a chess-board—you could have plowed it four ways without touching a hill." [4] It is doubted that all fields were so tidy.

[4] Dr. C. W. Larison, *Fisher Family,* p. 74.

For forage the pioneer depended on native grasses to a considerable extent, but eventually English grasses including timothy and white clover were introduced. The yield of hay, either native grass or timothy, was sometimes increased by the practice of "watering the meadows," a simple method of irrigation by which the water from springs and small streams was diverted so as to flow over low grasslands. Kalm found the practice widely used in the Philadelphia area in 1748, but apparently it was not common in West Jersey. Hunterdon County farmers declared that the ditching required too much work, and one of Joseph Capner's native neighbors "laughed heartily" when Joseph turned water on one of his meadows. The large yield from this croft and from a meadow watered by Isaac Passand a few years later convinced the doubters, and some of them took up the practice. This is not to imply that the Capners introduced the custom to Hunterdon County, for it was already known. A few farmers here and there continued to water low meadows for a time in the early nineteenth century, but the practice was eventually abandoned, probably because of the high cost of labor.

The harvesting of the various forage and small grain crops had certain points of similarity. Hay crops were cut while green with sickle or scythe, and grain in the same way as soon as ripe. After curing, hay was raked into piles by use of a hand rake with wooden teeth, then piled into cocks with a fork. Wheat and rye were bound into bundles at cutting time, and later placed together into shocks. The bundles could then be handled with the same pitching fork which was used for hay. Oats were seldom bound, apparently because of fear of spoilage. They were laid in swaths and picked up by means of a special contrivance, the "oats fork." The next job was that of drawing or hauling the crop to a place of storage. The haymows, handy to the stalls of the animals and to the threshing floor, were filled, and the remainder put into nearby stacks or under the roofs of barracks.

An aged man of an adjoining county, recalling his youth, writes of "those scenes and sounds; the early start in the dewy grass; the falling into swath of the lithe lines of clover and timothy over the reiterated swishes of the curved blades in the zigzag rows of the stooped mowers; the 'wetting the banter' by a peculiar clip of the rifle or whetstone, wherein a hidden challenge was given the rest; the mortification of the luckless wight who was 'mowed around' at the risk of amputation of one or more legs . . . ; the raking round; the hauling in of the sunbaked hay; the pitching off; the group of tired and sweat-stained men around the supper table." [5] The day-by-day accounts of actual farmers, giving as they do a description of the whole laborious process of harvest, are less romantic. They emphasize the vagaries of the weather, the difficulty of securing a sufficient labor force at the right moment, and other serious problems. Even after the harvest was over, there was still the job of threshing to be done during some slack season.

It is probable that in early days threshing was sometimes done by the trampling of animals, or by use of a "tumbling Tom," a device consisting of a grooved or spiked log fastened by an axle at one end and drawn by an ox or a horse. The most common method, however, was by use of the flail, a very simple tool consisting of a wooden bar or swingle fastened by rawhide or eelskin to a wooden handle. When a threshing day arrived, the threshing floor of earth or oak planks was swept clean and the bundles of grain laid side by side in rows. The head ends were beaten with a flail, then the bundles were turned over and beaten again. Finally the grain was unbound and rethreshed in order to prevent any loss. The rebound "battons" of rye and wheat straw were regular items of commerce. The best rye straw was used in making bands for tying fodder, and the best wheat straw for filling bed ticks. More common uses for both were as bedding in stables and as feed for dry cattle.

Of the major crops corn was the one requiring the most work. As soon as it began to peep through the ground, and perhaps again later on, it was harrowed lightly. As the crop grew larger, it was tended with the same plow that had been used to break the ground or with a lighter one, the soil being alternately plowed away from and toward the row. Blackwell recorded that the effective killing of weeds still required occasional hoe work. The harvesting of corn was laborious, but could be spread over a longer period than was the case with other crops. In early days corn was topped and the blades removed for fodder, but the practice gradually gave way to the one of cutting the whole stalks, which were then bound into bundles with rye straw bands and placed in shocks. Later the ears were husked and the fodder stacked for winter use. Enough husks were saved to make corn shuck mattresses and to braid into horse collars. In the 1840's Blackwell regularly marketed his husks, receiving 60¢ per hundred pounds from a small manufacturer in Clinton. The demand for this by-product continued for several decades thereafter.

The coming of labor-saving implements and machines in the nine-

5 *Papers of Bucks Co. Hist. Soc.*, III, 374.

teenth century slowly revolutionized farming methods. Perhaps fortunately, the development did not come suddenly or for all farm processes at the same time. For example, the fanning mill, which eliminated the farmer's dependence on the wind when threshing grain, was already in common use before the American Revolution. From the employment of that simple machine to the use of the power thresher or the combine had been a long journey, marked by a series of improvements, additions, and combinations. Similar progress, now halting, now rapid, has marked the evolution of labor-saving devices of all kinds. The new tool or machine hailed as the last word in the performance of its particular function during one period was in the next made obsolete by new developments.

Changes in the equipment used for preparing the soil for seeding came slowly. Traditions report that wooden plows were used in parts of the county for several decades of the nineteenth century. However, after the invention of the Newbold plow at Burlington, New Jersey, in 1797, many iron plows came onto the market. The blacksmiths who made plows occasionally stumbled on a design which was superior, and were honored by having their ideas adopted by others. Sometimes the invention was patented and the rights of manufacture sold. Thus the Peacock plow, the Smith plow, and others had a considerable vogue in these parts during the early nineteenth century.

In 1828 John Deats, a blacksmith near Stockton who had probably put together the ideas of various men, patented the Deats plow. This plow, made by three generations of manufacturers in the Deats family, was popular locally for nearly a century. The letters of patent, which bear the signatures of John Quincy Adams, President, and Henry Clay, Secretary of State, list numerous advantages of the Deats plow; but its most important feature was that it had a moldboard "better calculated to scour than any other hitherto made." John Deats died while peddling the rights of manufacture in Ohio, but his son Hiram built up an extensive business at a shop near Quakertown beginning about 1831. During the 1840's he moved part of his business to Pittstown where he would have water power the year round. In 1852 he built a machine shop there, and established another plant at Stockton. In 1859 a foundry was built at Pittstown, and the Quakertown plant was abandoned. The Deats factories made not only plows but various farm machines, cook stoves, and other iron and steel goods. The Deats plow was made in many models, the light ones with interchangeable moldboards so as to be adaptable for such purposes as plowing corn. Blackwell had in 1858 two

"No. 5," one "No. 8," and one "No. 9" Deats plows. These implements were still popular enough to be listed by name in advertisements of public sales in the third decade of the twentieth century.

The Deats plow was not without rivals even in its home territory. The editor of the *Hunterdon Democrat* in 1843 mentioned the Whitney plow as turning sod with "elegance and apparent ease." Other plows used were the Fisher, the Wiard, the Bowers, the New York, the Champion, the Syracuse, and the Oliver. Special types of plows, as the subsoil plow, the "double plow," and the corn plow, were made by each of the different manufacturers. After the middle of the century the increasing use of steel in place of iron permitted the making of plows which were lighter and easier to handle. The riding plow apparently did not come into the county until about the turn of the century and, if vendue lists are an indication, was never very common. The most revolutionary change in plowing methods came with the introduction of the tractor.

The harrow in one form or another remained the chief pulverizing tool. The old "A" harrow gradually gave way to the two-horse hinged harrow; and as time went on, this implement became lighter, more adjustable, and finally of all steel construction. Some time before 1880 the spring-tooth harrow made its appearance. Homemade drags and rollers had probably been used from earliest days by some farmers to supplement the harrow. The drag was made from squared timbers, the roller from a log selected for roundness. The manufacture of rollers was soon taken up by blacksmiths and later by the manufacturers of agricultural equipment. Though logs continued to be used for homemade implements, the manufactured "plank land roller" became more common. Blackwell sold the one which he owned for $16 at his vendue in 1858. In more recent times the makers of farm machinery have devised new contrivances. The somewhat costly disc harrows were looked upon as "horse killers" in this region, and were not used to any great extent until the day of the tractor. The same was true of the "pulverator" and "cultipacker."

The corn cultivator was introduced into the United States about 1826 and considerably improved by 1840. This implement, mounted on skids or runners and pulled by a team of horses, cultivated the soil on each side of the row with two or more small shovels. The name of "corn plow" was eventually appropriated by the cultivator after the true corn plow had disappeared from the scene, which did not happen until fairly late. Blackwell still used the latter on occasion in the 1850's, and it was a common implement in the Readington area as late as 1867. Two of the most popular cultivators in this area were invented by residents of the county. The one invented by Oliver Kugler of Three Bridges was patented in 1873 and that by Reuben K. Niece of Frenchtown in 1878. Patent rights on the Kugler cultivator were sold to various manufacturers as far west as Illinois. Wheel cultivators appeared in the 1860's, but were not predominant in Hunterdon County until near the turn of the century. They, too, were sometimes called "corn plows," but the name was more commonly applied to the skid cultivator, which was a common item at farm auctions for a half century. Wheel cultivators were equipped with a variety of devices for stirring the soil and were usually made in both riding and walking types. Because of their superiority they eventually replaced the older type. The next development of any moment was that of the cultivator attached to a tractor.

There were no changes in methods of seeding grain until nearly 1850. Among the Blackwell papers is a broadside dated June 19, 1856, in which a man at Readington advertised himself as agent for sale of "Lee's Patent Premium Grain Drills." Blackwell, however, did not buy one; and though several other makes were soon on the market, grain drills were not adopted very rapidly in Hunterdon County. Interest was sufficient by 1874 for the Hunterdon County Agricultural Society to offer a prize for the best drill with a grass sowing attachment, and by 1878 one of the entries at the County Fair had a fertilizer spreading device as well. Vendue offerings and dealers' advertisements in the county papers after 1880 point to an increasing number of grain drills. The broadcasting of seed, however, persisted into the twentieth century.

The development of the corn planter paralleled that of the grain drill. A letter in the *Hunterdon Gazette* in 1825 described a new "corn drill" as simpler than others then in use. It is probable that this was a machine to push by hand, as horse drawn planters were hardly a success for another twenty years. One of the prizes at the first Agricultural Society fair, held in 1856, was for a horse corn planter. An advertisement of the following year described a one-horse machine which cost $16 or $18, depending on whether it would plant one or two rows. The corn was dropped in checked hills along with fertilizer, and pumpkins could be planted at the same time. An illustration of a two-row machine in the *Lambertville Press* in 1858 shows two simple wheel planters fastened side by side on a frame. Despite the saving of labor which could be made, farmers of this region were slower than those of the Middle West to discard the old ways. The Deats factory at Pittstown was still making furrowing sleds in 1895, and at least one farmer was using the old method of planting as late as 1909. However, the newer types of wheel corn planters which came on the market in the 1890's were well received by most farmers, and planting by machine eventually superceded older methods.

The season of most backaches for farmers in the old days was that of the harvest, when forage and grain crops were gathered in close sequence. Homemade and experimental mowing machines began to appear in the 1820's, but the few farmers who bought them had so many troubles that they resumed the old methods for another generation. The McCormick and other reapers, already common in the West by 1850, did not appear here in any numbers until some time later. Most early reapers could be changed into mowers by removal of the grain platform. In 1854 Blackwell went to Raritan and paid $140 for such a combination machine. His chief reason was the scarcity of harvest hands, the wage scale at the time being from $2.50 to $3.00 per day. Although it was in the middle of the harvest season, other farmers quit work and gathered on the Blackwell farm to see the new machine in operation. Within two days Blackwell's wheat crop was cut. The machine, transformed into a mower, then made short work of cutting his oats and grass crops. During the following years his "mower and reaper" needed frequent repairs, but Blackwell apparently considered it a good investment.

Many combination mowers and reapers appeared on the market during the following years, each with its claims of superiority. The self-rake for the reaper appeared in the 1850's, and was probably common by the 1860's. It saved the labor of the man who had formerly sat on the table of the machine to throw off the grain in bunches. In a reaping and mowing contest held near Flemington in 1858, the machines of seven different manufacturers competed and an eighth was kept away by a railroad accident. All did well in reaping, some not so well in cutting tangled clover. The biggest handicap, however, was the curious crowd which had come to watch. Under the auspices of the Agricultural Society such contests were held each year and always drew many spectators. The price of the combined reaper and mower was $135 in 1860 and $200 in greenbacks in 1868.

After 1870 the advertisements in the county papers show an increasing emphasis on reapers and mowers as separate implements, though the combined machine remained on the market for a time. In 1878 the Agricultural Society offered separate prizes at the County Fair for mowers, for reapers, and for combination machines. Among the entries that year was a twine binder, a machine which was to prove a truly revolutionary development. There were some objections to it at first as a "lazy man's contraption" and complaints as to the quality of work and the number of breakdowns, but the binder came in more rapidly than had the reaper, perhaps because its advantages were so obvious. Bonnell was well pleased with the way the machine which he bought in 1884 speeded up operations under ordinary conditions. A Readington farmer, using relays of horses, claimed to have cut twenty-three acres of grain between noon one day and the same time the following day.

The success of the mower and the binder was so complete by 1894 that the use of scythes to any extent caused comment. As time passed, both were perfected mechanically and were made with longer cutting bars; but the fundamental principles have not been changed greatly, even with the coming of the tractor. The small combine which cuts and threshes grain at a single operation and works well in small, uneven fields is a recent development of great consequence. The first such machine in Hunterdon County appeared in 1928 in the Ringoes neighborhood. Within a few years combines were accepted as a success, and during the past decade their number has constantly increased.

Labor-saving apparatus for the handling of hay has been gradually evolved during two centuries. The hand rake was replaced first by a kind of rake which was dragged across the field by a man. Later this implement was enlarged and horse power employed. One development of the horse rake was the "whoa-back-rake," so called because the driver had to back the horse each time in order to empty the machine. Appearing about 1820, it was still in general use after 1850. The revolving hay rake was a decided improvement in that the horse did not have to stop at each windrow. On July 18, 1855, Blackwell wrote, "Brought home a new revolving horse rake made by J. S. Skillman of Flemington." When he retired from farming three years later, this implement brought $5.50 at public auction. The revolving rake, often called the "flip-flop rake" or "tumbler," had its day until the 1870's, even though the sulky rake, mounted on wheels, had been invented much earlier. Equipped at first with a hand dump, then with a self-dump, the sulky eventually gained popular favor.

The hay tedder, a supplementary machine used to stir up the mown hay so that it would dry better, was first shown at the County Fair in 1870.[6] In 1883 the editor of the *Hunterdon County Democrat* wrote, "Throughout the county mowers and tedders and horse-rakes and hay-

[6] *Hunt. Rep.*, Oct. 6, 1870.

wagons give the rural landscape an animated appearance." In the 1890's the hay loader and the side delivery rake came into use, but because of their cost were accepted somewhat slowly. The pressing and baling of hay was another development of the nineteenth century. Stationary power presses to which the farmers hauled their hay appeared before 1840. The portable, horse-powered baler, sometimes owned as part of the equipment of an individual farmer but usually moving from farm to farm to do custom work, came in before 1890. A decade or so later a baler driven by a gasoline motor made its appearance. In recent years the pickup baler, which bales hay directly from the windrow, has become very popular because of its speed and saving of labor. The harpoon hay fork, very useful in elevating loose hay into haylofts, was invented about the middle of the last century.

The coming of corn harvesting machinery was much behind that for grain and hay crops, perhaps because the need was not so great. Corn ripens so slowly that a farmer using a corn knife could usually cut sufficient fodder for winter use. Sled corn harvesters were first manufactured about 1886, and a few ingenious farmers, even recently, have made homemade devices of a similar nature. About 1895 the corn harvester, a low-slung, four-wheeled device pulled by one horse, appeared on the market, but was not widely adopted in Hunterdon County or neighboring areas. The corn binder was invented about 1890, but not until the 1920's, with increasing use of corn ensilage, were there many in the county. The mechanical corn picker is the newest development in corn harvesting machinery, and has achieved sudden popularity during the past few years. Corn shellers, hand and power, constantly developed both in speed and efficiency during the latter part of the nineteenth century. Those made locally by the Deats firm were very popular for many years.

Though there were successful machines for threshing grain in Europe before 1800, few were used in America until the second quarter of the nineteenth century. In 1830 a demonstration of a newly patented thresher was given in Flemington. The *Hunterdon Gazette* of August 4th reported, "It may be worked by hand, or by horse power, and is applicable to the thrashing [*sic*] of wheat, rye, oats, flax, etc., which it performs very expeditiously, and takes the grain clear from the straw." Within the next few years a number of threshers were put into operation in the county. These were "lever power" machines, run by a team of horses attached to a long sweep or shaft. For several years E. H. Conover manufactured threshers of his own patent at Headquarters. Another type, made at Ringoes, could thresh two hundred bushels of grain in a day. This implement, though not light, could be carried about in a farm wagon, thus making it possible to do custom threshing. Tolls charged were usually one tenth or one eighth of the grain threshed. The thresher operated by tread power probably came into use in this area slightly before 1850. In his diary Blackwell wrote on December 22, 1853, "George T. Gray brought home, and set up our thrashing [*sic*] machine. We put on Old Mike and run through one run of oats and it appears to run well—Price of Machine and Power $100 0/00 for which we gave our note payable 3 months after date—." His chief trouble during the following days was that Old Mike did not "keep up a steady motion." Soon tread power threshers were manufactured locally at a number of places. In 1856 the one made by Deats and Pierson took the blue ribbon at the first fair of the Agricultural Society.

As time went on, the thresher was refined and new features were added so that it became more efficient and faster. One important development was the combining of the thresher and separator soon after 1850. The new machine gradually replaced the tread power thresher, though a few of the latter are in use to this day. On December 13, 1888, Bonnell recorded in his diary, "Went in afternoon to Pittstown and bought a thresher and cleaner of Hiram Deats Jr. We agreed to pay him $195.00 and he allows us $20.00 for old machine and 10.00 off for cash." On December 22 he wrote, "The machine gained speed all the time. Threshed 100 sheaves of oats in 14 minutes [or] about one bu. of oats to the minute." After about 1890 large grain separators powered by portable steam engines were doing custom work for whole neighborhoods. Soon after 1900 gasoline engines were being employed for power, and within a decade gasoline tractors were appearing. The flail still had a use on the farm until about 1890, when cheap tarred twine made it no longer necessary to thresh out rye by the old method in order to obtain rye straw for bands. Another tedious task of the old days, treading out clover seed on the threshing floor with horses, was eliminated with the invention of clover hullers. The first ones were driven by water and usually were located at mills. The lever power threshing machines hulled clover after a fashion, and later a huller was devised to use with the treadmill. In the twentieth century the clover huller was adapted to fit new kinds of power.

Simple machines for slicing up root crops, pumpkins, etc., probably appeared quite early. In 1841 Philip G. Reading, who lived near Flemington, made such a device, had it patented, and exhibited it at the local tavern. In its early days the Agricultural Society gave prizes for root cutters. Another device was the straw and fodder cutter, in common use in the 1850's and probably much earlier. Run by hand or attached to a tread power, this machine was most useful in preparing feed and bedding. It has remained a common farm implement to this day, though going through various changes to fit it to modern power devices. A more recent development, one which came to Hunterdon County just before 1900, was the corn shredder. The first machines handled only husked fodder, but later ones were equipped with a husking apparatus. The first owners had regular runs for custom work, but later many farmers bought small shredders for their own work.

A modern development of great consequence was that of the silo and the ensilage cutter. The underground or pit silo was first developed in Europe, then copied in the Middle West. The *Hunterdon County Democrat* in 1881 did not know of a single one in Hunterdon County, though there were several in Pennsylvania. By the turn of the century, however, silos became more common, and most of them were of the type with which we are familiar today. Most of the early "patent silos" were of stave construction, those of tile not coming in until just before the World War. As these specialized buildings rose higher into the air, larger ensilage cutters with stronger blowers were needed. A silo near Pittstown in 1912 was forty feet tall and held the green corn from seven acres of ground. With an increase in dairying in this region the number of silos multiplied. Many farmers have in the last decade learned to supplement them with temporary storage devices of snow fence or woven wire and sisalcraft paper. In recent years there has been a sudden turning toward grass silage. Soybeans, Sudan grass, alfalfa, red clover, and other hay crops are put into the silo along with a certain proportion of molasses or other preservative. The new process bids fair to have a considerable effect on the whole dairying industry. The newest type of ensilage cutter is attached to the mowing machine and chops up the ensilage in the field. It is then hauled to the silo, into which it is put by means of a blower.

During the course of the nineteenth century machines appeared for a multiplicity of uses besides those already mentioned. In early days water was drawn from open wells by means of a rope or chain and bucket, sometimes with the assistance of a well sweep, a pulley, or a windlass. The first pumps, made of oak or hickory logs bored hollow with a pod augur, were common at the end of the eighteenth century. In 1838 two Flemington men advertised that they held patent rights for Hunterdon County for "Maxwell's Patent Cast Iron Double Lifting Pump." Dr. Bowne, always progressive, was one of the first to try it. The iron pitcher pump was probably in use even earlier than this. Another device which saw constant improvement but no great evolution was the cider press. The latest model was on exhibit at the first county fair in 1856. The horse-powered "grist mill" for grinding stock feed was an accepted appliance in the 1850's, but never became very common. Lime spreaders, at first homemade, were manufactured commercially by the 1850's. Manure spreaders did not appear on the scene until about 1880, but by the turn of the century were quite numerous. One of the features of the Farmers' Picnic in 1904 was "a manure spreader in operation." Wood saws, popularly called "buzz saws," were evidently not very common until the day of the gasoline engine. Improved household devices received attention at the various county fairs. In 1858 a churn operated by a dog on a treadmill was exhibited. Sausage cutters, apple parers, and washing machines were regularly entered for prizes. In 1862 a Hunterdon County native, Benjamin Horn, patented a washing machine which he put on the market for $12, and in 1878 a clothes wringer was exhibited at the County Fair. The incubator, a great labor-saving device, came into use in the 1880's. The milking machine has been common only during the past two decades.

Most of the agricultural machines of the nineteenth century either used man's strength more effectively or else substituted that of horses. One of the greatest changes in farm operations in recent times has been the application of new kinds of power. The portable steam engine was made available for farm operations by the middle of the last century. A small type on exhibit at the County Fair of 1857 sold for $12. Not many individual farmers bought steam engines this early; but threshing machines, clover hullers, ensilage cutters, and other equipment were operated by steam at a later time. Steam tractors were in general use for custom work by the 1890's. Of even greater importance for the farmer was the invention of the internal combusion engine. By 1900 gasoline engines were being used for threshing, grinding feed, pumping water, sawing wood, and other laborious tasks. In 1904 John Case of Ringoes replaced his steam engine with a gasoline engine to drive his threshing machine.

The true victory of the gasoline engine on the farm came with its application to the tractor. The first gasoline tractors were used about 1895, but a decade or so of improvement was necessary before they were a practical source of power. The first ones in this region appeared just before 1910. In that year a sales agency was established across the river at New Hope, Pennsylvania. Tractors were demonstrated at the county fairs and received increasing attention. Scarcity of labor and the high price of horses during the World War caused a great increase in their use. The tractor of two decades ago, with its wide driving wheels and big cleats, was pretty clumsy by present standards and not nearly so useful as the one of today, but each year brought new and improved models. The advertisements of implement dealers and lists of articles for sale at farm auctions show how steadily the use of the tractor has mounted. The same has been true of the farm truck, which has itself made considerable difference in farm methods. The type of power most recently exploited has been electricity, which has been equally important as a source of light. After the World War Delco-Light and other home plants became numerous. The light and power companies eventually realized the possibilities of the rural market and made efforts to tap it. More and more farms have put in electricity, and the electric motor is rapidly replacing the gasoline engine for many kinds of farm uses. Farmers would be badly hurt, perhaps dairy farmers worst of all, if they were suddenly deprived of this new kind of energy.

It is readily apparent that not all farmers accepted new and progressive ways of doing things at once. The Hunterdon County Agricultural Society was offering prizes for the best grain cradles and the best ox carts as late as 1878. There were still revolving rakes in use in the hayfields in 1908, and in 1909 a man at Rowland's Mills cut six acres of oats with a cradle and put it into a barrack. Old methods of making fences, of planting and caring for corn, of harvesting grain and hay, of raising poultry, and of caring for cattle have persisted long after the introduction of labor-saving devices. Nevertheless, the wheels of progress have turned, and although the average farmer even yet does a lot of hard manual work, much less of it is sheer drudgery than was formerly the case. Furthermore, the productivity of the individual has been multiplied several times, while costs of production have been materially lowered. Sometimes we are inclined to decry the passing of the old ways, but we should always remember that at best they were laborious ways and far less productive than those of today.

CHAPTER VII

Field and Orchard Crops

DESPITE THE CHANGES in farming methods, the crops grown in Hunterdon County have not varied greatly since pioneer days. Thomas Budd and Gabriel Thomas in their descriptions of West Jersey spoke of wheat, rye, barley, oats, Indian corn, buckwheat, Indian beans, potatoes, melons, squash, pumpkins, and hemp, besides many of the vegetables and flowers which are there today, though often in changed form. Four or five of the plants mentioned were originally of American origin, but the rest were brought in from Europe. An account of crops raised by the Dutch in New York gives practically the same list, with the addition of tobacco and gourds, the latter being important for making dippers and other utensils. Pioneers coming into Hunterdon County by way of the Delaware and Raritan routes brought these various crops with them. Inventories among the earliest probate records for the county mention "Indian Corne" and fodder, "wheat in the barn," "woats in the stack," "bawley," rye, flax, and hemp. There is no evidence, however, that tobacco was ever raised successfully in Hunterdon County.

Indian corn, or maize, was an important crop on nearly every American frontier. Because of its hill cultivation, it could be planted among girdled trees and stumps, and with little care could compete successfully with forest weeds. Its growing season was short, and its yield was large in proportion to the amount of seed used. The grain could be made into a variety of foods for man; it was an excellent feed for stock; and any surplus could be made into corn whiskey, a product whose market price was sufficient to warrant transportation over primitive trails. New Jersey pork from corn-fattened hogs was an important item of export.

It would be interesting to know more about the long course of selection and development which brought into being the many standard varieties of corn. In early days the ear was much longer and stalks much taller than today. An effort to get types with a shorter growing period was evidently not made until quite late, and Blackwell in 1849 considered that June 9 was very late for planting. In 1851 a good portion of his crop was "soft corn," caught by the frost. Probably the amount of yield was the point considered most important in selecting seed, and progressive farmers tried out many varieties. Blackwell, over a period of fifteen years, grew Virginia Dent, Red Cob Flint, Red Cob Yellow, Red Cob Dent, Monmouth Dent, and ordinary white corn. At a Pomona Grange corn show at Ringoes in 1889 there were exhibits with such strange-sounding names as Bloody Butcher, Gourd Seed, and Eight-row Yellow. Down through the years numerous varieties have appeared and disappeared. Scientific testing by experiment stations and actual field tests by the farmers themselves under Agricultural Extension Service guidance have made it possible in recent times to avoid a great deal of waste in securing satisfactory types. The most recent development has been that of hybrid corn. In 1935 six farmers of the county tried it out and reported good results. Its use increased rapidly thereafter, and today most of the corn of the county is hybrid. By 1940 the annual corn demonstrations sponsored by the County Board of Agriculture were largely devoted to testing varieties of hybrid corn.

The following census figures shows how important corn has been in the agricultural economy of the county during the past century:

Year	*Acreage Harvested for Grain*	*Bushels*	*Average*
1849		954,788	
1859		1,085,711	
1869		1,021,251	
1879	42,343	1,252,598	29.6
1889	34,789	936,585	26.9
1899	34,985	1,151,930	32.9
1909	32,156	1,083,816	33.7
1919	29,244	1,046,125	35.7
1929	18,110	661,542	36.2
1939	19,488	716,522	36.7

Besides a few hundred acres pastured each year, there has been an increasing acreage given over to production of ensilage. In 1919 there were 1705 acres so used; in 1929, 3325; in 1939, 6196. Because of the decreasing farm acreage of the county, better farming practices, and the need of raising more forage crops, the total corn acreage has decreased markedly in recent times.

The growing of wheat was of primary importance in the Middle Atlantic Colonies from an early period, probably because of the prominence of the crop in foreign trade. Wheat seems to have been successful from the first in this region, and in many areas usurped the place usually held by Indian corn. In Hunterdon County the evidence shows that both wheat and corn were important. In early days when clearings were small, the amount of wheat stored on farms was seldom large, but it increased as cleared areas grew larger in size. The wheat bins of Cornelius Van Horn of Readington Township contained 251 bushels in 1744. In 1765 Samuel Smith wrote concerning Hunterdon County, "The land is generally good for tillage; wheat the staple; their flour is carried to New-York and Philadelphia markets."

The continued growing of wheat, combined with other poor agricultural practices, eventually wore out many lands. John Rutherford sadly wrote in 1786: "Our whole attention seems to be taken up with raising Wheat, tho' it appears we are very little skilled in it." In a letter of November 5, 1792, John Hall spoke of Joseph Capner's yield of fifteen bushels per acre as unusually high, and three years later stated that much land in the Flemington area had been reduced "below bearing wheat." Better land care in the early nineteenth century brought up the yield for individual farmers, however. In 1841 the first prize in wheat production for the county went to William P. Emery, who raised 39½ bushels on one and one-fifth acres. In a diary entry of February 3, 1852, Blackwell figured that his average crop for 1851 had been 19.66 bushels per acre, which he characterized as "not a bad crop." In 1883 it was reported that some farmers were getting as much as thirty bushels per acre, which was over twice the state average for that year, and twice as much as Joseph Capner was able to raise in 1792.

Unfortunately there are no county-wide statistics as to crop yields prior to 1849. Census figures for wheat production in the county since that year are given below:

Year	*Acreage*	*Bushels*	*Average*
1849		188,828	
1859		241,805	
1869		340,178	
1879	21,365	234,795	11.0
1889	18,527	234,627	12.7
1899	22,404	284,150	12.7
1909	19,693	305,918	15.5
1919	18,200	260,166	14.3
1929	13,615	241,787	17.8
1939	13,079	288,272	22.0

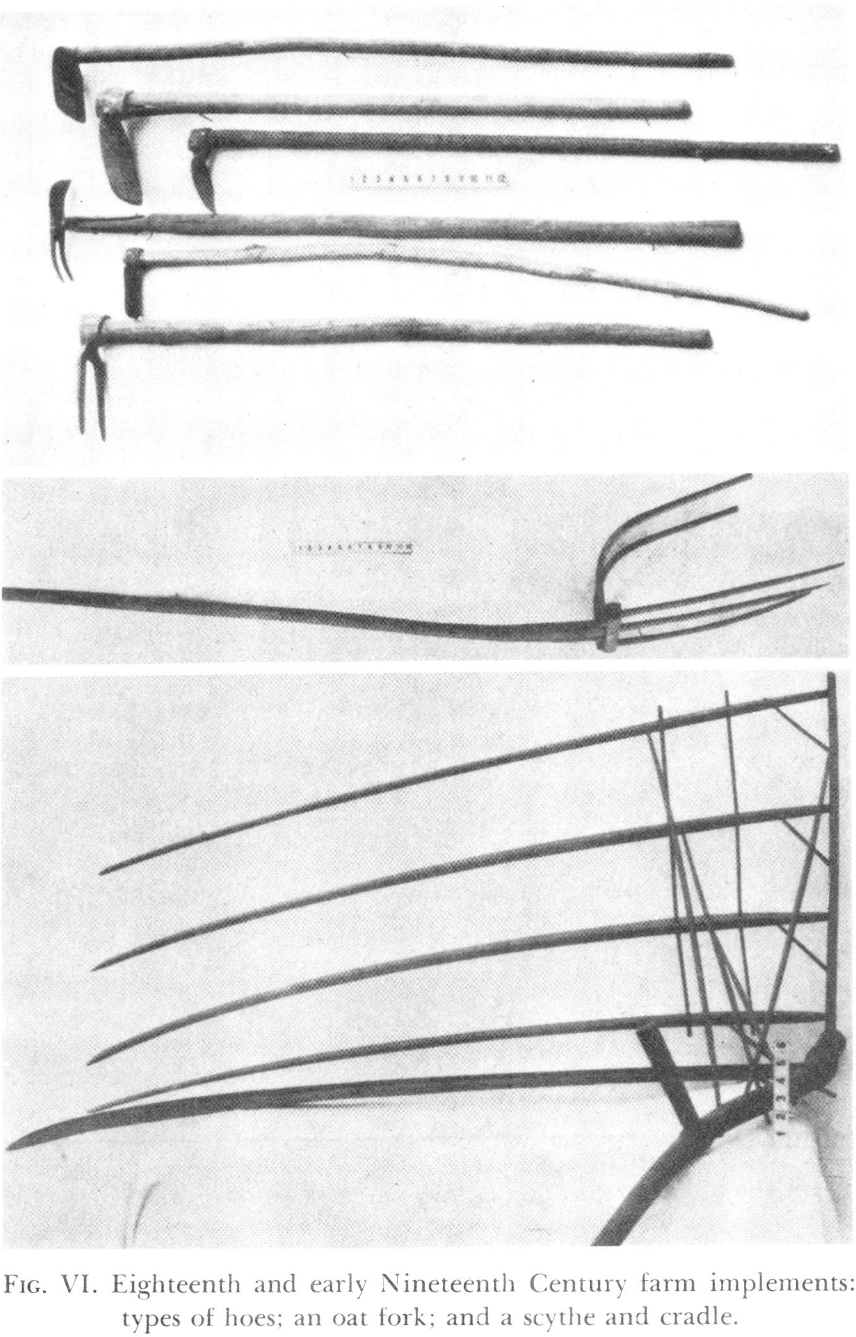

FIG. VI. Eighteenth and early Nineteenth Century farm implements: types of hoes; an oat fork; and a scythe and cradle.

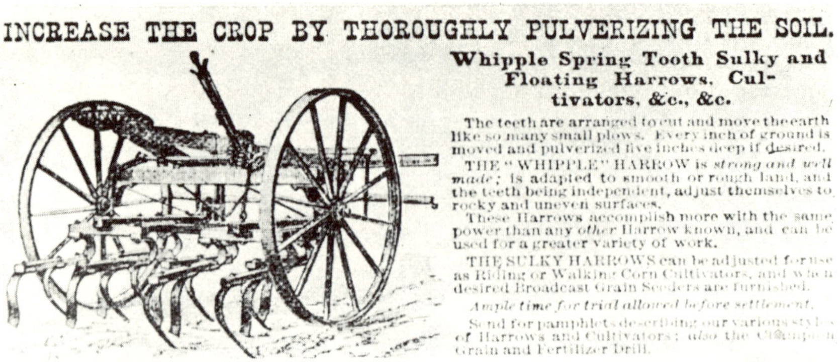

FIG. VII. Corn cultivator advertisement in the *Democrat-Advertiser,* May 2, 1884; wooden moldboard plows; an early iron moldboard plow.

The area devoted to wheat, fairly constant for several decades after 1879, was about one tenth of the improved acreage of the county. From the smallness of the average yield, it would seem that growing wheat on poor lands was never very profitable. Use of wheat as a nurse crop for various grasses and legumes probably accounts to some extent for its persistence. The decrease in the amount of wheat sown since 1919 has been accompanied by a much higher average yield. This increase is due to better crop practices and to use of new varieties as well as to the withdrawal of submarginal lands from production.

Very little has come down to us concerning the methods by which the pioneers found the varieties of wheat and other crops best adapted to the climate and to particular soils. Evidently only winter wheat was grown in West Jersey, and this was probably because it was less subject to rust or "blast" than spring wheat. An old tradition in the Fairmount area tells of a farmer named Fox who brought in or developed a variety so superior that farmers came from miles around to buy seed from him. Fox Hill is said to have been named after this man. Later experimentation, of which we have some record, was probably illustrative of that practiced by progressive farmers from a fairly early time. In 1805 a member of the Hankinson family of Amwell Township recorded an experiment in which he sowed Red Chaff wheat and bearded wheat in neighboring fields. Later in the year he wrote that the returns from both were equally satisfactory. Philip Case recorded in 1822 that he had sowed three strips in "Cannaday," White Bearded, and Red Chaff wheat respectively. Unfortunately, we have no record of his results. In 1844 Blackwell experimented with White Flint seed from the Patent Office and at another time sowed white Genesee wheat. Each year he sowed several varieties then in common use and recorded their relative merits. By such experimentation farmers found which varieties were most satisfactory for their own farms and their own neighborhoods.

Hardiness was considered the most important characteristic of wheat in early days. But after the appearance of the Hessian fly in the late eighteenth century, resistance to that pest was the chief point of merit for any particular variety. Since this insect attacks the maturing stalk, the sturdiness of the stem of the plant was of prime importance. In the 1830's Mediterranean wheat was hailed as "exempt from the fly," and sold at premium prices. Blackwell tried this wheat but reported that it, too, was "much crinkled down" in some years. Other pests which bothered him were the chinch bug and wheat rust, but he made no efforts to find resistant varieties. Some farmers probably knew that much of the damage from the Hessian fly could be avoided by late sowing, but the knowledge was evidently not widespread until fairly late. With the coming of experiment stations and commercial seed concerns, the testing and developing of varieties has been taken out of the hands of individual farmers.

Rye is hardier and surer than wheat, and early inventories show that it was once quite important. Foreign demand during the Colonial period may partly account for this. Kalm in 1748 and Cazenove in 1794 spoke of rye as if it were of some consequence in areas near Hunterdon County. Since the average yield per acre in New Jersey is less than that for wheat, there has been little incentive in modern times to keep on sowing rye, except occasionally as a cover or pasture crop. The following table taken from census figures indicates how rye cultivation has decreased during the past century in Hunterdon County:

Year	*Acreage*	*Bushels*	*Average*
1849		133,662	
1859		120,741	
1869		26,799	
1879	11,057	87,418	7.9
1889	7,019	61,398	8.7
1899	5,682	64,540	11.2
1909	6,832	80,679	11.8
1919	7,226	77,137	10.7
1929	1,404	19,674	12.1
1939	226	3,690	16.3

Oats have been grown in this area from the days of earliest settlement, and throughout the eighteenth century probate papers mention large or small stores of this grain. In the nineteenth century it still remained an important crop, grown by nearly every farmer. Oats culture before the coming of the grain drill was very simple. The ground was plowed, then the seed broadcast by hand and harrowed in with a spike-toothed harrow. On April 4, 1851, Blackwell plowed a small field, sowed his oats, and harrowed the ground all in the same day. On the following day he sowed clover seed on the same field and again went over it with a harrow. The process was sometimes finished off with a roller. There seems to have been little experimentation as to varieties of oats until late. In 1860 there were at the county fair exhibits of black oats, white oats, Poland oats, and one-sided oats, the last evidently a great curiosity.

Beginning in the late 1860's, there was much excitement over the newly introduced Norway oats. This variety was vigorously pushed by the United States Patent Office, and stories of enormous yields were told. For a time it commanded the premium price of $3.00 per bushel, but its fall from public favor was as sudden as its rise. In 1878 the kinds of oats receiving most attention at the County Fair were the Waterloo, Schurman, and White Dutch varieties. A few years later the Excelsior variety became very popular.

Seed companies eventually multiplied the number of kinds of oats from which to choose, and the experiment stations and Agricultural Extension Service have relieved the farmer of much of the hit-or-miss experimentation which was necessary earlier. The following table gives oats production in Hunterdon County during the past century:

Year	*Acreage Cut for Threshing*	*Bushels*	*Average*
1849		679,539	
1859		830,653	
1869		902,737	
1879	29,584	854,852	28.9
1889	26,034	581,341	22.3
1899	22,802	486,740	21.4
1909	22,772	459,418	20.2
1919	23,390	437,994	18.7
1929	11,395	288,463	25.3
1939	12,674	333,131	26.3

Besides the acreage given above, there has always been a small amount of oats cut for hay. The decrease in importance of the crop during the past twenty years parallels the great decline in the horse population of the county and of the nation. A factor in the decline of oats growing has been the increasing use of corn as a feed for dairy animals.

The usual Hunterdon County breakfast in 1787, according to Mrs. Capner, included buckwheat cakes. In winter these cakes were leavened, in summer unleavened. As far as we can tell, the first statement would apply at any time during two centuries, and to some extent even yet. In early days nearly every farmer had a patch of buckwheat. In 1748 Kalm found that the first two weeks of July were the season for sowing. Any surplus not used for human consumption could be fed to poultry and swine. The straw, considered worthless, was spread on the fields and orchards. Isaac Passand wrote in 1793 that buckwheat was often used as a nurse crop for timothy and clover. Naturally this grain was a common article of commerce. An action in 1782 before Peter Brunner, a Readington Township justice of the peace, was for forty bushels of buckwheat. Stores usually kept a few bushels on hand for retail trade, and buckwheat by the bushel or "in the ground" was often sold at farm auctions. Blackwell always raised some buckwheat for family use; and on August 5, 1854, he recorded in his diary, "We had a nice little buckwheat shower last night." In 1879 the local census taker reported that 144 of the 281 farms of Readington Township had grown buckwheat during the previous year. Their average crop was about forty-five bushels per farm.

But from the time of the Civil War to the present, the acreage devoted to this crop has constantly dwindled, probably because of competition of more productive regions. By 1882 the Hunterdon County Agricultural Society had withdrawn all prizes for buckwheat entries because of the decline in importance of the crop. A Grange exhibit at Ringoes in 1889, however, still included an entry of silverhull buckwheat, and unsuccessful efforts were made to revive interest in its production. The following table shows how production in the county has declined since 1849 and especially since 1909:

Year	*Acreage Threshed*	*Bushels*	*Average*
1849		103,271	
1859		91,835	
1869		41,527	
1879	5,741	82,099	14.3
1889	2,911	16,692	5.7
1899	3,291	51,280	15.6
1909	4,367	78,679	18.0
1919	2,296	34,722	15.1
1929	565	11,104	19.6
1939	299	5,587	18.6

Barley is a cereal which has been raised in the county since early days, but never in significant quantities. Stray references through the years show some interest in this and neighboring regions, but for some reason the crop was not tried by most farmers. Blackwell grew a few acres each year in the early 1840's, but stopped after a time. In 1849 only an occasional farmer reported having a small barley field. From that time until recently the yield of the county was never as much as five thousand bushels in a census year. Under Extension Service auspices, and because of the need of finding a substitute for wheat under the Agricultural Adjustment Administration program, a slight movement toward increasing this crop occurred in the late 1930's. The acreage in 1939 was 527 and the yield 13,343 bushels, more than twice that of rye.

The securing each year of an adequate hay supply to carry his animals through the next winter has been a perennial problem for every farmer. The pioneers of West Jersey found native grasses in abundance and proceeded to use them for both pasture and hay crops. Some of the native lowland grasses survived and were apparently for a long time important sources of hay, and to this day occasional farmers have small "bottom meadows" of native American plants. Unfortunately, most of the American plants were annuals and easily killed out by close cropping. For that reason English perennials were soon imported. Various sources speak of English grasses used in Hunterdon County but do not mention actual types during the early period. Probably the chief ones grown for hay were timothy and red clover. In the 1780's the Capners and their friends wrote back to their native Leicestershire for seeds of timothy and various clovers. Sometimes they wished to get better seeds, at other times they wanted varieties not found here.

Timothy, sometimes called herd's-grass, was often sown with wheat and was the most common of the English grasses. The *Pennsylvania Farmer* in 1804 emphasized it as a common crop and mentioned as other hay grasses redtop, orchard grass, burnet, and sainfoin. Unfortunately, we have no means of knowing to what extent or in what combinations these various crops were used until a comparatively late period. Hay by the ton became a standard article of commerce in Hunterdon County, but those buying and selling did not usually feel it necessary to specify the kind of hay in their records. When they did, it was usually timothy; and it is probable that this crop, which grows well on most of the soils of the county, was the most important one for hay. Locally produced seed was given preference in order to avoid importing new weeds with the timothy.

Clovers of various kinds were known early in America, but were not grown to any great extent until after the Revolution. Just when clovers were successfully introduced into Hunterdon county is not known. According to a tradition at Wertsville, an itinerant Scotch teacher showed the farmers of that locality how to raise red clover. Clover seed at a shilling per pound was an article of commerce at Flemington in 1790. In 1793 John Hall wrote that farmers were beginning to raise various clovers and he was hopeful that their use would spread; and at another time he suggested that his relatives bring trifoil seed with them to America. In 1795 Mary Choyce wrote to relatives in England, "Mr. Case begins to mow clover tomorrow June 1st and thats more than you will do." Dr. John Bowne, at least from 1802 on, had one large field or more of red clover each year. The *Pennslyvania Farmer* in 1804 deplored the growing of clover and timothy together, as the two have a different maturing time. This advice was the exact reverse of that given by the New Jersey Experiment Station in 1925, in which the common practice of over a century was approved. The production of clover seed was a laborious job at best, even after the coming of the tread power huller, and farmers seldom had much surplus for sale. Blackwell charged his neighbors one-fourth toll when they brought their crop to him for hulling. The competition of areas with a less humid climate caused farmers of this region to stop hulling clover before 1900. One consequence of this has been the importation of dodder, a bad meadow weed, in the clover seed which was brought in from other parts of the country.

Other clovers have been introduced from time to time, but red clover remained the most popular. Crimson clover was introduced in the early 1890's, and alsike probably a few years later. Just when sweet clover, never very important, was first grown is not known. All of these have usually been grown as companion crops with timothy. The following table, taken from census records, gives some indication of clover and timothy acreage in the county during recent years:

Year	*Clover Acreage*	*Timothy Acreage*	*Mixed Timothy and Clover Acreage*
1899	3,209	(not given)	(not given)
1909	953	17,499	21,781
1919	1,315	12,899	18,062
1929	1,417	(not given)	24,922
1939	(not given)	(not given)	20,068 (including clover alone).

Only in recent years has the predominance of these forage crops been at all threatened. Farmers have on occasion, especially in dry years, turned to millet and Sudan grass, but the acreage has never been large. Two recent contenders among forage crops are alfalfa and the soybean. In 1795 John Hall reported that farmers about Flemington had been trying lucerne, as alfalfa was then called, but the experiment did not work out well. Over a century later the New Jersey Experiment Station and the Hunterdon County Board of Agriculture were having difficulty in getting Hunterdon County farmers even to try this crop again, and there were only fifteen acres of alfalfa in the county in 1899. In 1906 and 1907 a few farmers were cooperating by planting two-acre plots for observation; but although the experiments were successful, the difficulties of getting a crop started deterred most farmers from trying it. The modern practice of inoculating seed has changed the situation markedly. In 1919 there were nearly two thousand acres of alfalfa in the county, in 1929 over three thousand, and by 1939 over seven thousand. The growing of soybeans to any great extent is of very recent date. In 1925 only 110 acres were grown in the whole county. Four years later eighty-seven farmers reported 526 acres of soybeans grown alone and 83 acres in which they were mixed with other crops. The success of the new crop plus the necessity under the Agricultural Adjustment Administration of finding substitute and soil-building crops led to a great increase in acreage in recent years. In 1939 Hunterdon County farmers grew 3230 acres of soybeans for hay and grain, besides a large but unspecified amount plowed under for green manure.

A food crop unknown to the Hunterdon County pioneers was the potato. The plant originated in America, but so far as is known was not cultivated by the Indians of the Eastern United States. Apparently it was introduced into the English colonies by Scotch-Irish immigrants early in the eighteenth century, and slowly taken up as a garden vegetable. Potatoes were not a common crop in New Jersey before 1750, and Peter Kalm merely mentioned them along with other garden crops. Inventories in the Hunterdon County probate records seldom mention the crop at all prior to the Revolution, which probably indicates that the small patches grown for family use were reckoned of little monetary value. A tradition at Sandy Ridge says that an immigrant school teacher taught the skeptical natives the advantage of cutting their seed potatoes rather than planting whole ones.[1] Yet by the end of the eighteenth century potatoes were an essential part of every meal, and the production of a crop was a matter of importance for every family. The sudden growth in popularity, common to much of America, may be partly explained by the ease of growing and storing potatoes for winter use. Since they were not ordinarily an article of commerce before 1800, it is probable that most people who ate them grew their own. By the 1820's, however, they were as standard in country trade as were butter and eggs.

The earliest mention of a potato variety name found for this area is

1 L. A. Groff, "Van Dolah's School."

in the records of Solomon Holcombe, who was selling Foxite potatoes at his store in Mount Airy in 1824. A quarter of a century later the Foxite was still a favorite, but had rivals in the white, blue, and common Mercer potatoes. At the county fair in 1858 prizes were given for the white Mercer, common Mercer, Foxite, Pink Eye, Pink Eye Kidney, Sand Lake, Scotch, and common white varieties. By 1860 the Peach Blow, Buckeye, and Mexican potatoes had appeared, and within a few years many new varieties, such as the Prince Albert, the Early Goodrich, and the Early Rose were being ably extolled by their proponents. Apparently the claims made for the Early Rose were justified, for it became and remained for a long time a popular potato. In the 1870's a Flemington merchant, John S. Emery, sought to boost production—and his sales of seed potatoes—by offering each year several hundred dollars in prizes for large yields. The names of the varieties which he introduced have long since disappeared, as have most of those which appeared at a later period. For example, the "Queen of the Valley" was a favorite for several years during the 1880's. Continuous selection by seed growers and experiment stations to secure better keeping and handling qualities brought into being the modern commercial varieties, and older types have been gradually abandoned. On March 24, 1903, the editor of the *Hunterdon County Democrat* mourned the passing of "those little kidney-shaped potatoes that were ready to dig early and when cooked with green peas made the most toothsome of dinner vegetables."

The production of potatoes was of importance on the Blackwell farm in the 1840's and 1850's, and several varieties were always grown. Digging, sorting, and burying the yield were regular tasks each autumn, and from time to time during the winter the "potato holes" were opened for sorting out potatoes which had not kept well. Except for the substantial amounts put into the cellar for home consumption and those kept for seed, the crop was hauled to nearby village stores for sale. In 1849 Blackwell sold over three hundred bushels. Potatoes were definitely a commercial crop for him and for his neighbors, and the local census returns taken in 1849 indicate that such was the case for many farmers. As other regions more adapted to potato growing came into fuller production two or three decades later, Hunterdon County farmers turned to more remunerative ventures. A few, particularly those in the Landsdowne soil areas, have kept on and have even expanded their operations. A leading producer for the past forty years has been Clifford E. Snyder of near Pittstown, and a more recent one is his neighbor Robert Crane.

But the production of the county as a whole has declined rapidly, as is shown by the following table:

Year	*Acreage*	*Bushels*
1849		78,526
1859		92,985
1869		86,807
1879	1,718	115,823
1889	1,138	52,708
1899	1,112	88,073
1909	950	66,663
1919	842	48,685
1929	297	27,733
1939	172	26,331

Broomcorn was a crop produced on a small scale probably from pioneer days. It was easily grown, its culture being like that of corn. In the fall the heads were gathered and stored, and when time permitted, the seeds were stripped off for poultry feed. The straws were later bound into brooms by amateur home broom makers, by itinerant professionals, or by artisans in small shops, like the ones which functioned at Flemington and Frenchtown. The coming of machine production and the rivalry of areas more adapted to the plant caused both broom manufacture and broom corn cultivation to die out in this area. The census of 1890 gave one acre of broom corn in the state, none in Hunterdon County.

A most interesting example of how distant happenings could affect the economy and daily lives of the farm folk of this region was the Civil War development of sorghum cane. To go with the standard breakfast of buckwheat cakes and sausage, it had long been considered necessary to have available a full pitcher of New Orleans molasses. This article was one of the most important items at every country store. When the Civil War broke out, shortage of this old stand-by created a dire emergency, though it would be going too far to say that this explained the opposition to the war within the county. However, there were men ready to rush into the breach. In nearly every neighborhood a farmer or villager put up a horse-power sorghum mill and saw to it that his neighbors had an opportunity to get "Chinese sugar cane" seed. Farmers usually put out patches of from a quarter to a half acre, enough to produce fifty or more gallons of syrup. Before frost the stalks were stripped while standing, then cut down and carted to the mill. The sorghum maker collected from 40¢ to 50¢ per gallon or one half the sorghum as toll. This molasses was quite satisfactory with wheat cakes for breakfast or with mush and buttermilk at supper, but cooks complained much concerning its baking qualities. When New Orleans molasses returned to grocery shelves, the growing of cane declined at once. The last season for most sorghum mills was in 1864. In 1869 only 2527 gallons were produced in the county, and by 1889 the crop had almost entirely disappeared.

Flax was a crop of some consequence, though of small acreage, during much of the county's history. Eighteenth-century inventories mention partially manufactured flax and the various tools and implements used in its preparation. Kalm in 1748 and Burnaby in 1779 reported that flax was a common crop in West Jersey. Bounties offered by the English government in 1764 for flax shipped to England probably did not have much effect on production, but bounties paid by the Province of New Jersey between 1765 and 1772 may have had a temporary effect. However, little incentive was needed to cause farmers to grow sufficient flax for home use. Local store, mill, and personal accounts show how general was cultivation of the crop in Hunterdon County. By the terms of a lease in 1811, the renter was to give his landlady one third of the flax, but was allowed to keep all the seed. Each farmer apparently raised enough of this crop in early days to keep busy the spinning wheels of his own household. If possible the crop was sowed on Good Friday. An old superstition was that if the owner saw a tall, well-dressed woman on St. Valentine's Day, his flax would be of good length and well coated.[2] The patches of tall green plants with their delicate blue flowers and waxlike balls were definitely ornamental to every homestead.

As manufactured cotton cloth became cheaper, home manufactures of flax declined, and by 1849 the process had gone so far that relatively few households except in the northern part of the county had home products of significant value. For a time, however, farmers continued to grow it for seed, and the number of oil mills increased. The most famous of these was that at the King Mills near Quakertown. In 1870 Joseph King advertised in the *Hunterdon Republican* that the manufacture of linseed oil at that place began in 1814. His oil mill is said to have operated until 1886, that at Frenchtown until the early seventies, and that at Spring Mills until the sixties. Apparently it was the fact that the farmers quit raising the crop that killed the linseed oil industry

[2] J. W. Lequear, "Flax Culture and Manufacture."

in this area. Flaxseed, so long a familiar item of trade at every store, disappeared as an article of everyday commerce.

Only one side of the story has been presented. As home industry declined, utilization of the straw had become a problem for those still raising flax. Apparently many farmers who were producing flaxseed threw away the straw. Where long hauls were not necessary, growers often sold it to rope and twine factories or to storekeepers and millers who acted as their agents. From the 1830's on there were such factories at Lambertville and Elizabeth, and others later operated for a time at Frenchtown, Milford, and Hamden. Pulled flax, harvested and cured by the old methods, was worth far more than cut flax or flax straw. In 1839 John W. Snyder of Clinton offered from $5.00 to $8.00 in trade per long ton for "cradled flax, if rotted," and $17 for "full rotted flax in the stem." [8] If flax was pulled, however, the seed, being immature at pulling time, was of less value. Moreover, pulling flax by hand was a very laborious job. In the late 1830's William Brittain and John Silvers of Amwell Township invented and patented a flax pulling machine, which did custom work for at least three seasons, but eventually dropped out of the picture. By this time it was more usual to cut flax than to pull it.

Apparently it was only by a judicious utilization of all parts of the plant that the growing of flax could be made to pay. Areas far from market were already dropping out of production in 1850. In Bethlehem Township only sixteen out of 212 farmers grew it. Yet in districts where the straw could be sold, the crop evidently paid for a time. The following census figures show the rise and fall of flax as a commercial crop:

	Pounds of Fiber	*Bushels of Flaxseed*	*Tons of Straw*
1849	122,153	9,563	
1859	44,589	2,674	
1869	226,936	5,916	
1879	46,401	5,127	759
1889		8	1

The low figure for 1859 indicated a downward trend temporarily stopped by the Civil War, which shut off much of the cotton supply. During that time flax fiber production received an impetus still reflected in 1869.

During the 1870's flax declined greatly in importance, and the only cordage factory which remained in operation in New Jersey was that

[8] *Hunt. Gaz.*, Aug. 20, 1839.

at Lambertville. In 1880 the state legislature tried to save the day by offering bounties of $7.00 per ton for flax and 3½¢ per pound for dressed fiber. The following year, however, it hastily amended the act to exclude mowed or cradled flax. The bounty caused immediate action. In Hunterdon County a flax factory was started at Frenchtown by a Philadelphia group, and farmers were offered $20 per ton for pulled flax and $15 for mowed straw. Under various owners this concern operated for at least three years. Probably the cordage factory at Lambertville flourished during the same period also. Farmers in areas which had long since given up the crop rushed to buy flaxseed, and soon patches of two, four, and even eight acres dotted the countryside. One Kingwood Township farmer was reported to have grossed $90 per acre. But when the artificial stimulus of the bounty was removed, the whole industry collapsed. By 1889 there were but two acres of flax in the county. Competition of other areas producing flaxseed and the appearance of cheaper fibers help to explain its decline. Other factors were the great labor involved, the belief that flax exhausted the soil, and the fact that it did poorly on recently limed land.

Garden and truck crops have never been of great commercial importance in this area, though home gardens have always been a mainstay. European travelers spoke deprecatingly of American gardens, but Rutherford reported in 1786 that the women made such good gardens that they were "half the support of their families." Lists of the names of common vegetables of this and neighboring areas at various times during the past two and one-half centuries contain a great many of the favorites of today, and few that we no longer have. Sweet potatoes, then called "Bermudian potatoes," came into West Jersey just before 1750. Although there have been efforts to make them a commercial crop, the production in Hunterdon County has never been of great consequence except for home use. Rhubarb, or pie plant, did not come in until about 1835, and it has remained a crop for home gardens.

For a time around 1840, sugar beet production was encouraged in this area, but the crop did not win public favor. Turnips were of some importance for a long period of time, being stored for the winter use of both man and beast. Their popularity has declined, though some farmers now raise the rutabaga type. Hops were grown on a small scale during the middle of the past century. Cabbage was probably much more important in the days of homemade sauerkraut than it is today, though in recent years there has been a small amount of commercial production. The much-esteemed watermelon was produced more widely in earlier times than now, probably because soils were newer and transportation of melons from other areas impracticable. Pumpkins and squashes, grown from earliest days, were commonly fed to hogs and cattle. Many farmers thought it was necessary to remove stems and seeds lest the cattle choke themselves. Farmers long vied with one another as to who could raise the biggest pumpkins. In 1869 the prize winner at the County Fair weighed 114 pounds officially. This crop is still grown in the cornfields of the county, but probably not as much as formerly.

The tomato or "love apple" was long thought to be poisonous, but was fully accepted in the vegetable gardens of Hunterdon County by 1847. Since that time it has always been popular for home use. As a commercial crop, however, it has had its ups and downs in this region. For over thirty years, beginning about 1863, the Butterfoss cannery at Lambertville supplied a market for tomato growers, using annually in later years the product of some three hundred acres. Other canneries outside the county, particularly those at Pennington and Hopewell, New Jersey, and Titusville, Pennsylvania, also bought some Hunterdon County tomatoes. In 1890 a catsup plant was built at Stockton, and in 1893 and 1895 canning establishments were started at Ringoes and Bloomsbury. The canneries put up small amounts of corn, beets, and apples, but their chief concern was with tomatoes. Prices paid the farmers averaged from $7.00 to $8.00 per ton. For one cause or another these plants ceased operations soon after the turn of the century. Some were purchased and closed by competitors, while others failed in business. Tomato growing naturally declined as a result.

The growing of tomatoes for the New York market began about 1900, with the center of production in the Cherryville-Quakertown-Pittstown area. In 1927 there were shipped from Flemington 61,253 crates. The crop was always something of a gamble, bringing handsome returns in some years and only losses in others. In recent years tomato growing in the county has staged a comeback because of the operation of large canning establishments at Trenton and elsewhere. In 1936 there were 140 acres in Hunterdon County under contracts with these companies, by 1939 over 300 acres, and by 1944 over 1000.

There has been during the past two or three decades a small amount of commercial production of other vegetable crops. Each year a few hundred acres of sweet corn and a few fields of string beans and other vegetables have been grown. The latest development is the growing of peas and lima beans under contract to canning companies. The first year for this kind of arrangement was 1939. In that year eight farmers in the Pittstown area grew forty-six acres of peas and forty of lima beans. In 1940 the county acreage nearly tripled that of the year before, and considerable interest was shown in the new project. Much credit for this development goes to Robert Crane, chairman of the vegetable committee of the Hunterdon County Board of Agriculture. Approximately one hundred acres of peas and 225 acres of lima beans were contracted for in 1944.

In this discussion of crops we have said little about the farmers' disposition of their lands. In the era of general farming, each had his arable divided among various grain and forage crops. Because of the necessity of using nurse crops for grasses and the advisability of planting corn occasionally on each field in order to retard weed growth, there has always been a certain amount of rotation of crops, though a too heavy emphasis on the production of grains. Rutherford in 1786 wrote:

> These lands are in general laid out in Farms from fifty Acres to four hundred Acres, and of each Farm the arable Land is commonly laid out into four Fields, and the Course of Crops is 1st year Summer Crops of Indian Corn, Oats, Flax, and Buckwheat, 2d year fallowed, 3d year in Winter Grain, and 4th year in Grass. [B]y this frequent tillage, and from the small stock of cattle having little help by manure, the Lands are very much exhausted. . . . [B]esides the arable each Farm has commonly an Orchard, a piece of natural meadow ground, and a proportion of one-fourth of timber Land for fencing stuff, Buildings, Farming Utensils, and Firewood.

Farm advertisements all through the nineteenth century show similar division of farm land as to function. The number of fields on a farm varied with the topography of the land, but three or four main categories of crops were almost always given. A Bethlehem Township farm advertised in the *Hunterdon Gazette* in January, 1839, is a fair sample. Part of the farm was "seeded down in the best manner with clover and timothy." Another portion was in grain "ready for seeding with grass in the spring." The remainder of the tillable land was to "come in for oats and corn in the spring." From necessity farmers eventually showed more concern about overcropping. Better rotations were practiced and green crops plowed under. In 1825 the *Hunterdon Gazette,* quoting from an agricultural magazine, suggested a rotation which called for plowing under clover once each five years. The average farm in this area has always had a diversity of crops, though there has been a gradual shift away from grain to hay and forage.

Gabriel Thomas, one of the earliest writers of West Jersey, reported with great pleasure that, "befriended by the Sun's hot and glorious Beams," fruit trees here bore earlier and more prolifically than in England. English settlers in Burlington County before 1700 were growing apples, peaches, quinces, apricots, plums, cherries, grapes, gooseberries, currants, raspberries, blackberries, and strawberries. As the area of settlement spread, pioneers naturally took these plants with them. The West Jersey Society required that its lessees in Hunterdon County put out one apple tree for each acre leased. Kalm in 1748 wrote of large orchards of peaches, apples, and cherries near Princeton, very near to Hunterdon County. Fruit trees in New Jersey, particularly peaches, tended to bloom before the last frosts, and it had already been discovered that because of air drainage the best places for orchards were on hillsides.

A very good eyewitness description of the orchards and wild fruit of the central part of Hunterdon County was written by Thomas Capner in a letter of November 17, 1787:

> The planted fruits are apples, plums, cherries, currants, etc. Almost every person has an orchard which is thought small except it is 9 or 10 acres. The trees are planted in rows about 6 or 7 yards apart for the convenience of plowing them and cherries are in proportion as apples. Almost every farmer plants a number by the roadside, so that any person that choose may gather what they want in a good year and likewise peaches, but they are not quite so plentiful here as cherries, though they are more so in some places. Plums are not so plentiful as in England, except wild plums. Wild fruits are grapes, wild cherries, strawberries, raspberries, blackberries, chestnuts, hickory nuts, walnuts, hazelnuts, butternuts, persimmons. There are three kinds of grapes, 1. Fox Grapes. They are very large and sweet, only have a disagreeable [odor] while eating, though a many people are very fond of them. They make excellent pies. . . . 2 are a small grape not quite so sweet as the English. They would make very good pies were it not for the number of seeds they contain. 3 are a white grape which eat exceedingly well. Wild cherries grow in bunches like currants, eat something like a sloe only bitterer, but not quite so sour. Strawberries in some fields grow nearly as thick as [?]. By places we gathered a considerable quantity. Raspberries and black berries grow very thick on bushy land, but are not quite so large nor so good as English, but in some places people cultivate them and there they are preferable.

Orchards contributed considerably to the diet, both solid and liquid, of their owners. Cider, both sweet and fermented, was a favorite beverage from pioneer days. The inventory of John Reading's estate, the first settled within the present county, mentioned a cider press. As the market for brandy grew, larger orchards were planted. In 1802 Hugh Exton sent twelve hundred bushels to a distillery and kept back three hundred bushels for his own supply. At a later time the production of "Jersey lightning" by church members and even by ministers was to prove a bone of contention in many churches. But before temperance days such things were taken for granted. However, there were other and more sustaining uses for fruit. Apple and cherry pie, apple dumplings, cherry pudding, and other fruit dishes were much extolled; and apple butter and "peach leather" were mainstays of the winter. Before the Revolution Crévecœur spoke of the universal American custom of drying various fruits for winter use. The dried peaches which John Hall took from Hunterdon County to England with him in 1790 were there pronounced delicious by critical relatives. Dried apples, peaches, cherries, raisins, fresh apples by the bushel, other fruits in smaller quantities, cider by the gallon or barrel, apple brandy, and other fruit products were standard items of commerce over a long period of the county's history.

The English settlers near Flemington after the Revolution approved of the varieties of fruit which the Americans grew, but Isaac Passand, who owned a nursery near Philadelphia, imported the choicest of English types and planted some on his Hunterdon County farm. In the days when all farmers were orchardists, there was widespread interest in every new variety. It was long a countryside tradition at Ringoes that all the "Fallawater" apples in that section were descended from a single scion brought from Shamokin County, Pennsylvania.[4] Farm advertisements of the early nineteenth century emphasized grafted fruit of many kinds. A Readington farmer in 1845 advertised for sale young peach trees "inoculated from the choicest variety . . . that are sold in the New York market."[5] At about the same time Blackwell had in his orchards Crawford, Old Mixon, Freestone Heath, and other peaches, Golden Fall Pippin, Bow, Bellflower, Rhode Island Green, July Sour, and other apples, Bartlett pears, and at least two varieties of plums. He ran his own farm nursery and made considerable effort to obtain new varieties. A Mount Airy Negro brought him cuttings from a choice wild grape, and Blackwell christened the new vine the "Robert" after the donor.

Other fruit growers were as enthusiastic about their orchards as was Blackwell. At the county fair in 1860 one farmer of the county exhibited sixty-one varieties of apples, while a New York nursery, in-

4 Dr. C. W. Larison, *Fisher Family,* p. 62.
5 *Hunt. Gaz.,* Oct. 29, 1845.

terested in getting customers in the locality, showed thirty-five kinds of apples, seventy-two of pears and thirty-five of plums. A Copper Hill nursery in 1857 sold its customers fifteen types of peaches, and one at Sand Brook in 1892 advertised thirty-five kinds. In the spring of 1890 Elmer Bonnell planted seventeen varieties of peaches in a new orchard. An orchardist, by choosing his varieties carefully, could have his fruit ripening at a constant rate over the season. Certain kinds were stand-bys over a long period and were much mourned in passing. The editor of the *Hunterdon County Democrat* on November 21, 1882, wrote, "The true old-fashioned bell-flower apple is as scarce in these days as honest politicians." Now that numerous new commercial varieties have replaced former types, old men recall with nostalgia the Sweet and Pippin apples, the Harvest pear, and the Honest John, Reeves' Favorite, and Crawford peaches.

One of the most interesting developments in the county's economic history was the half-century-long commercial production of peaches. Long afterwards there were a number of stories current as to where and when peach growing here got its start. As a matter of fact, this fruit had been grown since early days, and expansion to commercial proportions was doubtless gradual and not ascribable to the leadership of any one person. A most important factor was the development of fast transportation facilities to large markets. In the 1830's the area about Trenton grew many peaches, and expansion of commercial growing a few miles north into Hunterdon County was to be expected. In 1852 Dr. George H. Larison planted an orchard of three thousand trees near Sergeantsville and started a nursery. His neighbors at the time laughed at his experiment, but a few years later he was hailed as "an experienced peach grower," and an account spoke of the "unnumbered thousands of trees in the county." [6] By this time Flemington, Ringoes, and Lambertville were sending consignments of peaches regularly to the New York market by rail. A new factor in farm values was the ability of the land to produce peaches.

In 1862 orchardists of five Hunterdon County townships and of Hopewell Township in Mercer County met with representatives of the Delaware and Belvidere Railroad at Ringoes. It was arranged that each day a special train was to start at Flemington, picking up loaded cars of peaches at various points. At Lambertville cars from Stockton would be attached, and at Bordentown the train would be rerouted to South Amboy, where special boats would carry the fruit to the market at the

[6] *Lambertville Press*, Mar. 24, 1859.

foot of Dye and Barclay streets in New York. Peaches leaving Flemington at 4:45 A.M. one day were at the market by 3:30 A.M. the next. In the following year the South Branch Railroad was built from Somerville to Flemington and provided a shorter route. The daily departure of the peach train from Flemington on this line was quite an occasion during the season. Various points on the New Jersey Central, particularly Whitehouse station, were also important shipping points. After the building of the Easton and Amboy (Lehigh Valley) in 1875, that line, too, provided facilities. On one day in 1882 the South Branch hauled twenty-one carloads from Flemington, the Lehigh Valley picked up thirty-three cars from various county stations, and the old Delaware and Belvidere, by this time part of the Pennsylvania system, gathered up ten cars. From New York Hunterdon County peaches went on to Boston, Albany, and even to Montreal. The question of freight rates during the years resulted in frequent meetings of protest, in yearly dickering with the railroad companies, in Farmers' Alliance activities, and even in Interstate Commerce Commission hearings and special state legislation. It is to be guessed that during the heyday of the industry the hauling of peaches was a very lucrative business for the railroads.

A number of subsidiary industries thrived as a result of peach growing in this area. Nurseries were of course a necessary adjunct and labored mightily to keep the orchardists supplied with young trees. Another dependent occupation was the making of peach baskets. In the early days of commercial peach growing, basket makers were located in the south part of the county, at Copper Hill, Wertsville, and other points. But as peach orchards spread, the industry moved northward. In 1882 the sturdy Beecher baskets, some of which still turn up at farm auctions, were first made in Flemington. In 1883 a rival factory was started in Flemington and, in short order, others appeared at Frenchtown, Whitehouse Station, Landsdown, West Portal, Bloomsbury, Glen Gardner, Pattenburg, Clinton, Pittstown, Califon, and other places. These factories gave employment to hundreds of men and women during a portion of each year, except when the peach crop failed. The early baskets were substantial affairs, made of oak staves and furnished with linen covers. They cost 20¢ apiece in quantity lots in 1874, and 12½¢ a decade later. Arrangements with the railroads, at least until 1888, provided for the return of the baskets from New York free of freight charges. As time went on, baskets became flimsier and new machines speeded up their construction. In 1891 Elmer Bonnell had to pay only four cents each through his local Farmers' Alliance With the decline of peach raising in the county, which will be discussed later, the basket factories fell by the wayside one after another, except for a few in the heart of the small peach district remaining.

Besides the men who made baskets, the peach growing industry gave employment each year to hundreds of orchard laborers. Bonnell's diary mentions at some length the varied tasks in a peach orchard. The trees had to be set out, and later, if they were seedlings, they had to be budded. Each year the trees had to wormed and trimmed, and the brush had to be hauled away. The ground must be regularly plowed and harrowed, and that next to the trees hoed. Weeds were pulled, and ashes put around the young trees. Baskets must be hauled, wagons made, and peach houses built. During picking season the demand for labor was often greater than the supply.

Many farmers dealt directly with commission merchants. Bonnell ordinarily shipped his peaches to the New York market, consigning them to any one of over a dozen dealers with whom he had connections. Others took the simpler course and sold to the agents and speculators of various stripe who rushed in each season. In 1887 there were buyers from New York, Newark, Syracuse, Buffalo, Boston, Cleveland, and Cincinnati at the peach auctions, and many individual agents combed the county to dicker directly with the orchardists. Too often they kept their contracts only when they could do so profitably. In order to cope with these shrewd gentlemen, and with the railroads and basket makers, growers belonging to Farmers' Alliances established cooperative peach exchanges at Flemington, Whitehouse, Lambertville, Bloomsbury, Clinton, and Oldwick in the late 1880's and early 1890's. These marketing agencies, some of which were still functioning after World War I, sold millions of baskets of peaches and other fruit and doubtless helped the marketing position of the peach grower. But they were unable to control the market or secure good prices when the market was glutted with this highly perishable fruit. A small part of the crop might be dried or a fraction of it sold to peach canneries and distilleries, but growers in peak crop years were often faced with the choice of selling below the cost of production or letting the peaches rot. In 1891 Elmer Bonnell sold most of his crop at from 20¢ to 40¢ per half-bushel basket. After one sale he commented, "18¢ and glad to get that." Yet the year before, when prices were high because of a short crop, he had not sold a single basket.

Since the peach crop varied so much from year to year, it is difficult to say whether there was an overexpansion of peach growing. In 1872 some members of the Union Farmers' Club maintained that the day of profit from peaches was over, but others thought that no other means of land utilization was nearly so profitable. The same division of opinion would probably have characterized any farm group during the next thirty years. A farmer who lost his crop by frost one year and sold below cost of production the next soon lost faith. Naturally the sources of information emphasize those farmers who made enormous profits in particular years, and older persons now recall best the cases in which the returns from an orchard paid off the mortgage on the whole farm or enabled its owner to purchase more land. They do not remember as well the disappointments and failures. However, the constant expansion until 1890 indicates that the prevailing opinion was that peaches paid in the long run. Since insects and diseases were not important, spraying was unnecessary and costs of production were much less than they are today.

For a time after the introduction of commercial peach growing into the county, the center of production was in the south. It moved rapidly northward, however, and by the late 1870's production in the southern townships was falling at the very time that it was rising rapidly in the central part of the county. Continuous orchards near Quakertown in 1882 totaled 23,000 trees. By the late 1880's, areas about Frenchtown, Pattenburg, Cokesbury, Califon, and Whitehouse were coming into full development. In 1889 there were in the county over two million peach trees of bearing age, and nearly a million baskets were marketed in that year. The crop of 1890 was a total failure, and the large ones of 1891, 1892, and 1893 caused ridiculously low prices. Many orchardists were discouraged and did not set out new trees. From 1895 on there were reports from scattered sections of the county that peach orchards were falling under the ax. The northeast part of the county, with its newer orchards and improved transportation system, was still producing heavily. The years of 1897 and 1899 brought good profits, but the heyday of peach raising was definitely over. By 1899 the number of bearing trees in the county had fallen to just over a million. General opinion to the contrary, the peach industry of Hunterdon County was on the down grade before the appearance of the San Jose scale.

The San Jose scale is a tiny but very destructive insect, so-called because it entered the United States through San Jose, California. It first appeared in Hunterdon County orchards just before 1900. In that year several experimenters killed their trees by using an oil spray. By 1903 losses from the scale were considerable, and orchardists were frantically spraying with a lime, salt, and sulphur mixture or with one of the numerous sprays which appeared on the market. However, nothing was able to do the magic of completely eradicating the pest. Many pinned their faith on the development of the parasite ladybird beetle, but of course this insect could not increase rapidly enough to save existing orchards. The New Jersey Agricultural Experiment Station and the County Board of Agriculture kept the farmers informed as to new developments. But the scale, sometimes hopefully reported to be on the wane, always returned in full force whenever precautions were eased. Other peach pests and diseases also appeared to add to the orchardists' troubles. By 1909 the number of bearing trees of Hunterdon County had fallen to just over 300,000, and by 1919 to 180,255. Vigilant, careful growers still operate profitably, but peach growing has become a specialized industry attempted only by a relative few. Only 29,062 bearing trees were reported in 1940.

No other fruit crop of the county has ever been so important commercially as the peach in its halycon days, though others have been of some consequence in a less spectacular way. Before 1850 the apple was the most important fruit because of its keeping qualities and because it served as the source of an important alcoholic beverage. Any surplus might be turned into cash by hauling apples or cider to a distillery. In advertisements of farms before 1850, it was the apple trees of the farm orchard which were stressed. A shortage in the crop of 1849 was considered important enough by the local census enumerators to receive special mention. Since there were no serious pests to bother apples at that time, most farms had small orchards. The falling off in the number of orchard trees since it has become necessary to spray is evident from the following table:

Year	*Bearing Trees*	*Bushels Produced*
1889	140,741	68,793
1899	189,721	401,634
1909	80,961	103,785
1919	82,081	92,808
1929	68,441	69,615
1939	48,015	142,612

It will be noticed that apple orchards were being expanded during the 1890's, at a time when peach orchards were beginning to decline. Had it not been for the scale and other pests which attacked either the trees or their fruit, the apple might have replaced the peach as a market product in this area. The change in conditions not only prevented any such development but caused a decline in home orchards. Apples today are chiefly raised by specialists, and most farmers prefer to buy the fruit rather than incur the expense and bother of spraying their trees repeatedly each spring.

Pears were never grown to any great extent here, but the average farm of early days probably had a few trees. In the 1880's the number of trees increased rapidly, and there was some expansion until 1919, after which the number declined rapidly. The story for cherries is similar. For a time after the turn of the century there was a small commercial development in the Stockton and Sand Brook areas, but the number of trees decreased greatly between 1909 and 1919, and by 1939 was very small. As for grapes, production for home use is still common, despite the ravages of the Japanese beetle and other insects. One arbor, that of H. E. Deats at Flemington Junction, produces well after a century of constant bearing. In 1859 the county produced 4328 gallons of wine, but since that time production has been negligible. In recent years there has been some growing of grapes for market. Strawberries, always raised for home use, became of some commerical importance around 1880. The first statistics, those for 1889, show seventy-one acres of strawberries in the county. Since then the production has fallen off markedly, and in 1939 only fourteen acres were reported. Over one hundred acres of raspberries were reported in 1899, but the acreage has fallen rapidly since. High labor costs, a tendency toward specialization in other lines, competition of other regions, and the increase in plant diseases and pests have prevented any large development of small fruits in Hunterdon County.

CHAPTER VIII

Animal Husbandry

THE DOMESTIC ANIMALS of Hunterdon County, like the crops, have not changed as to kind since the days of early settlement. The results of developments through selective breeding have been greater, however, and the changes in emphasis in recent years even more marked. The pioneers of this area brought livestock with them from older settlements or, as was especially true for the German immigrants, secured animals from their neighbors after arriving. In early days when every man was concerned with the problems of erecting buildings and clearing fields, animals were given a great deal of freedom. Early probate inventories often contained such entries as "jades running at large." For over a century it was customary for farmers to turn cattle or sheep into vacant lands and to let swine feed freely on the mast of the woods in the fall. Fences were constructed more usually to protect growing crops than to confine livestock within bounds, and the owner of a growing crop must by law enclose it with a fence of specific height and type if he expected to collect damages for depredations by estrays. Various devices were used to keep animals from wandering too far and from breaking through fences into cultivated fields. Cattle, sheep, horses, hogs, and even geese wore yokes. A cow sometimes had her head tied to a front foot or had both front legs bound together. Bulls were restrained by having heavy logs dragging from their horns. Chain hobbles or "hopples" were often used for horses. But such methods did not entirely solve the problem, and the matter of straying animals was of such importance as to be regulated by law, the administration thereof being left to the various townships.

In each township a public pound, often run in connection with the poor farm, was maintained. Estrays were taken in, a full description entered in the records, and arrangements made for feeding. When the owner appeared, he could reclaim his stock by paying the board bill. If the owner could not be located by advertising, he forfeited all rights. To make identification possible, each farmer had specific methods of marking his animals, and the marks were registered in a township record book. Horses and cows were branded, and sometimes had their ears marked. Other animals had their ears cropped or slitted in various ways. Two ewes in 1742 were described thus in the Hopewell Township book of estrays, "Marked as follows, one hath a Crop on each Ear and a Half penny Under the Under Side of ye Near Ear—the other hath a Crop on the Near Ear and a Half penny Under the off Ear." Six years later a stray sheep had "a Hole in Each Ear and a Slitt in each Ear and a Crop in the off Ear." As more land came under fence, the use of brands and other markings was gradually discontinued.

The matter of estrays remained of some importance, however, down to the day of barbed wire fences. As the use of marks declined, it became more difficult to identify animals, and the question of ownership caused many lawsuits. After the appearance of the *Hunterdon Gazette* in 1825, advertisements by those finding animals show that wandering stock was still a problem. A typical advertisement was one of December 18, 1838: "Came into the enclosure of the subscriber near Flemington . . . a red Cow with her horns bored. The owner is requested to prove property, pay charges, and take her away." Sometimes it was the owner who advertised hoping to find his lost stock. Estrays were still a menace to growing crops. A lease agreement of 1818 in the Case records required the renter "to take care of the adjoining Land of Philip Case from Creatures Trespassing on the same."

During the general farming period and until fairly recently it was to a farmer's advantage to have many kinds of animals on his farm. Cattle were necessary for dairy products, meat and leather, and oxen were used as work animals by some farmers and lumbermen till comparatively late. Sheep were kept primarily for wool, swine furnished most of the winter meat supply, and poultry gave both eggs and meat to the family larder. Horses were used for travel and as draft animals. A tax list of 1722 shows that nearly every farmer of this area had cattle and horses and nearly half of them had sheep.[1] Inventories of early estates give similar results. Samuel Coates in 1723 had nine horses, four oxen, ten other cattle, twenty sheep, and some swine. Duncan Oliphant in 1734 owned sixteen horses, fifteen cattle, sixteen sheep, and sixteen hogs. Emmanuel Coryell, a farmer on a fairly large scale, in 1748 possessed eight oxen, thirty-nine other cattle, nine working horses, ten other horses, nineteen hogs, and forty-eight sheep. John Rutherford, speaking of the average farm in 1786, said, "Each Farm keeps from 20 to 40 Sheep, from 3 to 10 Cows, besides young Stock, from 3 to 8 Horses, unfortunately few

1 Snell, *op. cit.*, pp. 185–86.

Oxen being used for Draught." The probate records indicate that his estimate was too high for sheep, but otherwise about right.

It was not until farming became more specialized over a century later that the situation changed to any extent. The following table, the figures of which are taken more or less at random from the local schedules of the 1850 census, gives a fair sample of the diversification common during the nineteenth century:

Name and Township	*Improved Acres*	*Horses and Mules*	*Milk Cows*	*Working Oxen*	*Other Cows*	*Sheep*	*Hogs*
M. Robinson Kingwood	50	3	5		3		2
Samuel Bellis Kingwood	62	3	4		2	8	8
Theodore Young E. & W. Amwell	80	2	4	2	1	14	9
Peter Stout Alexandria	85	4	4		7	17	20
John Barber Delaware	90	6	6		5	20	9
John Bird Franklin	90	4	6		2	12	7
Charles Rockafellow Alexandria	90	4	5		4		10
Daniel Reading Raritan	94	4	4		5	13	9
Daniel Opdycke Kingwood	100	4	5		3		2
Conrad Apgar Tewksbury	102	4	5		3	16	6
Uriah Bonheim Kingwood	110	5	6	2	1		9
Daniel Anderson Bethlehem	127	6	7		8	33	21
Philip Hoff Kingwood	260	9	11		5	17	27
A. F. Williamson E. & W. Amwell	280	10	10	6	62	50	12
G. Conover Bethlehem	300	11	6	2	30	150	35
Thomas Hill Raritan	302	9	10	2	28	55	27

At that time the 281 farmers of Readington Township, owning 24,826 acres of improved land, had together 875 horses and mules, 1216 milk cows, 68 oxen, 697 other cattle, 1752 sheep and 2029 hogs.

The earliest settlers in West Jersey were pleased that pasturage was so plentiful for cattle and the feed for swine in the woods in such great quantity. Because of the difficulty of harvesting sufficient hay and providing sufficient stabling, many settlers slipped into the habit of wintering most or all of their stock outside. Kalm in 1749 reported that in New Jersey only the Dutch and Germans stabled their animals, and Schoepf, three decades later, said that the Germans of Pennsylvania treated their livestock better in winter than did others. Even horses were often left out of doors. The brand of a stray mare caught in Hopewell Township in 1737 did not show up until "After the Shedding of her Coat." As barns were made larger, more of the animals were put inside, but sheep and young cattle were often wintered in outside sheds.

Eventually more attention was also paid to feeding. Hay and grain crops were increased, and although the average farmer was hard pressed at times to grow enough feed for the winter, livestock in general fared better than in early days. Many farmers supplemented their feed supplies by purchasing the by-products of the flour mills, oil mills, and distilleries. Dr. Bowne accepted "mixed feed" from one of his patients in lieu of money in 1802. During the early decades of the nineteenth century the Cases near Flemington regularly bought shorts, bran, ground corn, and mixed horse feed. Blackwell in 1845 fed his beef cattle a diet of corn meal, buckwheat bran, "corn ears ground," and pumpkins as long as they lasted. He supplemented his home supplies by purchases when necessary. The coming of the railroads had by this time made the importation of feedstuffs from distant regions possible. Elmer Bonnell in 1890 regularly bought malt, wheat bran, oil cake meal, and grist for his animals. With the expansion of the dairy industry in the twentieth century and a greater emphasis on balanced rationing, an increasingly larger part of feed for livestock has come from faraway places. In 1931 the United States Department of Agriculture made surveys of the Flemington and Clinton areas. In the former farmers grew only 52.9 per cent of their dairy feed and 65.2 per cent of their poultry feed. In the latter the percentages were still lower, 42.8 per cent and 57.9 per cent respectively.

Farmers have always taken a keen interest in methods of treatment of sick and disabled animals. The practices of this region in early days were probably in no way superior to those of English farmers of the same period. The earliest work on veterinary science found in Hunterdon County was printed in London in 1785 and bears the title of *Farriery Improved.* It is a queer hodgepodge of pseudo science and folklore. The author had no greater understanding of the fundamental causes of disease than was common at the time, and remedies were often queer concoctions of very doubtful value. Some "cures" were more reasonable, however and were evidently the result of trial and error. More interesting are some of the remedies in the commonplace book of the Hankinson family of Amwell Township. About 1788 an entry was made giving a cure for the "dum stagers" in horses. "Assifidity" or "devil's dung" was soaked in vinegar, then "stued." Pieces of cloth were saturated in the solution and fastened in the ears of the suffering horse. Three cures were suggested for "Niagary horse Disorder." The sick animal might be fed boiled sweet fern or drenched with either a linseed oil-saltpeter mixture or a brew resulting from boiling together several kinds of bark. Breathing the smoke of burning leather or "any other stinking thing" would also help.

The *American Farrier,* printed in Philadelphia in 1832, was evidently a stand-by of David Bellis of Copper Hill, in whose library a copy was found. It is more modern in tone than *Farriery Improved,* but the remedies are no more scientific. Bleeding, giving of simple physic, and light feeding were the chief suggestions for internal disorders. Soapsuds, greases, and oils were used for scratches and sores. Simple herbs, such as poke root and plantain leaf, were among the commonest remedies and given for a variety of ailments. Mr. Bellis bought other books on the subject, and cut out clippings from newspapers and magazines. Other farmers, of course, had access to the same sources of information. The county newspapers often copied items from agricultural papers and books. The *Hunterdon Gazette* in 1826 recommended a cure for spavin which included angleworms fried in butter. In recent times the germ theory of disease and the giving of scientific training to veterinarians have done much to remedy the situation. The use of serums and vaccines and the control of diseased livestock by legislative action are modern trends in animal care.

American domestic animals are the descendants of those brought from Europe. The voyage was long and hard in early days, and the number transported was comparatively small; but the natural increase, carefully husbanded, soon supplied the new land. Indiscriminate breeding and poor care, however, caused a progressive deterioration in American live stock at a time when that of Europe was being improved. Eventually, gentleman farmers here became interested in what was happening in agriculture abroad, and soon were importing blooded sires and experimenting, not always wisely, with selective breeding. The *Pennsylvania Farmer* in 1804 said, "The first great error in improving of livestock in America, is in their passion for the largest kinds." Despite this failing the net result of their work was an improvement of farm animals and a dissemination of knowledge as to breeding principles. In the nineteenth century that which had been the hobby of a few men became a widespread practice.

The rearing of livestock could nearly always be done cheaply in a new country. Abundant forage and free grazing gave the frontier an advantage over older lands. The time eventually came in Hunterdon County when it seemed more economical to buy animals brought from the West than it was to raise them. In early years the driving of horses, cattle, sheep, hogs, and even geese to market at Philadelphia was a common practice in this area, so the driving in of livestock to replenish farms was no great break with tradition. The practice probably started soon after 1800. There is no evidence concerning hogs, but the number of sheep and cattle involved was large. Many local farmers became professional drovers, bringing in herds of grass-fed cattle from New York and sheep from Ohio. As wool production decreased in the East, fewer sheep were imported, but the long, slow trips with cattle were continued until the 1890's. When the sources of supply moved farther west, cattle drovers turned to the use of railroads.

One of the earliest accounts of West Jersey speaks of the large size to which swine grew when turned loose in the forest. Although corn-fed hogs gave the "sweetest" pork, the great abundance of mast caused most swine to be put into the woods in the fall. Each owner marked his pigs by cropping and piercing their ears so that he could identify his property when the herds were rounded up at slaughtering time. In the meantime they were protected by law unless they broke through a fence into someone's crops. A stray hog in Hopewell Township in 1729 was described as "a white Sow having a Crop on the Near Ear and a hole in the off Ear." The farmer using those markings could claim the animal. Swine on range were often sold "sight unseen." An item in the inventory of the estate of James Stout of Amwell Township was "one Old Sow in the woods," valued at four shillings. As vacant lands diminished, farmers often enclosed their own woods and orchards for pasturage. In order to keep them within the fences, hogs were yoked and had rings put in their snouts. A farmer with a good hog pen to use on occasion had an important asset. The fact that male hogs were seldom castrated probably helps to explain the difficulty of keeping them from wandering.

The chief purpose in raising hogs was the provision of a family winter meat supply. In early days there was probably little surplus pork; but as herds increased and roads improved, it became possible to send pork packed in brine to distant markets. Country and village stores acted as middlemen. In 1787 Peter Case killed twenty-seven fat hogs on one butchering day. He sold one to a widow in the nearby village of Flemington, stored sufficient pork for his family during the winter, and sold the remainder, over three thousand pounds, in barrels to the village storekeeper. Large scale butchering was an annual event for the Cases, as for many farmers. Pork was worth $24 per barrel at New Germantown (Oldwick) in 1817. However, the marketing of fresh and cured pork was usually merely supplemental to the supplying of family larders, as was the selling of hogs on the hoof to local buyers or to drovers. During most of the nineteenth century all farmers killed and cured their own pork, and butchering time remained one of the important occasions of the year; but toward the end the dominance of the Middle West in pork raising and the very successful marketing practices of the packing houses caused one family after another to stop doing their own slaughtering, and in many cases to quit raising hogs. The scalding barrel was allowed to disintegrate, and the butcher knives, the scrapers, the ham hooks, and the gambrels were put on the shelf or discarded.

In the latter part of the eighteenth century, as hogs were allowed to run less freely, more attention was given to care and breeding. In 1805 two sows butchered by one of the Cases dressed out 329 and 370 pounds respectively, while eight pigs born that spring yielded from 131 to 179 pounds of meat each. Hogs reported each year to the *Hunterdon Gazette* were considerably larger than this. Many weighed five, six, and even seven hundred pounds when dressed, and the largest tipped the scales at 917 pounds. It was obvious that size was being overstressed, and already by 1840 a new emphasis on quality rather than large size was noticeable. Better farmers were acquiring the improved breeds brought from Europe, and especially from England. Berkshires were exhibited at the first state fair in 1840, and in 1856 the Hunterdon County Agricultural Society gave prizes for Suffolk, Byfield, and "common" hogs. During the next two decades stray references were made to improved breeds of both English and American origin. In 1881 the Agricultural Society had five entries of Essex hogs, eleven of Berkshire, ten of Poland China and Magie, fourteen of Chester White, and six of Yorkshire, but none for the famous Jersey Red. The Poland China and Chester White are breeds developed in America. The Magie was apparently a strain of the Poland China. Farm advertisements from that time to this show that these breeds have remained popular, particularly the Berkshire, Poland China, and Chester White, all of the fat or lard hog type.

The following table gives the number of swine on the farms of the county at dates indicated, as reported by various Federal censuses:

Year	*Number of Swine*
1850	24,613
1860	17,829
1870	15,311
1880	25,056
1890	30,539
1900	23,611
1910	20,539
1920	16,917
1930	3,628
1940	4,161

Only 760 farmers reported having hogs in 1940. The marked decline of the previous twenty years was due to the rivalry of more advantageously located regions and the decline of home slaughtering. A sudden betterment of the ratio of hog prices to feed costs led to a noticeable increase in the hog population of the county in 1941 and 1942, but a reversal in the ratio led to a rapid decline in 1943.

Forty-nine of the 122 tax-paying farmers of Hunterdon County as then constituted owned sheep in 1722. It is possible that the number was not greater because of the danger from wolves. If the number of estrays in Hopewell Township is any indication, the number of sheep in the area increased rapidly after 1735. Wandering sheep, like cattle, often broke through fences, and if the fences fulfilled legal requirements, the owner was liable for damages. Sheep were kept primarily to provide wool for spinning, though surplus animals might be killed for meat. Mrs. Capner in 1787 wrote that it was easy to sell sheep by the quarter late in the summer. Her nearest neighbors, the Cases, had twenty-eight head. Some farmers had a prejudice against sheep on the score that they pulled up the grass. Isaac Passand wrote in 1793 that his renter would not accept sheep as a gift. Apparently, however, the dislike was not general.

The Capners and their relatives, with their English experience, believed in the value of sheep, but thought that the scrubs of their neighbors needed improving badly. Joseph Capner, though he put most of his efforts into dairying, gradually acquired a flock of sixty sheep. Although other breeders in New Jersey were importing Southdown rams, Joseph Capner and his son after him specialized in Bakewell Leicester or Dishley sheep. It is not clear how or when they first acquired this breed. The son, Hugh Capner, when an old man, said that they had been smuggled in, but his meaning is not clear. Possibly they purchased them from the Captain Beans who is said to have imported Bakewells into New Jersey in 1806. Robert Bakewell, originator of the strain, had been a neighbor of the Capners and Extons in England, and this probably explains their interest in Bakewell sheep. Just when the Extons first acquired this breed is not known either. In 1828, after years of negotiation, Hugh Exton sent his son Joseph back to his native Leicestershire to purchase two fine rams, but he had evidently acquired other Bakewells much earlier than this. The Extons for several decades were well known as breeders of both Bakewell and Cotswold sheep. There are no records to show how extensive was the Spanish Merino craze in this area. An acquaintance of the Capners near New Brunswick imported a Merino ram soon after 1800, but Joseph thought Merinos were too expensive. A small fulling and weaving shop in Flemington about 1810 had only one Merino fleece brought in during its two or three years of existence. However, a movement so widespread undoubtedly had some effect in Hunterdon County.

The practice of improving native sheep by use of blooded rams became common after about 1820. Full- and half-blooded Tunis sheep were much advertised by 1830, but the various English breeds met with most favor. A flock near Flemington in 1827 contained seventy Southdown and Bakewell animals though it is doubted that all were pure-bred. The state fair in 1841 gave prizes for Leicestershire Bakewell, Lincolnshire, Southdown, Saxony, and mixed breeds. The first Agricultural Society fairs at Flemington in the 1850's offered prizes for Southdown, Bakewell, Cotswold, and Merino sheep, and sometimes for "native or cross bred." The classes by 1871 were Leicestershire, Longwool, Bakewell, Southdown, Southwell, and native. The English breeds, some of which were much alike, retained their popularity, the Southdowns being especially liked.

The first complete statistics for sheep in Hunterdon County were those of the 1850 census. The local schedules show that nearly every farmer had sheep, and flocks of twenty-five and even fifty were not unusual. The spinning wheels of the county had for the most part been discarded before this time, but there was a ready market for wool. In 1848 Blackwell sold his wool at a small factory in Clinton. Sometimes he would slaughter a sheep in the fall for family use. Apparently the keeping of sheep paid fairly well for three decades after the Civil War.

Large droves from Ohio, Kentucky, and Indiana were brought in annually and sold at the villages of the county. Later, sheep and lambs were imported by rail also. Farmers bought Western sheep, kept them for a year or two, fattened them, and then sold them. Besides the profit on the sheep themselves, they received the increase of lambs and a shearing or two of wool. But in the early 1890's the market collapsed, and a decline set in which has continued to this day. The following table summarizes census figures for sheep raising in Hunterdon County since 1850:

Year	*Number of Sheep*
1850	24,541
1860	19,320
1870	22,790
1880	23,888
1890	10,556
1900	10,688
1910	9,384
1920	2,553
1930	1,640
1940	959

Only ninety-one farmers reported owning sheep in 1940. A few still maintain that there is profit in sheep raising, but most farmers today concede that they cannot compete with more favorably located regions.

The first horses imported into West Jersey, according to Thomas Budd, were "good serviceable Horses, fit for both Draught and Saddle." Like other farm animals horses were allowed to graze freely in the woods, but were given little other feed. They were usually branded and in early days often had their ears marked. The estrays reported in Hopewell Township were a motley lot. The following are examples of descriptions between 1729 and 1735: "a dark Brown Horse . . . [with] a small bell," "Mare of a Darke Sorrel or Chestnut Cullored," "a Yellow Duncoulored Mare," "Bright Bay Horse aged about 13 or 14 Year old," "a Black Mare with a Smale Star and a narow Blaze Down her face . . . and hath been trimmed to a comb mane," "a white Mare," "a Mouse Coulered Yearling mare Colt," "a Yellowish Bay Mare . . . with a Blackish Main and Tail," "Light Gray Horse . . . a Racking pacer," "Brown Bay Horse . . . Short tail undocked," "a Dirty collored Grey Horse Somewhat flea Bitton or red spotted, aged about 7 or 8 years old, with a halfpenny on the uper sid of the near Ear and a halfpenny on the under side of the off Ear." Many estrays wore yokes and most of them were

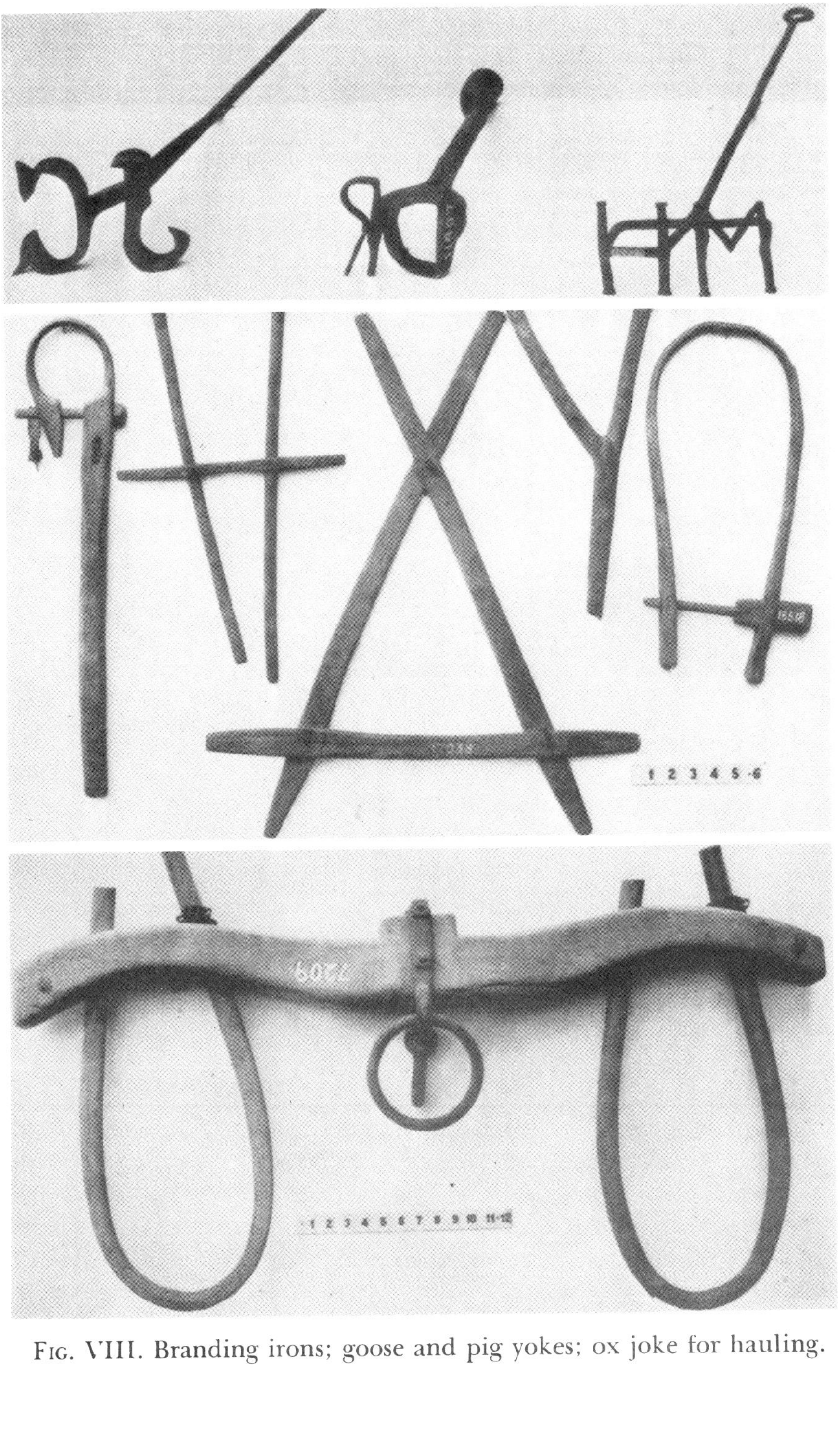

FIG. VIII. Branding irons; goose and pig yokes; ox joke for hauling.

Fig. IX. Lower and upper sections of a grist mill of the type commonly called tub mill.

small, only two or three in the course of the eighteenth century being over thirteen hands or fifty-two inches in height. Occasional horses had a "Dutch comb," and quite a number had docked tails. Only a few were shod, but many had saddle marks. Riding equipment and "horse geers" in the probate inventories show how important the horse was both for travel and for work. Good horses were valued highly at all times, and the birth of a good colt was cause for elation by the owner. Two entries on the flyleaf of John Lambert's memorandum book read as follows: "The one Eyd. mare folded a valient mare Colt June the 14th 1774," "The one Eyd mare folded a 2nd valint mare Colt June 23: 1775."

Efforts at improvement of horses through breeding became noticeable earlier than for other livestock. By castrating all male colts except one or two of the best, a farmer was practicing selective breeding to a certain degree, though often his best was none too good. By New Jersey law unconfined stallions were required to be at least fourteen hands or fifty-six inches high, but it is doubted that the law was generally upheld. American stallions were not rated very highly by observers from Europe even late in the eighteenth century, but those who had comparatively good ones could and did receive fancy prices for their use. Demarr Ford of Readington Township charged three pounds for use of a stallion named Farmer in 1782. In 1805 Jonathan Moorhead of near Flemington charged $10 to "warrant a fold" and William Philips asked $8.00 "the single leap" for Expedition. In 1810 Cornelius Vanhorn had a stallion called Trafalgar which he hired at $12 "the season" or $16 to "insure."

Interest in good horseflesh became so great in the early nineteenth century that exhibitions were held in Flemington and Lambertville, with prizes for best teams and best single horses. The numerous stud advertisements show an increasing interest in good animals, but the great emphasis was on fast road horses rather than on draft animals. Stallions with fancy names were periodically quartered at every village, and were advertised as "celebrated running horses," "full blooded horses," "imported horses," or "turf horses," and their pedigrees, real or imagined, played up in great detail. Only an occasional one was called a "field and road horse." Some interest was eventually shown in good work horses, however. The first state fair in 1840 emphasized racing horses, but special mention was made of some imported brood mares of a draft breed. The first Hunterdon County fair in 1856 offered prizes for the best work horses shown, and the practice was continued thereafter. Stallion advertisements from the Civil War onward included those of blooded draft animals, and these were evidently well patronized by farmers trying to improve their stock. Several stock farms in Hunterdon County in the 1890's and 1900's specialized in racing stock. Lynne Bel of East Lynne Stock Farm near Flemington ran first at Belmont Park and second in the Kentucky Derby in 1895. For the next six or seven years this horse was the darling of the county as he piled up record after record.

When it became more economical to buy horses reared in lands farther west than to raise replacements, farmers turned to the horses brought in by drovers. Just when horses were first imported cannot be discovered, but Joseph Case had two mares from Ohio already in 1836, and sixty Kentucky mules were sold at Vansyckle's Inn in 1842. In the 1850's tavern yards at Lambertville, Clover Hill, Flemington, Clinton, Readington, Lebanon, Three Bridges, and Frenchtown were the scenes of regular horse auctions. The number of such sales increased greatly after the Civil War. By the 1870's horses were also being imported by rail, though drovers were slow to relinquish their trade. Animals from western Pennsylvania, West Virginia, Kentucky, Quebec, Ohio, Indiana, Illinois, Michigan, Missouri, Iowa, Nebraska, and regions farther west were sold to farmers of Hunterdon County, as also were "car horses" released by the electrification of street cars in the cities. Prices fluctuated, but demand kept them rather high. A drove which were sold in Flemington in 1872 averaged over $180 in greenbacks and another averaged $192.60. "French horses" brought only $120 in 1877, and Ohio horses slightly more in 1880, but the slump did not last long, and in May, 1883, the average price at one sale was $191.25. "Indian ponies" and "western ranch horses" were naturally much cheaper.

By 1890 the demand for horses was very great, despite the fact that the agriculture of the area was badly depressed. William S. Riley of Flemington, an auctioneer to whom reference has already been made, reported to the *Hunterdon County Democrat* that year a sale at Lebanon in which brood mares in foal brought up to $365 each, young fillies up to $200, one- and two-year-olds up to $180. During the next few years prices declined somewhat, then began an upward trend which did not stop until 1914. The slump even after that date was not marked until tractors became numerous in the 1920's. Mr. Riley, who for over two decades cried sales for E. E. Opdycke of Frenchtown and A. S. Case of Flemington and Three Bridges, the two leading horse dealers of the county, had a flair for statistics and reported regularly to the *Hunterdon County Democrat.* The average price of one good group of "Western horses" in 1897 was $112.11, of another in 1898 was $139.45. The best carload lot reported in 1901 averaged $187.62; that of 1902, $191.15; that of 1903, $203.25; that of 1906, $231.04; that of 1909, $251.13; that of 1912, $277.55; that of 1915, $250.43; that of 1918, $266.40; and that of 1921, $241.67. There were always cheaper horses from various parts of the West and from the cities, and in 1918 huge lots previously bought for military purposes were turned loose on the market. But it was a very poor horse which brought less than $100 at any time during the period.

During the 1920's the number of sales decreased rapidly as the horse was displaced in agriculture, but occasional carloads of horses from Minnesota and other places are still imported. In general western horses have proven fairly satisfactory for the needs of this area. Some animals have always been lost as a result of the hard trip and the change in climate, and most horses have required special treatment for a few weeks while becoming acclimated and used to heavy work. Some attempts have been made to revive horse breeding in the county in recent times, particularly of Percheron, Suffolk, and other blooded draft animals. The following figures give the horse and mule population of the farms of the county at census periods since 1850. The effect of the rivalry of the tractor and the motor truck is obvious:

Year	*Number of Horses*	*Number of Mules*
1850	7,793	182
1860	9,421	463
1870	9,520	689
1880	9,673	740
1890	10,075	599
1900	8,661	202
1910	8,961	152
1920	8,058	247
1930	4,121	157
1940	2,949	114

From the days of earliest settlement, cattle played an important part in the agricultural economy of this region. Every farmer needed cattle for dairy products and meat, and often desired working oxen in addition. The earliest settlers in West Jersey allowed their cattle to roam the woods and reported that the animals put on weight. But Peter Kalm in 1749 wrote that lack of care and particularly the custom of leaving the cattle out all winter had resulted in a small, ill-favored breed. He reported that pastures were poor, milk production small, and the flavor of milk, butter, cheese, and meat inferior. The marked deterioration of the breed was probably due to lack of selection in breeding even more than to the factors mentioned by Kalm. The stray cattle described in the township records of Hopewell were certainly a nondescript lot as to color, being red, red and white, brown and white, black, black and white, brindle, and various other combinations, but the descriptions as to size, such as "small" and "midling," do not tell us much. The Capners in the 1780's and the Extons in the 1790's reported that American cattle were smaller and less productive than the English cattle with which they were familiar. However, they found that they responded to good treatment rather quickly.

The earliest specific description of dairy practices in Hunterdon County is that found in the letters of the Capners and their friends. Cows were pastured on natural vegetation and sometimes on sown crops such as rye and clover. During the winter they were given the hay which had been put in barracks for them. There was little grain in their ration, except that corn was given to those that were being fattened for slaughter. A home supply of dairy products and meat was the chief objective. Each family produced milk and butter and sometimes cheese for themselves, and small surpluses were sold through the village stores to nonfarmers. Bull calves were often butchered when a month or two old, part or all of the meat being sold to neighbors. Others were castrated and raised as beeves, though a few were trained as draft animals. Poorer milk cows and especially old ones were fattened and butchered. Some of the meat was eaten fresh or sold, and the remainder pickled in brine for later use. Barrels of beef were sometimes sold to the village store. A few industrious farmers had expanded their herds so that they sold considerable amounts of butter, cheese, and beef.

Joseph Capner had been a dairyman in Leicestershire and, after looking over the situation in Hunterdon County, decided to become one again in America, but on a larger scale than any of his neighbors. He planned to market his milk, as in England, in the form of cheese. Neighbors pronounced the first cheese good, but Joseph's mother was much concerned about the taste. She and her helpers soon mastered the situation, however, except that during the hot New Jersey summers cheeses sometimes "went mad" or "heaved" in the press. During the second year Joseph began making occasional trips to Philadelphia to sell cheese, and he exhibited his product at the Agricultural Society of that city. His uncle John Hall used his influence with his friends in high places, and Isaac Passand made the rounds of the taverns in order to build up trade. Soon the demand was greater than the Capners could supply. One day in 1790, as Mrs. Capner turned a cheese out of a vat, she said, "There is a dollar." Passand, always a wit, spoke up, "Send word to England. Your pot kittles dollars."

In 1791 the Capners sold over three thousand pounds of cheese from nine cows. This meant an average of from three to four thousand pounds of milk per animal per year, high for that day. The following year Isaac Passand wrote:

> It seems but a little wile [*sic*] since I said 200 mite [*sic*] be made of a cow and was laughed at but now I am more confident that 400 ma [*sic*] be made, that comes to £12 Sterling for each cow. This foretelling of profits will seem very wide to an Englishman but if he was to see our poor little cows and their short tho' sweet paster [*sic*] he would be more Astonished at wat [*sic*] is past than wat I say ma be done.

The Capners increasingly specialized in cheeses not common in America, and some of their net cheeses were personally examined and bought by Martha Washington. Joseph Capner expanded his herd to twenty milkers in 1793 and acquired an additional five the following year, though he was at the same time experimenting with the breeding of horses and sheep. Besides cheese the Capners made some butter, which they sold locally for the most part. When Hugh Exton settled at Union Farm in the late 1790's, he took up dairying and cheese making on an even larger scale than his brother-in-law, and kept it up until his death over three decades later. Thomas Capner at one period of his life also produced cheese on a farm near Trenton.

The English immigrants, no matter how interesting their story, did not change to any great extent the development of dairying in this region. Their neighbors were interested in their methods and doubtless learned something from them. Some of them may have expanded their own operations as a result. But it was too early for an intensive development of dairying here and particularly of something so laborious as cheese making. Isaac Passand thought Americans lazy and told English friends that if they expected "hard work to be done they must either bring it in their bones or bring somebody with them to do it." [2] The Capners and Extons brought over children of relatives and friends to serve as dairymaids and helpers, but ran into difficulties as they grew up and married. Mrs. Hugh Exton in 1826, at the age of sixty-eight, wrote, "I . . . am enabled to not only attend my Dary [*sic*] of from 50 to 58 cows but perform much of the labour of cheese makeing myself—it is

[2] Mary Choyce to Mr. and Mrs. James Choyce, May 31, 1795, Capner Papers.

Impossible to procure people to do that kind of busness [*sic*] for us." [3] It is little wonder that the second-generation Capners and Extons turned gradually to the raising of sheep and beef cattle. A third-generation Capner girl, however, took first prize for her cheeses at the county fair in 1859. For most farmers dairying remained a side line, though the money or store credit gained by the sale of surplus butter and meat no doubt was welcome. Dairying on a large scale was destined to wait until greater markets developed with the growth of large cities. As for the production of cheese, it gradually fell off in this area until by 1850 only a very few families made it for sale.

The average farmer of the first part of the nineteenth century had a few milk cows in various stages of production and some young stock of his own rearing. Nearly all cattle were "native," and would be today classed as scrubs. Bulls described in advertisements of estrays were of all colors and descriptions. One in Tewksbury Township in 1843 was "of mouse color with white in the face and white on the rump." [4] Blackwell's cows were of every variety and combination of color. Since there was not much incentive for greater milk production, farmers did little selective breeding to bring up the average yields of their herds. As a market grew for veal calves, steers, and fat cows, it seemed increasingly important to have heavy cattle; and those who brought in blooded bulls chose beef varieties. Hugh Exton imported from England in the 1820's a Durham Shorthorn bull and cow. During the following years his animals took prizes at various fairs, and many farmers brought their cows to his bull for service. Shorthorns were popular in the county for many years. In 1856 the Agricultural Society offered prizes for Durham, Devon, and Ayrshire breeds, and added Herefords and Alderneys the following year. The latter name was applied to both Jerseys and Guernseys. Although a few cows of dairy breeds were exhibited and a few prizes given for dairy products, the emphasis of the fairs at this time was on heavy breeds. As will be shown later, the total butter production of the county was not small, but to the average farmer the making of a few pounds of butter daily probably seemed trifling as compared with the marketing of a number of fat steers. Heavy breeds of cattle also produced the best working oxen, but not more than one farmer in ten owned oxen by 1850. Except for heavy jobs such as logging, oxen were definitely on the way out.

A number of occurrences during the last decades of the nineteenth

3 Letter of Mrs. Hugh Exton, 1826, *ibid.*
4 *Hunt. Dem.*, Mar. 29, 1843.

century had an important bearing on the future of cattle raising in this region. The transportation of beef from the West by refrigerator cars beginning in 1871 put beef production in the East at a decided disadvantage. At the same time the development of markets for fluid milk caused a new emphasis on dairying. A small part of the market was local, as milk wagon routes developed in the towns and villages. For example, daily delivery began in Flemington in 1867. At about the same time some farmers near Flemington began shipping fluid milk to New York City, and found the practice more remunerative than butter making. More important for the time being, however, was the sudden development of creameries. Sixteen were built in neighboring Bucks County, Pennsylvania, between 1878 and 1883. The first in Hunterdon county opened at Sergeantsville in April, 1881, capital being secured by the sale of 150 shares of stock at $25 each. Farmers were paid $1.40 per hundred for milk the first month, and by June the creamery was buying 10,000 pounds of milk daily from eighty farmers. Each day it made 240 pounds of butter, most of which went to city markets, and makers of ice cream in nearby towns bought 200 quarts of cream each week. In addition eighteen large cheeses were added each day to the stock in the ripening room. Skim milk was returned free to farmers to feed to their pigs.

Later in the same year similar creameries were started at Locktown, Little York, and Oak Summit and at Neshanic Station across the line in Somerset County. J. N. Pidcock built a creamery at Whitehouse Station in 1886, and others were built in the 1880's, either by local creamery associations or by private individuals, at High Bridge, Clinton, and Annandale. Spring Mills and Idell joined the procession in 1892 and Ringoes, Pittstown, and Flemington in 1893. Other creameries were started during the next twenty years at Clover Hill, Bloomsbury, Mt. Pleasant, King's Mills (near Quakertown), Sunnyside, Glen Garner, Reaville, Cherryville, Rosement, Everittstown, Three Bridges, Baptisttown, Oak Grove, Jutland, and Oldwick.

The coming of the creamery revolutionized the dairy industry. Many farmers continued to make butter, which they unloaded on the storekeepers with whom they traded, often to the financial loss of the latter. Lewis Runkle, local editor of the *Hunterdon County Democrat,* in an editorial entitled "A Chapter on Butter" in the issue of Mar. 4, 1890, pointed out that merchants, to keep the good will of their customers, often accepted poor butter in trade. As they could not send butter to markets often, much of it became rancid and did not bring good prices.

Most farmers, however, utilized the new market furnished by the creameries. Spring houses fell into decay, and churning equipment was discarded. Complete adjustment to the new situation did not come easily or quickly. Farmers objected to their treatment by state milk inspectors with Babcock testers, stockholders often grumbled at the small profits which the creameries returned, and inexperienced managers and workers sometimes turned out poor products. E. T. Bush later recalled that the cheese of the Oak Summit Creamery near Palmyra was "fit for decent burial, nothing more." [5] The prices of milk at times dropped to very low levels. For instance, the creamery at Locktown paid only 94¢ per hundred pounds of milk in October, 1896. At such times farmers were inclined to be exceedingly wroth, but few, if any, of them changed back to the home production of butter.

In the meantime the shipping of fluid milk to city markets by rail had been slowly developing. As mentioned previously, this began as early as 1869 at Flemington, and by 1878 daily milk shipments were also being made from Whitehouse Station. On July 5 the *Hunterdon County Democrat* reported that George B. Stothoff had "recently engaged in a new process of shipping milk in glass cans holding one quart each." He received four cents per quart where previously he had been paid two. Most milk destined for the city market, however, was sold by the farmers to dealers at milk stations. In 1899 they were receiving one and one-half cents per quart for milk, and city customers were paying six to eight cents. Some farmers of Hunterdon County participated in the attempt of the "Five States' Milk Producers' Association" to raise farm prices in that year, but little was accomplished. After 1900 the rivalry of the milk shipping stations began to hurt the creameries seriously, since their prices fluctuated less during the year, and often were considerably more than the creameries could pay. They and the one or two condenseries of the county found themselves on the defensive, and one after another was discontinued, sometimes simply stopping business, at other times changing over into milk shipping stations. The creamery at Three Bridges survived until 1927, that of Milford until 1929, those of Flemington and Whitehouse Station until the early 1930's. A small fraction of the milk of the county is still sold to creameries located outside its borders, and comparatively small quantities are purchased by cheese factories at Lebanon and Little York. Otherwise, all of the milk of the county goes to bottling plants.

Farmers, feeling at the mercy of the dealers, have organized in one

[5] *Hunt. Co. Dem.*, Nov. 24, 1932.

way or another. Since 1918 the Dairymen's League has attempted with varying success to control the prices of milk in the New York milkshed. Many Hunterdon County communities organized Dairymen's League Cooperative Associations, nine of which are functioning at this writing. The League's milk receiving station at Flemington is one of the largest of the county. In 1933 the Hunterdon County Milk Producers' Association was formed as a unit of the United Milk Producers' Association, which serves as a bargaining agency for its members. Over three hundred dairymen selling to independent dealers belonged to this organization in 1944. About fifty farmers in the southern part of the county are members of the Interstate Milk Producers' Association, a bargaining agency for the Philadelphia area. Their milk is sent to the secondary market at Trenton. Recently about twenty Hunterdon County dairymen joined the Consumers' Cooperative Association, a cooperative dairy operating at Belle Mead in Somerset County. This cooperative was formed through the efforts of the Farmers' Union, but is independent of it and markets its milk through a consumer cooperative group in New York City. In 1933, at the worst of the depression, milk prices were depressed to the lowest levels in half a century; and as a result of considerable agitation, in which this section shared, the New Jersey Milk Control Board was set up to cope with the situation by a system of price fixing within the state.

With the rapid development of a market for fluid milk came a change of emphasis in cattle raising. Cows came to be judged by the quantity and richness of their milk, and farmers practiced selective breeding with milk production in mind. Interest in dairy breeds was created, and a Holstein bull and two cows were imported by John T. Ellis in 1871. The Agricultural Society added one dairy breed after another to its list of prizes, and gradually eliminated the beef breeds. Herds of purebred Jerseys, Guernseys, Ayrshires, and particularly Holsteins appeared in the county. Greater interest in correct feeding was manifested, and farmers increasingly bought feed not produced on their own farms. To guide them in improving their herds, cow testing associations, now called dairy herd improvement associations, were formed. The first was organized near Ringoes in December, 1919, with Fred Totten as secretary. Others soon started in the Flemington and the Pittstown-Quakertown areas. In the 1930's associations were organized for the cooperative ownership of blooded sires of good pedigree.

A revolutionary change in breeding practices came in 1938. Under the leadership of E. J. Perry of the New Jersey Agricultural Experiment Station and the sponsorship of the Hunterdon County Board of Agriculture, farmers of Hunterdon County and nearby parts of Somerset and Warren counties organized the first cooperative artificial breeding unit for cattle in the United States. An expert from Denmark was imported to train local veterinarians in artificial insemination. The association was launched with about one hundred members, who together owned over one thousand cows. The first calf by this method was born on the farm of Richard S. Schomp near Stanton, February 15, 1939. At first only Holstein bulls were used, but later Guernseys, Jerseys, and Brown Swiss were added. In 1939 Mulhockaway Farm near Clinton was chosen as a central location for the building of bull barns and as a headquarters for operations. By April, 1940, over 3300 cows were enrolled in this project, and the first calves resulting from artificial insemination were themselves being bred artificially. Bull pens of the county have become vacant as more and more farmers have seen the advantage of the new method, which gives them the use of good proven sires at a cost of $5.00 per cow per year.

For over a century farmers of this area have had to balance the costs of raising their own calves against the price of cattle brought in from regions of lower costs. An advertisement in the *Hunterdon Gazette* of August 13, 1838, shows much about methods used in earlier days:

> There will be offered by the subscribers at Klinesville, about 150 head of cattle, such as Good Working Cattle, and from two to four year old Steers and Heifers, and Cows, all in good order. They are expected to be there on the 16th of August, inst., and remain until Monday the 20th.

Droves were brought from Ohio and Kentucky even after the Civil War, though dealers gradually turned to use of the railroad for long-distance transportation. New York cattle were driven in until fairly late, but finally they, too were shipped by rail. They were imported in especially large numbers from before 1890 until 1920. After that time they were superseded for the most part by cows from Wisconsin and other Midwestern states and from Canada. Prices have varied greatly from time to time. At a sale at Force's Hotel in Flemington in 1883 the average price was $53.50. In 1920 one herd of New York cows averaged $165, and in 1927 one from Wisconsin averaged $181. Eight head of good local cows averaged $296 at a sale near Flemington in 1930. Cattle prices slumped for a time during the depression of the 1930's, but at the present writing are again very high. A complicating factor in the importation of dairy animals has been state legislation regarding tuberculosis and, more recently, of Bangs disease. Losses through slaughter of diseased cows, as well as increasing milk prices, has increased the demand.

The following table gives various statistics of the dairy industry in Hunterdon County. It should be remembered that there has been a greater increase in the quality of dairy cattle than there has been in their number. Totals at various census times are not strictly comparable because of changes in classification. The figures for dairy products are as of the year preceding in each case:

Year	*Milk Cows on Farms*	*All Cattle on Farms*	*Butter Made in Pounds*	*Milk Sold in Gallons*
1850	10,691		975,679	
1860	12,037		1,010,674	
1870	12,983		965,243	594,568
1880	15,555	23,047	1,053,955	1,394,429
1890	15,722	21,182	1,171,181	(not given)
1900	15,150	23,226	805,147	4,077,169
1910	17,214	24,841	505,861	5,292,466
1920	17,186	22,937	204,630	6,344,728
1930	14,825	20,955	63,904	8,390,452
1940	17,754	24,001	12,209	12,179,624

Despite its importance today poultry raising has been merely a side issue during most of the history of the county. In early days fowl were allowed to run about freely, often roosting at night in trees about the farmyard. Geese were sometimes kept within rail fences by use of yokes. That not too great importance was attached to the keeping of poultry is shown by the scant mention in the records. Store accounts in the eighteenth century show no large amount of trade in eggs or poultry, since nearly every family kept enough chickens to supply meat and eggs for itself. In 1804 the Capners occasionally sold eggs to the village store, and their letters mention ducks and "Guinea birds" as well as chickens. Dr. Bowne in 1811 sold geese at 50¢ each and ducks at 60¢ a pair, and in 1816 he sold three turkeys for $2.00. As markets grew more important, prices increased. In early December, 1856, chickens and geese were worth 10¢ per pound in Flemington, turkeys and ducks 12¢ per pound, and eggs 24¢ per dozen. In January, 1869, Solomon Holcombe paid three cents each for eggs. The moderate size of the trade in live poultry even in 1870 is shown by a *Hunterdon County Democrat* item of September 28: "The largest consignment of poultry (1562 head) ever made in this section was by Park and Larew, one day last week, to George

A. Poulson, with Runyon and Applegate, New York." Poultry raising was not of much importance even this late, and chickens still did not get the care given to other farm animals. The country fairs for over thirty years showed a preponderance of prizes for fancy show birds rather than utilitarian breeds.

The first great technical advance in poultry raising was the development of the portable incubator, which came into popular use about 1880. Like all inventions the new machine was severely criticized at first. Sympathy was felt for the poor motherless chicks, and the flesh of incubator hatched chickens was pronounced insipid and thought by some to taste of kerosene. The success of individual farmers and the efforts of incubator salesmen soon broke down this resistance, however. The coming of the incubator corresponded with a great increase in the market for poultry products and consequent prosperity for poultry growers. Flocks increased in size both for chickens and other poultry. A market developed for eggs for setting, for young pullets, and soon for newborn chicks. Newspapers reflected the trend of the times by giving advice on the care and feeding of poultry, and Farmers' Institute and Grange lecturers talked on problems of poultry management. The results of work done at the New Jersey and other experiment stations were soon known by most farmers. A lively interest developed in the results of poultry exhibits at fairs and in improvements in incubators, brooders, and other equipment.

As was natural under the circumstances, an increasing specialization was soon evident. The "chicken business" became a specialized profession for many, who, even more than dairymen, dropped all pretense of general farming. The division of function was soon carried further. Some poultry farmers ran hatcheries, others fattened chickens for market, and still others kept laying hens. Those who raised chickens for meat turned to breeds which combined size with fair egg production, and particularly to Barred Plymouth Rocks, Rhode Island Reds, and New Hampshire Reds. Those interested only in egg production acquired Leghorns, and the White Leghorn eventually superseded all other chickens in popularity. Very recently crosses of various breeds have become quite popular both for eggs and meat. Chick hatching soon became a highly specialized industry in itself. D. C. R. Hoff, one of the pioneers in chick hatching in Hunterdon County, is still in the business at the present writing, though now located across the line at Neshanic Station in Somerset County. Hatcheries became increasingly important after 1910. In 1923 over three and one-half million chicks were sent by mail from the post office at Frenchtown, and since that time the industry has expanded greatly. Kerr Hatcheries, near Frenchtown, pioneered in two important recent developments, chick sexing and artificial insemination. The first has become very important for egg producers, as it permits the elimination of male chicks at hatching time.

In 1912 the Hunterdon County Poultry Association was organized. This organization, particularly in its more active period beginning in the 1920's, has sponsored a number of movements very helpful to poultry raisers. It encouraged entries in egg-laying contests outside of the county, and in 1929, with the blessing of the New Jersey College of Agriculture, backed the building of the Hunterdon County Egg Laying Contest plant near Flemington. The procedure and results there have pointed the way for flock improvement to farmers of the county. In their rapid expansion poultry farmers have had the guidance and advice of the experts of the New Jersey Experiment Station and New Jersey College of Agriculture, and the constant help of the Agricultural Extension Service. Another organizational development of the county was that of the Flemington Auction Market Cooperative Association, which has proven a very successful marketing device. Sponsored by the Poultry Association in the first place, it now has identical membership with it. Beginning with an auction of sixty cases of eggs in a Flemington basement in 1930, the Auction Market within five years passed the million dollar mark in gross annual sales. Since then it has continued to expand and is today one of the most successful auction markets of the United States. Since 1938 the Poultry Association and the Auction Market have sponsored annual egg and chick shows at Flemington.

As might be expected, poultry raising in general and egg production in particular have shown phenomenal gains in recent decades. Better equipment, balanced feeding, the use of lights in poultry houses, and, above all, a giant expansion of the industry have all played their parts. The laying ability of the average hen has been greatly increased. The average production of White Leghorns entered in the Hunterdon County Laying contest in 1929–30 was 161.5 eggs; in 1936–37 it was 187.9 eggs. The following table summarizes poultry production in Hunterdon County in the period for which census statistics are available. Although other poultry than chickens have not increased in numbers, their raising, too, has become specialized. Egg production is as of the year previous in each case:

Year	Chickens on Farms Apr. 1	Other Poultry	Eggs Produced in Dozens
1880	155,577	13,651	1,054,374
1890	223,402	27,109	1,071,828
1900	230,124	9,771	1,634,480
1910	327,729		2,090,043
1920	366,887	14,568	2,062,538
1930	508,431		4,080,015
1940	720,214	7,736	6,182,395

Honey bees were imported into West Jersey by its earliest settlers. "English flies," as the Indians called them, thrived in their new home. In the probate records of the eighteenth century, occasional hives of bees and a greater number of empty beehives are mentioned. Rutherford in 1786 wrote, "Bees are also reared in great abundance, a tenant with Care sometimes pays his Rent with this article." The usual beehive was simple, "a swelling cone, about one foot high, made of finger thick spiral strands of rye straw bound together with hickory thongs." [6] In order to get the honey, the bees of a hive were suffocated with sulphur smoke. A few specimens of these hives are still preserved in the county. They were used until about the middle of the nineteenth century, though many beekeepers knew earlier of improved types which did not necessitate the killing of the bees. After 1840 various patent beehives appeared on the market, and Blackwell reported good results from the use of one of these "palaces." In days when honey was more used as a sweetening agent than at present, beekeeping was more general, and honey and wax were standard articles of commerce at the stores of the county. With the increasing use of sugar honey became less important and beekeeping less general. The following table summarizes available statistics. Production is that of the year previous in all cases:

Census Year	No. of Hives	Honey Produced (pounds)
1850		14,171
1860		19,161
1870		8,585
1880		13,377
1890		21,480
1900	1,633	21,060
1910	1,799	32,777
1920	2,009	21,638
1930	1,406	23,673
1940	1,487	18,251

[6] Mercer, *Tools of the Nation Maker*, p. 11.

The beekeeper of today must have considerable technical knowledge, especially as to how to combat new diseases. The industry has not increased in recent decades, nor has it received a great deal of attention from anyone except the relatively few brave souls who keep bees.

CHAPTER IX

Transportation

THE DEVELOPMENT of means of transportation is nearly as important for the welfare of an agricultural community as the choosing of suitable farm crops and domesticated animals. In Hunterdon County one of the greatest debts owed by the pioneers to the Indians was the system of primitive thoroughfares which gave access to all parts of the county. Indian paths were designed for foot travel, but early surveyors and land agents were able to travel many of them on horseback. The first settlers used these trails to great advantage, and by a little ax work made them fit for wagon travel. Supplemental roads were added from time to time by the township overseers of highways, often as the result of petitions of the inhabitants. Old roads occasionally reverted to farm land. By a process of trial and error from earliest days to the present, roads have been located where most useful. The road system of the county, of which the Indian trails formed the skeleton, has been gradually filled out to its present form.

The most famous of the county's thoroughfares was the York Road, which crossed the southern part of the present county in a northeast-southwest direction. In 1711 the inhabitants of Pennsylvania built a road from Philadelphia to "John Reading's Landing," at present-day Stockton, with the understanding that it would be extended across West Jersey in the direction of New York. Such a road was opened, but when a ferry was established at Lambertville a few years later, the main route of travel changed to cross the river at that point. The term "York Road" came to mean the entire road from Philadelphia to New York by way of Lambertville. The villages of Mount Airy, Ringoes, Reaville, and Centerville later developed on the Hunterdon County portion of this trail. At Reaville the Amwell Road branched off from the York Road to lead to New Brunswick. An alternate route from Lambertville was the Hopewell Road, which ran across the extreme southern tip of the present county and later made a junction with the Trenton-New Brunswick road. The main north and south route of Hunterdon County was the one which led from Trenton through Flemington, Cherryville, Pittstown, and Hampton, thence on to its ultimate destination at the Delaware Water Gap. This road crossed the York Road at present-day Ringoes. A highway from New Brunswick to Easton crossed the north part of the county from east to west at an early date. In 1765 it was referred to as "the great Road . . . to Potterstown." [1] How early the road from Flemington to Frenchtown acquired importance is not known, but there was a ferry across the Delaware at present-day Frenchtown by 1741.

The very necessary secondary roads were of many types. There were "main roads" or "six rod roads," "four rod roads," "two rod roads," and "beaten paths." Travelers on even the main roads had to contend with mud, stumps, stones, and log corduroy. Roads were under the care of the townships which they crossed, and American aversion to taxation was too great to permit much improvement. Except for a few which were well drained, the condition of the roads of the county depended on the season, and during periods of thaw they were particularly bad. As a consequence poor roads were accepted as part of the American scene. Justice Ezekial Cole accepted "bad going" and "high water" as good reasons for not appearing in a justice suit at Whitehouse in the 1790's. One reason for building a Presbyterian Church at Flemington in 1791 was that bad roads in winter and high water in spring prevented regular attendance at the Amwell Church.

After more than a century of poor roads an effort was made by private initiative to step into the breach made by public neglect. Turnpike companies were given monopolies by the state in return for surfacing and maintaining roads. Their costs of construction were to be repaid and presumably a profit made by charging tolls for use of the road. Unfortunately, because of inadequate capitalization and insufficient tolls, most turnpike companies were unable to surface more than a fraction of their length and were often unable to keep repaired the damage done by heavy wagons. Hunterdon County, not on any main routes of travel, had few turnpikes. The New Jersey Turnpike, connecting New Brunswick, New Jersey, with Easton, Pennsylvania, was the most important. The company which constructed the route was incorporated in 1806. In Hunterdon County it took over the old road leading through Whitehouse, Potterstown, and Clinton. Despite the heavy stream of wagon traffic the turnpike was not a financial success. Farmers along the route, and especially those of Hunterdon County, forcibly resisted payment of the tolls. After thirty years of operation the company surrendered its

1 Petition, Emley Papers.

Hunterdon County portion to the various township overseers in 1838. The Georgetown and Franklin Turnpike, constructed in 1816, crossed the southern tip of Hunterdon County. Following the old Hopewell Road, it gave the region about Lambertville, which was then called Georgetown, a better road to New Brunswick. This highway was an important thoroughfare for Jersey and Conestoga wagons from both sides of the river. Wagons from Pennsylvania crossed at first by ferry, then by the new bridge which was built in 1814.

Two other turnpikes were more local in their nature. The New Germantown Turnpike Company was incorporated in 1813 with the avowed purpose of constructing an improved road which would connect the New Jersey Turnpike with the Washington Turnpike, another cross-Jersey route further north. This road took its beginning "between Bayles' mill and the white house" on the New Jersey Turnpike, ran through New Germantown (Oldwick), and struck the Washington Turnpike "at or near Netzer's mill in the county of Morris." [2] John Honeyman, a tavern keeper at New Germantown, was one of the chief promoters of the scheme. The impoverishment of Honeyman and other stockholders by the deflation in land values during and after the War of 1812 are said to have caused the failure of this company. The time for completion of the turnpike was extended in 1817 and again in 1820, but eventually the act of incorporation was repealed in 1832. The Spruce Run Turnpike was also intended to connect the New Jersey and Washington turnpikes, the Hunterdon County portion to run from Hunt's Mills (Clinton) northward up the narrow valley of the Spruce Run. It was never completed, though portions were apparently improved. On a journey in the early spring of 1828 Dr. John Blane traveled sections of the Georgetown, New Jersey, and Spruce Run turnpikes. He was distressed at the "passage . . . through the mountain" on the Spruce Run Turnpike, but otherwise found these roads were in better condition than the public roads. He described the main north and south road of the county, which was not a turnpike, as being nearly impassable. Attempts to revive turnpikes in later years failed for lack of public support.

In the nineteenth century the highway system filled out so that most farms were located on public roads. However, the old system of township control was retained, and most of them, at best, were fair-weather roads. A letter of 1831 summed up what was a common complaint. The road money of each township, never adequate, was parcelled out to

[2] *N.J. Session Laws,* Feb. 11, 1813.

various overseers. These men, being farmers, delayed action "until their corn [was] planted, the mud holes dried up, and the spring carting pretty well over." Then they fell to work and used up their appropriations as rapidly as possible during the slack season. The roads were liable to be bad again by the time of autumn hauling. During the long winter little was done to combat the mud and snowdrifts, everything being postponed until "after town meeting and corn planting." [3]

Year after year it was the same story. The town meeting of each township appropriated each spring several hundred dollars for the care of township roads. A taxpayer might, if he chose, work out a portion of his road tax. Neither the overseers nor their labor force would budge until their crops were safely in the ground. When the earth thawed out in the winter or spring, ruts were cut deeper and deeper. In May, 1868, the mire was so deep on one Raritan Township road that a horse broke its back. Haphazard work done by men whose chief concern was to work out their tax sometimes did more harm than good. Dr. C. W. Larison in his little magazine, *Ringos,* derisively called the method used the "one ditch system." Another writer angrily remarked that the usual way of "improving" the roads was "by throwing on them stones of the size of a man's head, covering them up with sod ploughed up from the gutter and turned upside down." He pronounced the road from Whitehouse to New Germantown (Oldwick) "the worst that any enemy of enjoyment in travelling could possibly devise, or, we believe, ever known outside of Virginia." [4] In the twentieth century most townships have turned to the more efficient plan of using supervisors, each responsible for the roads of a whole township. As late as 1933, however, the roads of Delaware Township were still in charge of some two dozen overseers.

The problem of snow in winter was a special one, usually inadequately handled. Snow was generally welcome, as it gave a fast method of travel, but when it drifted across the highways, sleds and sleighs had to take to the open fields. During the winter of 1835–36, snow fell intermittently for some weeks. One farmer wrote that "the People had to turn out with hourses [*sic*] and brake Rods [*sic*] in fields and every week or few days." [5] Similar difficulties were experienced in other years, the snowstorms of April, 1854, January, 1867, and December, 1872, being especially noteworthy. But it was the famous blizzard of March, 1888, which tied up traffic most completely. Elmer Bonnell reported drifts "as big as the white corn crib." It was two days before he could travel the short

[3] *Hunt. Gaz.,* Jan. 19, 1831.
[4] *Our Home,* I, 280, 377.
[5] Hankinson commonplace book.

distance to Clinton through the fields. The people of the countryside turned out in force to open avenues of travel. In some places drifts across the roads were so slow in melting that fields were badly cut up by traffic. In West Amwell Township a special town meeting was held to authorize the borrowing of money to hire shovel gangs. Other storms in January, 1895, January, 1905, January, 1910, and December, 1915, seriously incommoded travel. In the 1920's snow fences and motorized snow equipment began to appear. These, combined with centralized control and a public demand that roads be kept clear, have brought about a more adequate handling of the problem.

As the need for good roads became greater, there was an increasing demand for the construction of surfaced highways at public expense. State legislation of 1876 and succeeding years provided a means by which a county might take over, macadamize, and keep in repair certain roads. A proviso that the greater part of the cost of construction was to be assessed against landowners who benefited had a deterring effect, however, and no roads were macadamized in Hunterdon County at this time. Legislation of the early 1890's reduced the landowner's share to one tenth and introduced state aid up to one third of the cost, the county to pay the remainder and to maintain the roads after completion. Agitation in Hunterdon County increased, and a New Road League was formed. Plans for a surfaced highway from Flemington Junction to Flemington were defeated by the farmers along the route, however, and it was only through the willingness of John M. Davenport to pay more than his share of the cost that a strip of road near the fairground north of Flemington was macadamized. The forces opposed were able to prevent any further development for several years. The editor of the *Hunterdon County Democrat* thought that much of the demand came from bicyclists, and in a series of seven articles in 1902 urged that earth roads be improved rather than stone ones constructed. In 1901 Flemington macadamized its main street, and opponents were jubilant when heavy travel broke up the surface. By 1903, however, opposition was giving way, and the *Democrat* in its first issue of the year confessed that all indications were that Hunterdon County was "getting ready to embark in the stone road business." All newspapers of the county, with the exception of the *Clinton Democrat,* had fallen into line by 1907.

Much of the change in attitude was due to the new legislation which provided that the ten per cent formerly assessed against landowners should henceforth be paid by the townships. In 1904 special elections were held in East and West Amwell Townships, and the voters in each case almost unanimously favored the sale of township bonds to pay the township share of a macadam road from Lambertville to Ringoes. The county government then accepted its share of the burden, contracts were let in 1905, and the road was completed in 1906. By similar steps during the next four years stretches of macadam were built from Ringoes to Clinton, from Clinton to White House, and from Clinton to High Bridge. The whole system was tied to that of Mercer County by a macadam road from Ringoes to Woodsville, and short roads were also built from Asbury to Ludlow and from Spring Mills to Frenchtown. By this time the county had exceeded its allowable bonded indebtedness, and the only road built during the next two years was one from Clinton to Hampton, which gave the county a stone road for its full length.

These frail macadam ribbons soon proved inadequate for automobile traffic. It was necessary to apply oil each summer, and labor gangs were kept busy with repairs. Nevertheless, the desire to "get out of the mud" was very great. Local "good roads associations" and spontaneous booster meetings repeatedly sent delegates to the Board of Chosen Freeholders to urge the necessity of their particular projects, and the Board was forced to act on some of the demands in order to keep the good will of voters. Despite the high cost of maintenance, ways and means of building still more roads were found. By 1922 there were sixty miles of county macadam; by 1930, 150 miles; by 1944, over 200 miles. State aid to the townships, beginning in 1916, helped them to macadamize certain vital stretches not taken over by the county. The network has been filled in so that no part of the county is now far from a surfaced road. During these years methods of construction were adopted which enabled roads to stand up under wear and tear caused by motorized vehicles. The use of tarvia and other petroleum products as binders and for resurfacing older stone roads has proved very successful.

As costs increased, Hunterdon County felt itself at an increasing disadvantage. Its roads were expensive to construct, and there were no large cities to help bear the burden. The new highways were of undoubted benefit to people of other areas, and increasingly so as the numbers of automobiles increased. The fight to get a state road through the county was finally won in 1924 with passage of an act introduced into the legislature by David H. Agans, State Senator from Hunterdon. The state, with Federal aid, was to take over the road from Trenton to Buttsville, which included the main north and south road of Hunterdon County, and in addition the road from Ringoes to Lambertville and the one from Flemington to White House. In November, 1925, the stretch from Flemington to Ringoes was opened to traffic, the first concrete highway of the county. By legislation of the next few years the old New Jersey Turnpike, the road from Flemington to Frenchtown, and the road from Flemington to Somerville were also taken over by the state. These roads, with some changes to secure more direct routes and the addition of traffic circles, have since been paved, bringing the total of Hunterdon's concrete roads to about eighty miles. In the meantime the various boroughs and other municipalities of the county, sometimes with state or county aid, have paved or macadamized the more important of their streets.

The providing of means of passage across streams has always been a particular problem of transportation. In early days bridges of logs or hewn timbers were sometimes constructed by private persons or by overseers of highways. The whole county might be assessed for an important bridge, such as that over the Assanpink Creek at Trenton in 1716, but such occurrences were rare. In general, fords were far more common than bridges during the whole of the Colonial period. This meant that all travel except that by horseback was stopped by freshets. Thomas Capner, on his arrival in Flemington in 1787, wrote as follows to friends in England: "We crossed a [*sic*] many creeks or runlets, one in particular was so large that one of us were obliged to wade and carry the rest over, for there is no bridges, they all ride almost." Travel across the Delaware, until the building of bridges in the nineteenth century, depended on licensed ferrymen at a dozen or so points. A British officer captured by the Americans during the Revolution thus described the ferry at Frenchtown in 1778: "We crossed the river in scowls [*sic*], which are flat bottom boats, large enough to contain a waggon and horses; they are a safe conveyance and mostly used to cross the rivers in this country; they are rowed with oars on large rivers; but over a creek, . . . they pull them across by means of ropes fastened to either shore." [6] Another means of propulsion was the setting pole, used for pushing against the bottom of the stream.

In the nineteenth century more attention was given to the matter of bridges. Whereas roads were a township concern, important bridges were built by the county. By 1830 county expenditures for building and repairing such structures had advanced to the modest amount of $6,446.52 during the year, or nearly one third of the total county expenses. Nevertheless, even those bridges on the main stage roads were not always in a good state of repair. The amount spent constantly increased, and by the time of the Civil War was over $25,000 per year.

[6] Thomas Anburey, *Travels Through the Interior Parts of America*, p. 249.

Wooden bridges, often covered, were built at all of the old fords; these in turn gave way to bridges of stone, iron, and other durable materials. A Lambertville firm in the 1850's specialized in iron bridges, building among others those across the South Branch at Clinton, the Alexauken at Mount Airy, and Swan's Creek at Lambertville. The usual materials for bridges in the twentieth century are reinforced concrete and structural steel. A problem of modern days is to build bridges strong enough to carry giant motor trucks. Some of the picturesque old covered bridges lasted until recently, and one, "Green Sergeant's Bridge" on Wickecheoke Creek, is still in use at this writing.

The first bridges across the Delaware were built by private enterprise. In 1812 toll bridge companies were incorporated to build bridges at Coryell's Ferry (Lambertville) and Howell's Ferry (Stockton). The New-Hope-Delaware Bridge at Lambertville, opened to traffic on September 13, 1814, was a covered wooden structure of six spans, 1051 feet in length, and cost $67,936.37. The company was given extensive banking privileges, which later caused considerable difficulty. Ownership of the structure changed in 1863 and again in 1874. Floods in 1841, 1862, and 1903 tore out large sections, and in the latter year the whole bridge was rebuilt with a steel framework.

The company which built the Centre Bridge, as the one at Stockton was called, sold nearly five hundred shares of stock at fifty dollars per share. Most of the amount was sold locally, and former ferry owners on opposite sides of the river were among the purchasers. This bridge, which was also finished in 1814, was two hundred feet shorter than that at Lambertville. It had to be rebuilt in 1830 because of faulty construction, and again in 1841 when it was swept away by a flood. A ferry was probably established each time until a new bridge could be built, as was also the case at Lambertville in 1841. The bridge which was built in 1841 had fewer piers in order to facilitate the passage of rafts. It was well built, and the only bridge between Easton and Trenton to withstand the flood of 1862. The company paid small dividends until 1829, then none at all for forty-four years. After that time, and expecially after 1885, it paid the stockholders fair returns until its destruction by fire in 1923. According to Roberson the toll rates fixed by law in the beginning were as follows:

Coach, landau, carriage, phaeton or other "pleasurable carriage" with four wheels and four horses	75¢
Wagon with four horses	63¢
Carriage with two horses	50¢

Chaise, riding chair, sulky, cart, or other two-wheel carriage, sleigh or sled with two horses	38¢
Same with one horse	25¢
Horse and rider	12½¢
Foot passenger	3¢
Horned cattle	3¢
Sheep and swine	3¢

Certain concessions in rates were made from time to time, notably to shad fishermen, carters of lime, and people crossing the river to go to church. In 1817 wagons hauling construction materials for the Baptist Church at Sandy Ridge were allowed free passage. The bridge at Raven Rock, a few miles farther up the river, was built in 1835. It was a covered bridge 705 feet long.

After many years of agitation companies were formed and covered wooden bridges built at Riegelsville, Milford, and Frenchtown. The first, just outside the county, was finished in 1837, the second in 1842, the third in 1844. The Riegelsville bridge had three spans with two lanes and sidewalks. That at Milford, also built in three spans, had a single driveway flanked by sidewalks. The bridge at Frenchtown was considerably longer, 962 feet, and had six spans. The old Frenchtown and Milford bridges, repaired or rebuilt after the floods of 1862 and 1903, remained until 1931 and 1933 respectively, and were among the last of the covered bridges across the Delaware. The bridge from Point Pleasant, Pennsylvania, with its New Jersey end near Byram, was built in 1855. It was damaged by the flood of 1862 and entirely destroyed by fire in 1892. The new steel structure built in the latter year was partially swept away by the flood of 1903, necessitating considerable repair.

The sixteen or seventeen bridges across the Delaware, including those mentioned as serving Hunterdon County, had a long history as toll bridges. Early in the twentieth century, however, the agitation for free bridges gained volume on both sides of the river. Eventually both state legislatures were won over and a joint commission was created. The bridge at Trenton was made free in 1918. The first acts to affect Hunterdon County were the purchase of the Lambertville-New Hope and the Point Pleasant-Byram bridges by the commission in 1919, and that of Riegelsville in 1923. The site and piers of the Centre Bridge were acquired in 1925, the toll company receiving enough to pay back the full value of its stock. The piers were reinforced with concrete, and a modern steel bridge was built. The bridges at Raven Rock, Frenchtown, and Milford were purchased in 1928, and in 1931 and 1933 the picturesque old bridges at Frenchtown and Milford were replaced by modern concrete structures.

A considerable part of the travel of the eighteenth and early nineteenth centuries was by horseback. Everybody rode, young and old. Schoepf wrote in 1783, "To go afoot is an abomination to the American, no matter how poor or friendless." Mrs. Mary Capner in her practice of midwifery often rode long distances. In 1787 she wrote with characteristic good humor, "I was soon mounted on a green Velvet or Shag Hunting Side Saddle [and] away we ambled abreast for let Horses travel how [they] will they go abreast if there is two." Men and women often rode to church in large groups. It was a tradition in the Holcombe family that in December, 1783, Mrs. Richard Holcombe rode a mare to Buckingham, Pennsylvania, for her husband's funeral. On the return the mare, anxious to get home to her colt, swam the Delaware with her mistress. Women wore no riding costume, but sometimes carefully wrapped their outside skirts around their waists to keep them from being splashed with mud. Travel by horseback remained important until the day of the motor car.

Carts and wagons were the most common vehicles in the early settlements of Hunterdon County. Heavy wagons, pulled by either horses or oxen, were essential to every farm; and, as roads were improved, light wagons were developed for travel. In 1793 when she came to visit the Capners, Mary Choyce wrote to her parents, "We all came to Flemington . . . in a light waggon that is what they travail in in America. I think they are much pleasanter than in coaches. We left Brunswick by about nine in the morning and arrived in Flemington betwixt four and five in the afternoon." In the same letter she described a trip to Philadelphia in a covered wagon. In 1795 John Hall thus lectured relatives who were preparing to come to America:

> Emigrants must fall into the ways of the farmers here and go out with their own waggons, seldom more than two horses, and they ride and drive by lines, taking Hay and Provender for their horses and Crib for themselves with a blanket and lay in their waggons. [A]t all taverns there is a cover'd shed and manger to feed.—You then open your crib and dine or sup &c., having a glass or half glass of grog and the profits of that is to the tavern-keepers recompence.—[T]hose that can determine to fall into these methods and other ways habituate themselves to the country will feel themselves as happy as little kings in a few years.[7]

[7] John Hall to Mr. and Mrs. James Choyce, May 1, 1795, Capner Papers.

Wagon travel was made much more comfortable by an American invention, that of wooden springs on which was mounted a wooden seat.

Vehicles of other types were imported from abroad or developed in this country. The Capners reported two-wheeled carts, sulkies, and "chairs," usually pulled by one horse, chaises or "shays," which might have either two or four wheels and might be pulled by one or two horses, and occasional two-horse carriages. Jasper Smith of Amwell Township paid a tax on a phaeton, a light type of carriage, in 1790. Coaches were few and restricted to the wealthy. The Coxe coach, still preserved in 1870 according to "Traditions of Our Ancestors," was "heavy, lumbering, painted yellow, with the family coat of arms painted on each side." None of these vehicles provided comfortable transportation, but in some of them jolting was lessened by suspending the body of the vehicle from leather straps.

The pleasure wagon with lattice or panel body appeared early in the nineteenth century, and soon after it the trotting wagon with lamps of tin or glass for night travel. The elliptic spring, a great contribution to comfort, began to appear in the 1820's. A vehicle stolen in Flemington in 1825 was described in the *Hunterdon Gazette* as a new gig with cradled springs. It was painted a snuff color striped with black, and its interior was lined with blue cloth. The new harness was of breeching type, and the metal parts were heavily plated with nickel. Carriages and trotting wagons were soon equipped with the new springs. The latter, furnished with a top, became the buggy wagon, and a later development was the buggy. Tops made of tow, linen, or leather were used for protection from the elements. "India rubber cloth" was also used after 1840. The carriage shops of the county in the 1840's and 1850's advertised that they made trotting wagons, buggies, family carriages, Rockaway carriages, and fancy sulkies. All pleasure vehicles at that time were taxed. According to the township assessment book twelve carriages in Readington Township paid a tax of one dollar each in 1846. Pleasure wagons were divided into three classes. Those of the first class, of which there were 135, paid 75¢ each. There were 83 in the second class, which paid 50¢ each, and 15 in the third class, which paid only 30¢ each. Six gigs and sulkies were taxed at 40¢ apiece.

A marked characteristic of the second half of the century was an increasing diversity of vehicles. A new kind of wagon for road travel was the Dearborn, which had small front wheels and a high body so that it could be turned very sharply. Buggy wagons varied as to shape, and carriages as to number of seats. The two-seated family carriage was very popular. The type used in the 1860's was so short that the knees of the driver and his companion almost touched the dashboard, while those who sat in the back seat were equally crowded. A larger carriage, the carryall, could accommodate from six to nine persons. The Agricultrual Society in 1878 offered prizes for phaetons, buggies, sulkies, side-bar buggies, buggy wagons, phaeton buggies, family carriages, two-seated carriages, one-seated sleighs, and two-seated sleighs. Elmer Bonnell paid $140 for a new buggy wagon made to order in 1886. The canopy-topped surrey wagon or surrey, pulled by one or two horses, was popular from about 1880 until the day of the automobile. Among the items offered at a sale near Flemington in 1916 were the following: farm wagons, a "low down wagon," a spring wagon, a "store wagon," a platform spring wagon, jaggers, buggies, and surreys. It was at about this time that rubber tires were put on some horse-drawn vehicles.

The fastest transportation of early days was by sleigh and sled, but unfortunately the season of snow was often short. Many of the sleds and sleighs so common in the eighteenth century were probably homemade wooden affairs, though blacksmiths also made iron bound sleds for their patrons. In 1794 Joseph Capner bought a "new box sled with runners with Bowes and a cover over it like a carrier's wagon." The whole Capner family was disappointed when the winter passed without sufficient snow for a trip to Philadelphia. Family sleighs were often quite fancy. Those common in Readington Township in the 1860's were "long goose necked affairs with a very high seat in front and a low seat with a very high back for those who sat behind." [8]

The first snowfall of any season was quite an event. In December, 1787, Mrs. Capner wrote, "We have had such stareing [*sic*] and gaping at the Sleighs and Sleds runing [*sic*] both way along the road, not one waggon now let the business be what it will." Long journeys were often taken after a heavy snowfall. People ventured out to make distant visits, farmers took their produce to market, young people took pleasure jaunts known as "straw rides"; and throngs of people visited the village stores, attended church, and packed the courtroom at Flemington when court was in session. In February, 1793, the Capners drove from Flemington to New Brunswick in a little over four hours. At one place the sled upset and the whole family was spilled out. On the following day it rained, and they rushed homeward, hot bricks at their feet, extra clothing thrown over their laps, and an umbrella over their heads. Neighbors, caught farther from home, had to make the slow homeward trip to Flemington

[8] Stillwell, *op. cit.*, p. 26.

by wagon. Sleds were also used for carrying heavy or bulky goods. Joseph Capner employed a sled to haul his cheese to Philadelphia each winter when possible. In the *Hunterdon Gazette* of February 10, 1830, appeared the following item: "The Philadelphia mail stage came the whole way on runners on Monday."

Though the season of snow was never more than a few weeks each year, the use of sleigh and sled persisted until the coming of the automobile. After that time, however, the use of sleds decreased rapidly. On February 11, 1920, the *Hunterdon County Democrat* complained that only rattling car chains, not sleigh bells, were heard after a good snow, but in a later issue reported that some sleds and sleighs were being brought out of hiding. Since then the efficient handling of snow on the highways has made year round use of the automobile possible, and sleds have nearly disappeared.

Freighting by wagon was the chief method of exchanging products with other areas during a considerable portion of the county's history. Along the York-Amwell Road, the Easton Pike, and the Trenton Road long lines of heavy wagons carried loads of farm products to port and returned with loads of merchandise. Millers and storekeepers hired teamsters for regular trips to Raritan Landing, New Brunswick, and Trenton Landing. The Atkinson and Johnson store in Flemington in 1793 hired men regularly to haul flour, cider, barreled pork, butter, etc., to Trenton, returning with barrels of rum, tierces of fish, barrels of lime, kegs of raisins, bags of coffee, loaves of sugar, hogsheads of molasses, kegs of nails, iron rods, crates of hardware, and cases of general merchandise. At other times it employed them to haul lumber from the Delaware, and to do local hauling of gravel, stone, rails, and oak bark. The first court house at Flemington, built in 1791–92, required the carting of stone from a distance, of lumber from sawmills and from the Delaware, and of lime, paint, oils, and glass from New Brunswick. Plaster from New Brunswick, "bords from the river," "rale cuts out of the Swamp," and "hides from New Brunswick" were common items in the Case tannery records. Many farmers made periodic trips to Philadelphia, New Brunswick, and Trenton on marketing expeditions of their own.

The same pattern continued until the coming of canals and railroads. The Atkinson store at Flemington in 1820 and the Kingwood store in 1835 carted goods to and from both New Brunswick and Philadelphia. Dr. John Blane later recalled that when he settled in the north part of the county in 1828, carters carried farm products to New Brunswick or

Raritan Landing and brought back barrels of fish and molasses, sacks of salt, iron bars and rods, and all of the other many things necessary to the rural economy of the time. Often in the fall the farmers themselves would take loads of apples, potatoes, pork, lard, poultry, butter, apple brandy, nuts, and game to Philadelphia or Trenton. The trip, made by covered wagon, required several days, and it was necessary for some of the family to stay at home to do the daily chores. Those who went carried food in a wallet and lived in the wagon during the trip. It was an occasion for all, and gave the housewife a chance to buy clothes and household necessities. Merchants and farmers were not the only persons requiring the services of teamsters. In 1827 an Amwell Township man offered two months of work hauling hewn timbers to New Brunswick. The Hill pottery, the Case tannery, and the Deats manufactory regularly hired carters to transport their products, which had markets far beyond the confines of Hunterdon County. The Case Tannery paid a carter with team and wagon two dollars per day in 1835.

The coming of canals did not eliminate carting but did shorten trips. Many wagons formerly going to New Brunswick, Trenton, and Philadelphia now went to such canal points as Stockton and Lambertville on the feeder canal, to Millstone on the Raritan Canal, and to various points on the Morris Canal. The coming of railroads caused still further changes in teamsters' routes. In 1848 stores in Flemington hauled their merchandise, coal, and other goods from Lambertville and New Hope; a few years later they sent their carters to Whitehouse Station on the New Jersey Central Railroad. With the coming of a railroad to Flemington, most hauling except local draying was eliminated. A miller on Nishisakawick Creek who had previously sent his corn meal to Somerville or New Brunswick shifted to Uhlerstown, Pennsylvania, when the Delaware Division Canal was opened in 1834. When the New Jersey Central Railroad was built to Whitehouse Station in 1847, he again changed his route. In 1853 the building of the Belvidere and Delaware Railroad made it possible for him to utilize the railway station at nearby Frenchtown. The canals and particularly the railroads increased marketing facilities greatly for all farmers of the county.

The demands of passenger travel led to the establishment of stage wagon lines across New Jersey at an early date. The chief routes were from Perth Amboy to Burlington, from Perth Amboy to Bordentown, and from Trenton to New Brunswick. Stage wagons were heavy vehicles with straight sides and built-in seats. Large hoops held up a cover of canvas or heavy cloth. By the middle of the eighteenth century regular schedules were being worked out by rival companies. There may have been irregular stage service at an earlier date, but the first regular traffic in Hunterdon County came with the establishment of a new line from Philadelphia to New York by way of Coryell's Ferry in 1769. The promoters of this line advertised as follows in the *New York Gazette or Weekly Post Boy:*

A new stage is now erected to go from New York to Philadelphia by way of Powles Hook, from thence through Newark and Elizabethtown to Bound Brook, and the North Branch of the Raritan, to Coryell's ferry, the only ferry between Newark and Philadelphia noted for its shortness and conveniency over the river Delaware. This Road is known by the name of Old York Road through the finest, most pleasant and best inhabited part of New Jersey.[9]

Wagons were to start from New York and from Philadelphia each Tuesday morning, meeting at South Branch for an exchange of passengers. The complete trip, supposed to take only two days, cost twenty shillings. The first stage wagon crossed Coryell's Ferry September 26, 1769. This line like others was abandoned during the Revolution, but resumed service after the war was over. During the latter part of the eighteenth century the stage wagon gave way to the stage coach.

Thomas Capner in pursuance of his trade of millwright found it necessary to make many trips by stage coach. The account of one made in 1795 is interesting enough to merit quotation; apparently Flemington, though now the county seat, did not have a coach connection as yet:

The day I left you I took a passage in the Stage for York about 4 oclock—only 2 passengers besides myself, at night slept at Elizabethtown and arrived at York next day about 11 oclock on Sunday. [T]he Mondays Stage was full. I took a Seat in the Tuesdays Stage. . . . I got my business done . . . and Spent my time in seeing everything remarkable I could find. . . . I left York about 3 oclock on [T]uesday morning and arrived at New Heaven [*sic*] that night about dusk. . . . The lower part of York Island I passed without Seeing Much it being dark, the Middle part appeared Barren, in the upper part was some very green fields of Grass. . . . Wednesday the sixth of May left new Heaven at 3 oclock. . . . [T]he coach was crowded with Collegians from Yale. . . . [B]reakfasted at Gilford . . . Dined at New London—left new London at 4 oclock. . . . Thursday May 7 left Norwich about three oclock in the Morning. [W]e left all our collegians upon the road. . . . [W]e entered the State of Rhode Island about 10 or 11 oclock. . . . [W]e arrived at [P]rovidence about 3 or 4 oclock being detained by the breaking of the

9 Quoted in R. C. Holcombe, "The Old York Road."

> coach—changed Horses, dined and left between 4 & 5 and Rode to Boston by a little past 2 next Morning. . . . I felt much Fetiagued [*sic*] having been the last day near 24 hours in the Stage. . . . [M]y Spare time I spent seeing everything in town worth notice. I left Boston on May 14 postler packet and arrived the 15. [N]ow 40 miles to Hebron.[10]

Parts of this letter not quoted show that the writer was a shrewd judge of persons and a careful observer of the farming methods and ways of life of the people of the countryside through which he passed. Despite the hardships of stage coach travel, he enjoyed his trips as thoroughly as does the modern traveler.

Though stage coach service expanded considerably as roads were improved and bridges built during the early nineteenth century, the line through Lambertville remained the most important one in Hunterdon County. At an early date, perhaps before 1800, this route was shifted so as to include Flemington. It was spoken of as "the old Swiftsure route" in the 1820's, and had evidently been a mail stage for many years. The coaches crossed the bridge at Lambertville, followed the Old York Road to Ringoes, then took the road to Flemington, where they stopped for the night. The route then turned to the right, striking the Old York Road again at Centreville. The trip from Philadelphia to New York, made three times weekly, was supposed to take thirty hours. The through fare was $3.50, way passengers paying in proportion. New contractors in 1828 advertised that they would use "splendid Troy Coaches." Because of the poor state of the county's bridges and roads, there were several stage upsets early in 1832. Under various owners the Swiftsure line, bolstered by mail contracts, operated its complete route until railroads actually cut across the line of travel. When the railroad from New York reached Somerville, the stage route was shortened to end at that point, and after 1844 it ran only from Philadelphia to Flemington. After the railroad came to Lambertville in 1851, the line was discontinued, though local stages still ran between that point and Flemington.

At what date stage service between Flemington and Trenton was started is not known, but the day book of a farmer named Bergun Berkaw shows that there was such a stage in 1824. In 1829 a mail contract enabled a stage line to run from Trenton, through Ringoes, Flemington, and Pittstown, to Bloomsbury. This stage made connections with the steamboat from Philadelphia at Trenton. By 1830, if not sooner, the line was extended to Belvidere, and was still in operation in 1849. For a time at least there was some kind of stage service from New Hampton, prob-

10 Thomas Capner to Mrs. Mary Capner, May 16, 1795, Capner Papers.

ably connecting with the Trenton-Belvidere stage at Pittstown. It is probable, too, that portions of the New Jersey Turnpike had stage service at an early period. For at least half of a century the most important contacts with the outside world for people of the county were by means of the stage coach.

The immediate effect of the coming of railroads was an increase in the number of short stage lines, though the longer ones disappeared. The first new stage line of this area devised as a feeder to a railroad was the route from Flemington to New Brunswick. According to the announcement of its opening in the *Hunterdon Gazette* of August 1, 1838, it ran by way of Greenville (Reaville), Clover Hill, Shannock (Neshanic), Flagtown, and Millstone, and made connections with "the train of Cars from New York" at New Brunswick. The through fare was $1.50. When Millstone received a rail connection in 1855, the stage line was shortened to stop there. The line, which at least at times had a connection with Ringoes, ran for many years. As portions of the New Jersey Central were finished, other stage routes were started to carry passengers to various railheads. In 1844 the "Daily Line" connected Flemington with Somerville by way of Centreville and Readington. Another line came from Jugtown (West Portal), or perhaps from Bloomsbury, passed through Bricktown (Perryville) to Clinton, then followed the old New Jersey Turnpike to Somerville. When the New Jersey Central reached Whitehouse Station, that point became the stage line terminus. As soon as the Easton and Amboy (Lehigh Valley) Railroad reached Barton's Bridge (Flemington Junction) in 1875, a stage was started to connect that point with Flemington. Short lines ran at various times from Milford to Clinton, from Annandale to Clinton, from Pottersville to Whitehouse Station, and from New Germantown to Whitehouse Station. As late as 1914 a horse-drawn stage was still operating regularly between High Bridge and Clinton.

Water transportation was of great importance in West Jersey in early days when land travel was slow and laborious. In 1685 Thomas Budd wrote, "After great Rains, we may bring down great quantities of Goods in flat-bottom-Boats, built for that purpose, which will then come down, by reason of the Landfloods with speed." Flat boats, rafts, and dugout canoes were employed in flood time on even small streams. Farmers in the eastern part of Hunterdon County utilized the South Branch to transport their produce to Raritan Landing and New Brunswick. The Delaware, however, was the only river extensively used, and transportation on that river had to cope with serious hazards. Between long, quiet

Fig. X. Flemington Bicycle Club in the early Nineties.

Fig. XI. Lebanon peach exchange about 1904. The auctioneer was John Rockafellow and the clerk, John D. Hoff. Selling fees charged were one cent per basket.

FIG. XII. Engine 308, Flemington Branch, Lehigh Valley Railroad. Used in the 1890's.

FIG. XIII. U.S. mail coach that traveled between White House and Flemington at the turn of the century.

stretches of water there are rifts or rapids where the water rushes through tangles of rocks. The worst of these on Hunterdon's portion of the river is at Well's Falls near Lambertville, where in less than a mile the water drops twelve feet. Pilots navigating a boat or raft through the "Horse Race" at this place charged a considerable fee. The rapids at both Well's Falls and Howell's Falls (Stockton) were improved somewhat in 1793 by men hired by the state of Pennsylvania, but the Delaware remained navigable only in times of high water. The "spring fresh," caused by melting snows, lasted three or four weeks. The "pumpkin fresh" in the fall was of shorter duration.

Freighting on the Delaware depended on the evolution of craft which could navigate the rapids. The raft and the Durham boat were the two early solutions to the problem. The raft was used chiefly as a means of getting timber and lumber from the Lehigh Valley and upper Delaware to market. The first man to "run the river" is said to have floated from Port Jervis, New York, to Philadelphia in 1746. Rafts were long and narrow, and strings of them were fastened together. Some were composed entirely of logs, others were made of logs overlaid with lumber. Various places on both sides of the river were used as stopping places. Here rivermen bought and sold rafts, hired pilots, and fortified themselves with liquor and food for the hazardous journey downstream or the laborious journey home on foot. Rafting was a colorful though perilous occupation, and concerned the people of Hunterdon County in a number of ways. Inns and stores of the river towns profited from the river trade, some innkeepers being so industrious that they served the occupants of passing rafts by boat. Local pilots were paid good fees, and local speculators often purchased rafts and took them on to market. Local sawmills diverted a portion of the traffic, and teamsters carted the lumber which they sawed to villages of the whole county. The amount of timber taken down the Delaware each season was enormous. It was estimated that in 1828 over a thousand rafts, containing fifty million board feet of lumber, made the journey. The traffic was at its height for some years after 1840, but declined as the timber resources of the Lehigh Valley and upper Delaware were depleted during the following decades. There was a brief resurgence in the early 1880's, as the last stands of timber were cleared out. Very few rafts have used the river since 1900.

The Durham boat, supposed to have been devised about 1750 at Durham Furnace on the Pennsylvania side of the Delaware for transporting pig iron, probably evolved from a type of craft already common on the river. Smith in 1765 spoke of "long flat boats . . . made like troughs, square above the heads and sterns, sloping a little fore and aft, generally 40 or 50 feet long." These boats were used after freshets for hauling corn, wheat, flour, barreled pork, limestone, and iron. The Durham boat did not have square ends and was a little larger, but was not essentially different otherwise. It was sixty or more feet in length and about eight in breadth. It needed less than six inches of water to float when empty and less than three feet when fully loaded with 150 barrels of flour, 600 bushels of corn or wheat, or twenty tons or so of iron. A crew of three or four men was required, and long oars, a triangular sail, and setting poles were used for propulsion. The latter, operated from "walking boards" along the edge of the boat, were used where the going was tough, as when the wind failed on trips upstream. Similar boats were used on the Susquehanna, the Mohawk, and other rivers and are thought to have been copied from those of the Delaware.

Durham boats remained important for the greater part of the first half of the nineteenth century. Several hundred plied the river when there was sufficient water, and as many as one hundred would tie up above Well's Falls at night when the season was at its height. Many of them came from further up river, but most Hunterdon County river towns had small fleets or single boats of their own. According to tradition, a Durham boat operated by a man named Pidcock ran a regular packet service from the foot of Well's Falls to Philadelphia before the Revolution, and such a boat is known to have been in operation in the 1790's. It was this boat which lost the cider barrel of Senator John Lambert in 1809. Because he disliked the beer in Washington, he had sent home for a beverage more to his taste. Hunterdon County products, such as wheat, flour, corn, flax, pork, butter, and hickory nuts were sent to Philadelphia after each freshet.

The death blow to the Durham boat and to river trade generally was the construction of the canal paralleling the Delaware on the Pennsylvania side. Durham boats were not very well adapted to canal use, and few of them were used there. Many were turned to such humble employment as hauling paving stones gathered from the bottom of the river or sand scooped from sand bars. For several years a boat from Lambertville made regular trips by way of the Raritan Canal to New Brunswick for clams. An interesting journey was that made by one of the last of the Durham boats in 1863. This boat carried a load of ship's knees down the Delaware to Lambertville, then by way of the canal to New Brunswick. From there it was towed to Brooklyn Navy Yard, where its cargo was sold. When a naval officer insisted that the boat must have a name, the owner christened it the "Monitor." In 1865 this same boat made a last journey to Philadelphia and back, then was allowed to lie in the feeder canal until it disintegrated.

The "coal ark" was another river craft which was devised for a specific purpose. The first ark carried three hundred bushels of anthracite coal from Mauch Chunk to Philadelphia in 1806; later models carried over twenty tons. The ark was essentially a large box mounted on a flat boat, and often a half dozen or more were joined together in a train. They were devised for the carrying of coal, but were sometimes used for transporting building stone, roofing slate, or seasoned timber. Naturally, these boats could travel only in flood time. A few were successfully brought back up river, but most were torn apart and their materials sold in Philadelphia. Occasional arks and their loads were sold at Frenchtown, Milford, and Lambertville. With the construction of the canal, the day of the ark came to a sudden end.

The steamboat was not a success on the upper Delaware. After a number of attempts to establish steamboat routes, a sidewheeler, the "Major Barnet," actually made several trips between Lambertville and Easton. Travel was possible only during periods of high water, and the venture was never profitable. When the Delaware and Belvidere Railroad was completed in 1854, the Barnet was taken elsewhere. Except in its lower reaches the Delaware as an artery of commerce had many disadvantages. With a canal paralleling it on one side and, later, a railroad on the other, it gradually became a deserted stream.

The idea of a canal across New Jersey was already an old one at the end of the eighteenth century, but three decades of the nineteenth century were to pass before the idea bore fruit. Agitation for a canal was constant after 1815 and insistent after 1825. Three companies at one time or another secured charters, but none was successful in raising sufficient capital to start work. Every community of the state was selfishly interested in the matter, and local jealousies caused more than one deadlock in the state legislature. Those who favored state ownership were ably opposed by private interests at all times. Pennsylvania's opposition to the diverting of water from the Delaware was also a cause of difficulty until 1829, when that state, interested in a canal of her own, came to terms. The proponents of outright state ownership eventually lost out. The law of 1830 incorporated the fourth company to attempt a canal, and this one was destined to be successful.

Hunterdon County had not been behind other areas in its agitation, and most of the county favored the route eventually taken. People of the southern part of the county generally felt that they would benefit from the construction of a feeder canal through Lambertville. Those of the eastern and central parts of the county also supported the Raritan route, partly because it would be nearer than others proposed and partly because they hoped for a branch canal coming up the Raritan and South Branch as far as Clinton, with a short spur leading to Flemington. Citizens of these areas financed a survey from Clinton to Millstone and advertised the feasibility of the route, but nothing further was ever done. Hunterdon County was vociferous in its support of the Delaware and Raritan Canal. Public meetings in many places endorsed the plan, petitions favoring it were signed by many voters, delegates attended Internal Improvements Conventions, and lobbyists were sent to Trenton to influence the legislature. A grand jury in 1828 in a spirit of jest put the matter to a vote, and decided in favor of the Raritan Canal by a tally of fifteen to six. The same jury voted twelve to nine in favor of Jackson for President. All of Hunterdon's representatives in the lower house favored a canal bill in 1828. Only one, Isaac G. Farlee, opposed the bill of 1830, and that was because he, like the editor of the *Hunterdon Gazette,* favored state ownership. It is interesting to note that two prominent foreign-born citizens, Paul H. M. Prevost and Thomas Capner, were active proponents of the bill. The latter was much opposed to state ownership, in whole or in part, and was especially against investing the state educational fund in the canal.

The canal as finally provided for ran from New Brunswick to Trenton by way of the Raritan, Millstone, and Assanpink rivers. Water for the main canal was to be brought in by a navigable feeder running alongside the Delaware from Trenton to Bool's or Bull's Island, just below Saxtonville (Raven Rock). As was natural, Hunterdon County was especially interested in its portion of the feeder canal, on which work was begun in 1832. At Bool's Island, a low riprap stone dam was built to divert water into the canal, despite the objections of the shad fishermen. Since the construction of the canal was done by hand, a considerable labor force, native born and foreign, was needed. One observer, John Lequear, has left the following description:

> In the year 1832 the construction of the feeder was begun, and in 1833 more vigorously pushed. In that year I several times accompanied my father on trading expeditions with beef and potatoes from the head at Bull's Island to Centre Bridge. The line was thickly dotted with board shanties. . . . The excellent sand stone quarries in the neighborhood, which the company had

secured, afforded them fine material for the construction of their locks and bridges.[11]

The Delaware and Raritan was completed in 1834, and in 1838 an extension was made from Trenton to Bordentown.

The effects of the canal were immediate. The long wagon haul to New Brunswick was shortened to stop at Millstone, and much carting was diverted to Lambertville, Stockton, and Bool's Island. At the latter point rafts were taken into the feeder canal for use by the sawmills along its course. The feeder had sufficient flow to operate these sawmills and several flour mills located near Lambertville. In the 1840's this village had a tremendous boom, and became an entrepot for a considerable shipping business. Canal boats of one hundred tons burden plied the canal to Trenton and on to New Brunswick or down to Bordentown, from which place they might be shunted into the Delaware Division Canal. In 1843 one enterprising man advertised thus: "He intends to continue the boating business at Lambertville, being provided with a decked canal boat, freight can always be kept dry; his intention is to make one trip every week, conveying freight to and from New York. . . . A good store house to store goods at Lambertville, going and returning." [12] A canal freight line in 1858 was making two trips weekly to Trenton. It is little wonder that the sleepy village became an enterprising town, and that its commerce, mills, and factories supported an increasing population.

Another factor in the growth of Lambertville was its connection with the canal system of Pennsylvania. The Delaware Division Canal, paralleling the Delaware, was started in 1827, but because of engineering difficulties was not entirely completed until 1834. This canal influenced Hunterdon County more than is sometimes thought. The *Hunterdon Gazette* of February 3, 1847, quoted a letter which read as follows: "There are permanent bridges across the Delaware at each of these places [Milford and Frenchtown], over which is hauled a large quantity of Corn Meal and Flour to the canal on the Pennsylvania shore, manufactured at Little York, Mount Pleasant, and Everittstown, each of which are pleasant villages situated a few miles from the river." Boats at Lambertville were shunted across the river from one canal system to the other, and limestone, coal, and pig iron were brought across for Hunterdon County and regions beyond. In 1854 a better connection was made and an overhead ferry system installed.

11 Clipping, *Lambertville Record,* 1891.
12 *Hunt. Dem.,* Mar. 8, 1843.

Farmers of the northern part of the county were also benefited by the Morris Canal, which traversed the nearby counties of Morris and Warren. The canals ruined the river trade, except for rafting, and by efficient handling of bulky products, particularly coal, competed successfully with the railroads until a late date. This was particularly true of the Delaware and Raritan, because of its position. The lease of the canal to the Pennsylvania Railroad in 1871 was a mistake, though the fact was not immediately apparent. The canal was still active at the turn of the century, but from that time on its decline was rapid; and in 1933, just a century after its beginning, it was closed to traffic. In 1934, because of some hazy idea of creating a scenic waterway, it was acquired by the State of New Jersey.

Even before the main canals of New Jersey were constructed, the proponents of horse railroads were arguing that they were less expensive and used less horse power than canals. Forward-looking citizens of Hunterdon County were determined that their section must not be neglected; and when a railroad from Elizabethtown to Somerville was projected, public meetings at various points urged the practicability of an extension on to Flemington and Lambertville and finally on to Philadelphia. A spur to Clinton would serve the central part of the county. Petitions were signed, resolutions passed, and lobbyists, including Thomas Capner, sent to Trenton. It was some years before the agitation bore fruit, and by that time locomotives had displaced horses on most railroads. Eventually, however, in 1839 the Elizabethtown and Somerville began to crawl westward toward Hunterdon County. A stretch of road from Elizabethtown to Plainfield was opened in February, and by the end of the year the line had reached Bound Brook. A correspondent of the *Hunterdon Gazette* rightly predicted that a "great day" was coming. In 1841 the road was extended to Somerville in Somerset County.

Despite popular agitation the railroad came no nearer to Hunterdon County for several years, until in 1847 a new company, the Somerville and Easton, was incorporated; and before the year was out, the railroad was pushed to Whitehouse in Hunterdon County. After some delay it was extended to Clinton Station (Annandale), but did not reach Easton until 1852. In the meantime coach service from the railhead to Easton had been provided. A giant celebration marked the completion of the road. The directors and their guests made the complete route in a special train of eight "splendid" cars drawn by the "gigantic" engine Pennsylvania—all decorated with flags and streamers. Crowds gathered at the stations and along the right of way to signalize the event. In 1849 the Elizabethtown and Somerville and the Somerville and Easton railroads were amalgamated to form the Central Railroad of New Jersey, usually known as the New Jersey Central.

The early sections of what became the New Jersey Central used a type of rail consisting of a six-inch timber surmounted by an iron strap. Whenever a strap broke, it had the dangerous tendency to curl up through the car above. In 1843 a passenger was transfixed and instantly killed by such a "snake head." From Somerville onward, however, the new iron T-rail was used, and three tracks were laid so that cars of two different gauges could operate. The T-rail eventually replaced all other types throughout the United States. The arrangement for two different gauges of cars on the Central was retained until 1875, when one outer rail, used by wide-gauge cars, was taken up.

The New Jersey Central did much to develop the upper part of Hunterdon County. As the railroad crawled westward, stage lines and teamsters changed their routes to utilize each new railhead. Farmers now had a nearby outlet for their products, and manufactured goods from the outside could be imported much more easily. A Clinton firm in 1856 proudly advertised in the *Hunterdon Gazette* that it carried Michigan lumber. Trade flourished, population increased, and contacts with the outside were greatly augmented. A person might go to New York, spend six hours there, and return the same day. In 1868 an observer wrote:

> Any one riding along the N. J. Central R. R. will be astonished at the rapidity with which villages and towns are springing up, and at the increasing population of places which had hardly a name before the railroad was opened. New Hampton shows a most rapid development, . . . while at High Bridge and Spruce Run manufactories are going up and laborers by scores coming in and settling.[13]

The people of the county were proud of the Central, and noticed each improvement in facilities. One of the biggest locomotives bore the name of "Hunterdon." An engineering feat of the original builders had been a wooden bridge more than one hundred feet high and nearly a quarter of a mile in length across the South Branch. It was this bridge which gave the town of High Bridge its name. After a few years the bridge became shaky, and the company decided to make a giant fill about it; and for five years, including the period of the Civil War, the work went on. Blackwell, who went to see this wonder in 1861, thought that it would

[13] *New-Brunswick Daily Fredonian,* Jan. 22, 1868.

"take quite a pile of dirt," and he was right. In 1856 a junction was made with the Delaware and Lackawanna at Hampton Junction, now New Hampton, and both railroads kept shops there until 1889. In one day in 1870 between seven and eight hundred cars were shifted from one line to the other at this point. A branch line from Somerville to Flemington, built in 1863, was known as the South Branch or the Short Line, and soon became a busy one. Despite constant urging plans to extend it to the Delaware never matured. In 1876 the Central built a branch line from High Bridge which paralleled the South Branch and left the county above California (Califon).

Plans for another railroad across Hunterdon County, one leading from Trenton to Belvidere, had been made in the 1830's, but because of the Panic of 1837 the line was delayed for a decade. In 1847 the idea was revived, and received immediate public acclaim. Early in 1848 the state authorized such a road, but a second bill for a branch to Flemington was defeated. In 1851 the Belvidere and Delaware reached Lambertville, which benefited greatly from the railroad shops established there. Cars and locomotives were manufactured until 1871, when the Pennsylvania Railroad leased the line. A wood burning freight engine, the "Warren," completed in 1866, had four driving wheels, four pilot wheels, a wide-topped stack, and a huge kerosene headlight. The railroad reached Frenchtown and Milford in 1853, and its destination at Belvidere in 1854. Its arrival made a tremendous difference in the economic life of the Delaware Valley. Frenchtown certainly enjoyed its connections with the outside world. On one Saturday evening in 1869 eighty-three persons, male and female, met the train, though there were only four passengers.

After the state legislators refused to permit the Delaware and Belvidere to build a branch to Flemington in 1848, prominent citizens of that place and Lambertville organized the Flemington Railroad Company and pushed through a charter for a separate line. When the Flemington Railroad was finished in November, 1854, the elated editor of the *Hunterdon Gazette* wrote, "At last, the ears of our delighted population are greeted with the melodious sound of the steam whistle." On the first day of operation of the new railroad only two tickets to Ringoes and four to Lambertville were sold at the Flemington end. The passenger business soon increased, however, though it was never large. Freight traffic was important from the beginning, and the local stores now used teamsters only for carting from the railroad depot. On May 10, 1860, Blackwell wrote, "The railroad is doing a good rushing business in coal, lime, lumber, and procures besides merchandise—many of the country merchants have their freights sent here." The town now had rail connections with Trenton and Philadelphia and, less directly, with New Brunswick and New York. When the Short Line of the Central arrived in 1863, the interior of the county had rail connections in two directions. In 1912 John H. Steele, a grandson of Thomas Capner, wrote as follows concerning the early days of the Flemington Railroad, the one line which was strictly a Hunterdon County affair:

> The passenger cars were built with side doors and no steps at all, and the station platforms were all built on a level with the car floors, so that all the passengers had to do was simply step from the car to the platform on entering or leaving the train. The locomotive burned wood, and after sundown, when I was on the train, it was my great delight to watch the innumerable sparks which would come from the smokestacks of the engines.[14]

The Flemington Railroad Company paid a six per cent dividend in 1871, the only one which it ever paid, and in that year it was absorbed by the Pennsylvania system.

In 1872 the Easton and Amboy (Lehigh Valley) Railroad was begun. Since as direct a route as possible was chosen, many engineering problems were involved, most serious of which was the tunnel through Jugtown or Musconetcong Mountain at what is now called West Portal. Work on the tunnel began early in 1872 and lasted until 1875. The job proved a bigger one than had been expected, and the contractors had much trouble with their labor gangs. The locally famous "Pattenburg riots," in which several men lost their lives, have already been mentioned. A money stringency caused by the Panic of 1873 added to the difficulties. While the tunnel was being dug, local contractors graded and built sections of the road here and there, until finally in 1875, with the completion of the tunnel, the road was complete. A double track was soon laid, and eventually stations were built at Flemington Junction, Landsdown, Pattenburg, Bloomsbury, and other points for passenger trade, which was definitely a secondary consideration for the company. In 1881 a spur from Landsdown gave Clinton a railroad connection, in 1884 another from Flemington Junction reached Flemington, and in 1891 a third was built from Landsdown to Pittstown. In 1927 and 1928 the Lehigh Valley, because of its great volume of traffic, widened its tunnel at West Portal, making it the widest tunnel in the world. Engineering progress since 1875, when the first tunnel was finished, was reflected in the much greater speed with which the work was done.

A late development in this area was the building of the New Jersey

14 *Hunt. Co. Dem.*, Dec. 17, 1912.

and Pennsylvania, or Rockaway Valley Railroad. The company which constructed and operated it was organized at Somerville early in 1888, its leading figures being J. N. Pidcock, John F. Melick, and W. W. Sutton. The line was at first built from Whitehouse to New Germantown (Oldwick) and served as a feeder for the New Jersey Central, with whom a working arrangement had been made. In 1889 it was extended on to Pottersville, in 1890 a spur was run from New Germantown to Peapack in Somerset County, and in 1891 the main line was carried to Mendham in Morris County. The intent was to extend the line to Morristown at once, making a connection with the Delaware, Lackawanna, and Western; but because of inadequate capitalization and disagreements as to policy this was not done until 1900, when the company was reorganized and J. N. Pidcock was forced out. This short railroad opened up portions of Hunterdon, Somerset, and Morris counties badly in need of better transportation. An immediate drop in coal prices and better marketing facilities for peaches were results of the coming of the railroad.

The pay load of the Rockaway was never large, however, and the company was hurt at the beginning by the failure of the peach crop in 1890. The new company which was formed in 1900 was also unsuccessful, and the line was thrown into bankruptcy in 1910. The Rockaway Valley went rapidly from bad to worse from this time on, and intermittent service was dropped in 1914 when it lost its only good locomotive through failure to keep up installment payments. All attempts to bring it back to life were unsuccessful, and during the World War the whole line was sold for junk. Later the right of way was also sold. In 1934 the remains of an old locomotive, tender, and wrecking derrick still sat on fifty yards of brush-grown track left near Whitehouse Station. One writer says of this ill-starred railroad that it had been built on promises, and had died of "taxitis." [15]

Lambertville in 1905 secured a much-desired trolley connection with Trenton, but despite agitation at Stockton and other points, the line was never extended, the chief cause probably being the coming of the automobile. The success of motor transportation has had the same effect on the railroads of this area that it has had elsewhere in the nation. In 1898 fifty-four trains entered and left Flemington each day on its three railroads, yet within two decades the railroads were on the decline. Curtailment of freight and passenger service, the substitution of gasoline cars and busses for trains, and the discontinuance of station agents have

[15] *Ibid.*, Apr. 19, 1934.

marked an attempt at adjustment to the rivalry of motor vehicles. Railroads are still of great local importance for carrying freight, and in recent times have gained in passenger service because of an increase in the number of commuters. Wartime conditions have greatly increased their volume of business, but their position is still far from secure.

In the latest stage of American transportation much emphasis is again put on road-borne traffic. The humble predecessor of our modern motor-driven vehicles was the bicycle. The first bicycles, high-wheeled affairs which were not very usable on rough country roads, began to appear in Hunterdon County in considerable numbers about 1880; the sidewalks of villages were soon alive with them and young men ventured on long cross-country trips when roads were good. Bicycles naturally excited the mirth of those who did not use them. A Flemington editor told how a Lambertville man fell into a puddle when he tipped his hat, and how a Lambertville girl was so amused at a bicycle spill that she swallowed a pin. About 1890 the low-wheeled bicycle came into use, and during the next several years a bicycle craze swept the land. In Flemington a bicycle club, with very natty uniforms, was formed in 1890, and a similar club was formed in Frenchtown in 1897. Among the members were prominent citizens and church goers. Bicycle parades and bicycle races drew large crowds. By 1894 the craze had extended to the women, and dozens of ladies mounted on bicycles were a common sight. The fad collapsed by 1900, and the bicycle was largely relegated to utilitarian purposes, as indicated by the fact that in 1905 there were only fourteen members left in the Flemington Bicycle Club.

The first automobiles appeared just before the turn of the century. The toll bridge at Lambertville in 1899 charged a fee of 10¢ for one-seaters and 20¢ for two-seaters, and automobile races were held at the fair at Trenton in 1900. Horses had by now accepted the bicycle, but were outraged by these new contraptions. When William H. Fulper brought the "Red Devil" into Flemington in 1902, a gentle old dray horse promptly broke loose and scattered a load of chestnut coal up and down Main Street. Runaways became everyday occurrences, and Stockton women complained that Lambertville autoists made it unsafe for them to drive horses on the public highways. In 1905 the editor of the *Hunterdon County Democrat,* Anthony Killgore, favored legislation requiring that all cars be geared so as not to go over twenty miles per hour, and some people suggested that owners of automobiles be required to build their own roads.

The car driver was the butt of many a popular joke, and the most humorous sight of the times was an automobile being pulled by a horse. The following are some of the "automobilisms" of 1905: "Judge not an automobile by its smell; all comparisons are odorous." "It is a short road that has no mending." "All does not go that glitters." "Say not, 'We shall return at 5; ye may return at 6's and 7's." "Oil's well that ends well." "Though thou swear by thine auto seven times, the eighth thou will swear at it; that is Kismet." "The horse goeth not ten parasangs an hour, neither doth he explode." [16] Automobiles were blamed for their damage to roads, for the dust they stirred up, for their bad odor, for their danger to the public, and for the decrease in land values along highways. Editor Killgore wrote on April 17, 1906, "The autoists have forgotten that for every voter who rides in an automobile, there are, and always will be, a hundred who . . . will never own or have a chance to ride in one."

The opponents of the "devil machines" might as well have saved their breath, for each year saw an increase in the number on the roads. Thomas, Rambler, Dolson, Pope-Hartford, Stanley, Burrick, Flanders, Holsman, Overland, Liberty-Brush, Maxwell, Haines, Saxon, and Briscoe cars appeared in turn, along with those of more familiar names, the Reo, Packard, Pierce-Arrow, Hupmobile, Buick, Dodge, Studebaker, Hudson, and Oldsmobile. The Ford was common before 1910, and the Chevrolet excited much interest in 1916. The various marks of our motor age appeared rapidly. Licensing of automobiles began in 1903 and licensing of drivers in 1906. Oil preparations were applied on roads to keep down the summer dust. Automobile "stages," soon known as busses, appeared on the main roads of the county. Mail wagons gave way to automobiles on mail routes, dray wagons were replaced by delivery trucks, and fire companies bought motor-driven apparatus. The "road hog," the "hit-and-run driver," and the "hitch hiker" each received his share of criticism. The garage, the parking lot, and the roadside filling station became American institutions. In 1917 Editor Killgore commented sadly that a whole generation of young men had grown up who were unable to harness and hitch up a horse. With this remarkable change in means of transportation facilities, a new attitude developed toward good roads. Opposition to taxation and bonded indebtedness decreased as more and more families became car owners. The farmers generally have become increasingly dependent on good roads and on motor vehicles.

The supremacy of the automobile, the motor bus, and the truck now

[16] *Ibid.*, Aug. 15, 1905.

seems complete; but already there are signs that an era of air travel and transportation is beginning. When airplanes made forced landings in Hunterdon County in the 1920's, huge crowds gathered to see them, but now regular plane service over the county elicits no special interest. Just where air service will fit into the economy of such a rural area cannot now be predicted. Most people are sure that it will never be other than a supplemental means of travel. It might be well to point out to them, however, that the same was once said of the automobile. Anthony Killgore in an editorial of April 17, 1900, said, "The gentleman's roadster [a horse], however, will never give place to the automobile." Perhaps the day of the air freighter, the flivver plane, and the helicopter is not far distant.

CHAPTER X

Communication and Trade

THE DEVELOPMENT of communication and trade is closely allied to that of transportation. To a people now accustomed to the daily delivery of mail and to the use of the telephone and telegraph, the lack of facilities for communication in earlier times seems appalling. In pioneer days it was possible to send letters only by special messenger or by someone who happened to be making a trip in the right direction. When better roads developed, post riders and stages provided a means of sending and receiving mail, and service was fairly good for those living near post stations; but people who lived in the back country sometimes did not get their messages until months after they were sent. A post route across New Jersey was established soon after 1700 with post offices at Burlington and Perth Amboy, but since Hunterdon County was off the main line of travel, its inhabitants were not very well served.

The establishment of a post office at Trenton in the 1730's and the development of that town as an important depot for stage wagon and stage coach transport later in the century helped the situation somewhat, but much of the county was still isolated. With the establishment in 1769 of a stage line on the Old York Road, it became possible for people near it to send letters unofficially by stage drivers and their passengers, a common practice at all times and especially so for several years prior to and during the Revolution, when the entire American postal system was disorganized. After the Revolution it was some years before areas not on main post roads were well served. The Capners reported that mail service at Flemington in the late 1780's and early 1790's was most unsatisfactory. They and their friends often journeyed to Trenton to get letters from England, and sometimes they directed that communications be sent to relatives in New Brunswick or Philadelphia so that they would be sure to get them.

The general situation improved rapidly after passage of the Federal postal law of 1794. Many new post offices and post roads were soon established. The opening of a post office at Flemington, the new county seat, was a great boon to Hunterdon County, though the service was still so slow and expensive that letters were sent by other means where possible. Other post offices were set up at various points, and by the turn of the century there were six or seven within the limits of present-day Hunterdon County. Besides the one at Flemington, others were located at present-day Lambertville, Oldwick, New Hampton, Pittstown, Ringoes, and perhaps at Kingwood. There is some evidence, as well as a tradition, that another was established at Prallsville (Stockton), though probably not under that name, very shortly after 1800.

During the early part of the nineteenth century many store and tavern keepers in villages and at country crossroads became postmasters for their localities. Van Sickle's tavern in the High Bridge region was a post office by 1817, and Bonnell's Tavern at Hunt's Mills (Clinton) is said to have received mail once a week by sulky in 1818. In the 1820's stores or taverns at Baptisttown, Frenchtown, Clarksville (Glen Gardner), "Point Reading," and other places also became points of mail delivery. The county is said to have had twenty-six post offices in 1827. Many small ones were moved from time to time for political reasons or because it was difficult to find men willing to take on the job of postmaster for the small remuneration involved, while post offices in the larger villages shifted their personnel with each new President.

In the early days of stage coach and post rider mails were very slow, particularly at times of the year when roads were bad. Postmasters were sometimes accused of dilatory practices, especially as regards newspapers. Postal rates were high and paid at the destination until about the middle of the century. In 1831 it cost six cents to send a single-sheet letter thirty miles or less, ten cents for eighty miles, and somewhat in proportion for other distances. Double and triple letters cost two and three times the above rates. Because of excessive costs, as well as illegible handwriting and bad roads, post offices often had many unclaimed letters on hand. The general situation was improved by a number of developments: postage rates were reduced in 1845 and again in 1851, postage stamps were used after 1847, and prepayment of postage was introduced at about the same time. With the coming of railroads mail routes were changed, and the mail service was naturally speeded up. In 1877 there were in the county eleven contractors who carried mail from railway depots to parts of the county which had no rail service.

The next great change was the introduction of rural free delivery. The first routes in Hunterdon County were started in 1905, one running from Flemington and the other from Stockton, and others were started in short order. Post offices at Rosemont, Locktown, Quakertown, and Centreville were discontinued in 1906 and 1907, and others soon thereafter. When, at one fell swoop, Cokesbury, Mountainville, and Fairmount were changed to rural free delivery in 1915, someone truly remarked, "The abolishment of post offices to a certain extent takes a village off the map." [1] The process of change has continued since that time. Clover Hill lost its post office in 1930, Croton in 1935, and Reaville in 1937. A recent trend due to better roads and development of automobiles has been the elimination of small centers for rural routes. In 1932 the route from Three Bridges was discontinued and divided among the mail carriers of Flemington. The establishment of parcel post in 1913 brought an added public service, leading to an increase in mail orders, and markedly reducing the local express business.

The telegraph, one of the great time savers in communication, was established in this area at a very early date. The Magnetic Telegraph Company, which was formed in 1845 by the inventor Samuel F. B. Morse, had difficulty in securing a direct route from Philadelphia to Newark, and therefore built its line so as to cross the Delaware at Lambertville. Its Hunterdon County portion followed the Old York Road. Messages sent from Lambertville in January, 1846, were the first ever sent from Hunterdon County. In 1855 Flemington was given a connection, and during the Civil War the towns of the Delaware Valley received the benefit of a telegraph line which followed the Belvidere and Delaware Railroad. The New Jersey Central Railroad constructed the Banker's and Merchant's Telegraph from Whitehouse to Flemington in 1868 and later extended the line to Stockton. These lines and short ones which were built privately gave many of the villages of the county a telegraphic connection with the outside world. The short lines were later made unnecessary by the coming of the telephone, and the main ones were eventually acquired by Western Union and Postal Telegraph.

The cable connection with Europe was of some consequence even in Hunterdon County. In 1858, during a meeting of the Hunterdon County Bible Society at Readington, it was announced that the Atlantic Cable was completed; the message of Queen Victoria to President Buchanan and his reply to her were read; and a prayer of thanksgiving was offered up for "this great event of the age." Soon the papers of the county were featuring the latest news from Europe.

The telephone was of more direct importance to rural people than the telegraph. In 1878, only two years after its invention by Alexander

[1] *Hunt. Co. Dem.*, Sept. 29, 1915.

Graham Bell, a connection between a hardware store and a drug store in Frenchtown was reported as working "admirably." As an experiment this amateur telephone system was connected to the telegraph line and a conversation held with Lambertville sixteen miles away. All of the first telephone lines were local affairs, and telephones in village stores with their dreadful noises often scared the horses at the hitchracks outside. The first telephone exchange in the county was established at Lambertville in 1881 by an affiliate of the Bell Telephone Company, with the cost per subscriber $5.00 per month. The "Pennsy Bell," as the company was locally known, ran a line to Flemington in 1885. The *Hunterdon County Democrat* of September 15th announced its arrival thus: "The telephone hooked Flemington to New Hope, and now you can stand right here in Finch's store and talk to the New Hopers all day if you want to." A number of subscribers at once put in telephones, and a few months later a connection was made with Clinton.

The high rates charged by Bell gave local telephone companies their opportunity. About 1898 the various private lines in and near Flemington joined together to form a cooperative line with headquarters in the home of Asa B. Reading at Flemington Junction, and in 1900 Reading branched out as the operator of an independent local company under a franchise held by his neighbor H. E. Deats. The rates were only one dollar per month, and the company soon had several hundred subscribers. At about the same time another local concern, the Merchants and Farmers Telephone Company, was having a remarkable growth. Its beginning was a private line connecting the creamery at Sergeantsville with that at Ringoes. In a few years the system had subscribers in Flemington, Lambertville, Frenchtown, Clinton, Pattenburg, Little York, Stanton, and other towns nearby. A third group, the Lebanon Telephone Company, was started in 1902, gradually extended its lines in various directions from Lebanon, and reached Flemington in 1906. In 1911 the company became a Bell affiliate. All three lines ran into Flemington, and residents of that town could have complete service only by having three different telephones in their homes until the three companies were finally combined in 1931.

Other companies brought local telephone service to other parts of the county. Pittstown and contiguous territory were long served by the Franklin Telephone Company. The Stover Telephone Company, a Pennsylvania concern, had telephones in Frenchtown from 1906 to 1931, but the wires were not replaced when the new bridge was built in the latter year. The Delaware Telephone Company of Milford, organized in

1905, was later bought by the Riegel Paper Company. Another local company functioned for a time in the Stockton-Sandy Ridge area. As the result of a process of gradual absorption of the local lines, the New Jersey Telephone Company, a Bell affiliate, now owns nearly all of the telephones of Hunterdon County.

These revolutionary changes in communication have had a considerable effect on the life of the people of this region. Speedy mails and telegraph service changed business methods radically and proved of great convenience in other ways. In the heyday of peach orcharding a special telegraph operator was kept at Flemington each season to supply New York commission merchants with information as to the condition of the crop. Racing and other news from the county fairs were wired direct from the county fairgrounds. During the time of the Hauptmann trial in 1935 the details of the trial were constantly flashed to the outside world by telegraph. Telegraphic connections then and since have enabled farmers and dealers to keep in touch with market conditions. The telephone has given the isolated farmhouse and the quiet village immediate contact with the outside world, and to some extent has taken over former functions of the telegraph. A Norton correspondent of the *Hunterdon County Democrat* in 1910 admitted the convenience and saving of time, but felt that people had become too dependent on the telephone; limitation of service at night and on Sunday already seemed a serious hardship to people who had had no telephone service at all a few months previously. Sunday and night services were eventually brought to all parts of the county, and the degree of dependence is of course greater today than ever.

The radio was the third and latest of the inventions bringing rapid communication. The news of Peary's discovery of the North Pole in 1909 first reached Flemington by an amateur "wireless" telegraph set. A number of amateur receiving and sending stations were soon functioning, and "hams," as their operators call themselves, were still operating on certain wave lengths up to the time of the second World War. Radio, however, never became a means of sending messages for most people. Although chiefly a means of entertainment, programs such as market and weather reports, broadcasts of agricultural information, and news of the world happenings have utilitarian value.

Trade was of fundamental importance even during pioneer days. In West Jersey an attempt was made by the earliest settlers to introduce the European system of fairs and markets. At Burlington several fairs were held during each year for the convenience of merchants and farmers, and market days occurred quite often for the sale of foodstuffs and other farm produce; but fairs and markets were not practicable in areas of scattered population without towns of consequence. In Hunterdon County their functions were taken over, as in most of America, by the general store.

The early store accounts which are still extant indicate that the purchases of pioneer families were not numerous. A statement from one Lewis Moore to Henry Van Dolah of Sandy Ridge in 1731 shows that the latter visited the store of the former six times in as many months. His chief purchases were rum, molasses, powder and shot, sugar, salt, and pepper.[2] These articles were among the most important of those sold at the store of Janeway and Broughton in Somerset County, a store to which many Hunterdon County people came periodically in the 1730's and 1740's. Other important items sold from this store were as follows: woolens, calico, muslin, gingham, handkerchiefs, stockings, shoes, hats, dishes, pans, scythes, grindstones, other tools, nails, locks, wool cards, paper, spices, tea, indigo, soap, starch, redwood, alum, needles, scissors, buttons, buckles, pins, knitting needles, penknives, spectacles, mirrors, cutlery, combs, Bibles, jew's-harps, wine, wigs, and pigtails.

This store, with its regular wagon service to Raritan Landing, was doubtless far better stocked than were stores of Hunterdon County. However, as roads were opened and farm products for exchange increased, inventories were expanded. By the beginning of the second half of the eighteenth century, stores at Kingwood, New Germantown, Flemington, Alexandria (Frenchtown), and other places were doing a considerable business. The store at Alexandria had trade connections in Philadelphia and by employing Durham boats used the Delaware as a shipping route. The farm produce bought by these stores and others ultimately reached a seaport and was shipped to markets abroad. In return came European manufactured goods and other articles not produced locally, for which the local merchants acted as middlemen.

As a land of scattered wilderness homesteads developed into a settled countryside, the needs and wants of the people multiplied greatly. The accounts for a few months of 1771 and 1772 of a store which was probably located near Whitehouse show how much the situation has changed since pioneer days. They record articles of great variety: powder, shot, and flints; chisels, squares, gimlets, wool cards, cradling scythes, grind-

[2] Quoted in *ibid.*, Sept. 11, 1930.

stones, and shoemaker's knives; nails, locks, hinges, window panes, and rope; iron kettles, teakettles, porringers, knives and forks, wash tubs, and sad irons; soap, brimstone, candles, logwood, alum, and tar; calico, scarlet cloth, drill, linen, chintz, buckram, swanskin, broadcloth, flannel, lawn, taffeta, sheeting, cambric, buckskin, deerskin, and sheepskin; buttons, cotton, silk, wool, dye pots, indigo, whalebone, knee buckles, lace ribbons, and knitting pins; gloves, blankets, garters, ribbons, mittens, greatcoats, stockings, handkerchiefs, felt hats, and patterns for trousers; combs, pins, and looking glasses; salt, rice, bacon, butter, ginger, sugar, tea, nutmeg, allspice, molasses, mace, and pepper; bran and middlings; snuff, wine, and rum; rugs, chamber pots, whiting pots, baskets, saddles, bridles, and fishhooks. From farmers, during a period of less than nine months, the store bought over one thousand bushels of wheat and some corn.

Crévecœur at about this time wrote of the "multiplicity of shops with English goods" in America. The "irresistable temptations" and the "allurements of fineries" of which he spoke were certainly present in the stores of Hunterdon County. A farming population could gain the money to pay for imported articles only by increasing the sale of agricultural produce. John Lambert, a storekeeper and miller in the southern part of the county in the 1770's, purchased farm products and sent them by wagon to Trenton. Among the things sent were wheat, rye, corn, oats, buckwheat, flax, tow, butter, cider, vinegar, and feathers. Individual farmers by the sale of such products as these acquired the cash or, oftener, the store credit made necessary by their changing economy.

During the War of Independence paper money inflation led to a partial disruption of commerce, but after the war was over trade soon settled into its old groove. In 1786 John Rutherford wrote that the farmers of New Jersey were buying imported dry goods, tea, West Indies articles, and wine, and were selling abroad wheat, flour, rye, corn, pork, cattle, horses, flaxseed, butter, poultry, staves, and headings. Despite the imposing list of exports he complained that New Jersey generally had an unfavorable balance of trade, a condition true of most of the United States during the Confederation period. The stores of villages and crossroads still remained the chief agencies of local trade. Those in Hunterdon County bought beef, pork, grain, flour, butter, cheese, wool, honey, wax, bundles of rye straw, and other farm produce from their customers, and sold them exotic foodstuffs, woodenware, hardware, paints and dyes, cloth of all kinds, articles of clothing, tobacco and snuff, and alcoholic beverages. The stocks of goods were more varied than in earlier periods, the number of kinds of cloth in particular having increased greatly.

Not all buying and selling was done through the local stores. The Capners took much of their cheese to Philadelphia, and when they wanted the best of cloth they bought it in that city. The Cases sold their leather in New York City and other places, and many farmers carted their own grain to Trenton, Lamberton, Raritan Landing, New Brunswick, or Philadelphia. In 1803 Abram Hagaman took sixty-three bushels of corn to a merchant in New Brunswick. The merchant's statement showed that he bought tea, coffee, sugar, pepper, and calico, and took the balance due him in cash.[3] Livestock was often sold to drovers, who made regular trips to Philadelphia; nevertheless, the local storekeeper was usually concerned with most commercial transactions of a community. The accounts of Benjamin Guild, merchant and postmaster at Pittstown in the 1790's, have now disappeared, but a person who once examined them has left the following description:

> It appears that he dealt in nearly everything. Nothing was too small or too large to handle, if there were money in it. He dealt in grain and flour; pork, lard and hams; wagons, horses and cattle; cutlery of all sorts from pocket knives to grass and cradle scythes; brandy, whiskey and hymn books; paper, dry goods and groceries. . . . His teams were constantly on the go; but it appears from the records that they made more trips to . . . Lamberton, which is now South Trenton, than other places. These wagons were always heavily loaded in going and coming.[4]

Despite encroachments the general store remained of considerable importance far into the nineteenth century. A store opening in Flemington in May, 1831, advertised in the *Hunterdon Gazette* that it would carry "every variety of Goods usually kept in a Country Store." Such an advertisement promised a great deal, for the number of articles kept in stock, even by the stores of the smaller towns, was enormous. In the larger stores there was a tendency toward departmentilization, a division of the goods into such categories as dry goods, groceries, hardware, drugs and medicines, house furnishings, and liquors; and a store would often play up one type of merchandise or another in its advertising. Some departments were more remunerative than others. When a store in Flemington braved public opinion by discontinuing its grocery section in 1869, the *Hunterdon County Democrat* commented, "It is a well-known fact that little or no profit is realized on groceries." The large

[3] Larison, *Fisher Family*, p. 59.
[4] I. N. Leigh, "A Few Historical Facts. . . ."

stores often acted as wholesalers for smaller stores and taverns. In 1869 seventeen men of Hunterdon County paid federal taxes as members of wholesale firms, and four of them had had sales of over $50,000 each in a period of a year. The small stores were more modest in their operations, but endeavored to supply as nearly as possible the needs of their customers.

The account books of these general stores, with their minute itemization of every article sold, make fascinating reading. The buying habits of the people of a community, and hence much concerning their standard of living, are mirrored in the musty pages. Since every storekeeper had to face the rivalry of other stores of his own community, of those in nearby communities, and to some extent of those in Philadelphia, Trenton, New Brunswick, New York, and Easton, he made an effort to keep up with the trends and styles of the times. New kinds of cloth were constantly introduced among the "seasonable goods" in stock; for example, "jubilee cord" and "Grecian stripe" were fashionable in the spring of 1827. The cheapest muslins, the newest stoves, the best silver-plated cutlery, the most efficient plows, the best wine, the finest chinaware, the greatest variety of sugar and molasses, the richest cheese, and the most special of "fancy goods" were advertised by local stores. When buffalo robes became popular, a Flemington store introduced a good assortment at once. A store in 1828, though it carried plenty of plug and smoking tobacco, had no cigars; but a quarter of a century later the smoking habits of its customers caused another store to stock both domestic and part Cuban cigars. In the 1860's the sales of the new illuminant, kerosene, and of lamps and wicks for its utilization became more common than were sales of candles. Fruit cans for home preserving were coming into use at the same time. Ready-made clothing was becoming an important part of the stock of any general store, a fact which necessitated some observance of the prevailing styles.

Packaging of goods prior to sale was an interesting development of nineteenth century merchandising. In the early part of the century most goods except those in large containers were sold by weight or by measure, the weighing or measuring being done at the time of sale. When Solomon Holcombe opened his store at Mount Airy in 1825, nearly all articles were sold by the ounce, the pound, the quart, the gallon, or the yard. By the end of his career, forty-five years later, he was selling a surprising number of things in standardized packages or containers: lye, bluing, soda, pills, indigo, coffee essence, and matches were sold by the box; castor oil, soothing syrup, ink extracts, and whiskey by the bottle; tacks, needles, pins, corn starch, pepper, ginger, coffee, and saleratus in "papers"; and tobacco and salt in sacks. He still purchased a great many things, including sugar, coffee, tea, spice, flour, meal, and dried fruits in bulk and weighed or measured them out to his customers; but a long step had already been made in the direction of the modern custom of packaging.

During the course of the nineteenth century the functions of the general store were gradually usurped by shops specializing in certain types of goods. In the villages enterprising women handled millinery and women's clothing in their homes or small shops. Separate establishments developed for selling lumber and building materials. The hardware store took over a rather important portion of the business of the general store, selling all kinds of tools, utensils, and supplies for artisans. Paints, oil, glass, and associated materials were sometimes sold by hardware stores, sometimes by separate establishments. Apothecary shops became "drug stores" in the modern sense by first branching out into the handling of perfumery, brushes, and the like, and then rapidly taking over a large variety of goods which were easily handled. No hard and fast rules were observed as the newer types of stores developed, and there was naturally a great deal of encroaching on one another's fields. The development of new articles of commerce also caused many changes. For example, as new building materials, fuels, fertilizers, and household conveniences have come on the market, the modern "lumber company" has expanded till it sells nearly as diverse a stock as does the drug store. Increased specialization has created the dry goods store, the variety store, the bookstore, the furniture store, the machinery store, the grocery store, and the meat market. Such specialization has gone farthest in the larger towns, and of course is nowhere so complete as in more urban areas.

The general store maintained its place as a market for country produce for some time after 1800, but small local stores lost a portion of the business of their communities to the larger stores of Flemington and Lambertville, which in turn suffered from the rivalry of still larger markets in Philadelphia, Trenton, New Brunswick, and Easton. Current prices at these latter places were advertised in Hunterdon County newspapers, and local merchants were forced to keep in line. A farmer always had a choice between hauling his grain and other produce to distant markets or of selling to merchants and millers nearer at hand, but because of distances involved the local men had an advantage so long as their prices were satisfactory. In practice, therefore, they handled a great many kinds of produce. They bought and carted away a substantial portion of the grain of the county and purchased, barreled, and hauled to larger markets the apples, pork, eggs, butter, and hickory nuts of the countryside. Other articles bought or accepted in trade were wool, hides, live cattle, calves, lard, tallow, honey, poultry, game, white beans, turnips, potatoes, dried apples, dried peaches, dried cherries, cider, flax, hackled tow, cordwood, rails, hay, and old rags. Where the general store persisted into the present century, it remained a market for many kinds of country produce.

As mentioned previously, however, storekeepers of even the early period did not entirely monopolize the buying of country produce; and as time went on, their grasp slackened considerably. Animals were taken to market under their own power by professional drovers. Butter, milk, eggs, poultry, fresh meat, fruit, hay, grain, and wool were often sold to local customers by the producers themselves. Corn husk mattress manufacturers purchased considerable quantities of husks directly from the farmers. Tanners bought hides and skins from farmers and butchers. With the coming of canals and railroads, dealers in livestock, poultry, grain, baled hay, and other produce established themselves at various shipping points; and farmers themselves often shipped directly to commission merchants in New York City. When peaches became a commercial crop, they were, for the most part, carried to ultimate consumers by men from outside the county. The growth of markets for fluid milk has already been discussed in an earlier chapter. In recent times the coming of the motor truck has made the farmer even less dependent on local buyers than formerly. The development of the Auction Market at Flemington since 1930 has given the whole county a new outlet for poultry products and livestock. Buyers, most of them from outside the county, now come to the Auction Market and compete with one another at public sales.

An interesting though not too important kind of trade was that carried on by itinerant peddlers, already a problem in early Colonial days. Dr. Bowne's servants, white and black, in the early nineteenth century were regular customers of pack peddlers. County licenses were supposedly required, but few bothered to secure them. In 1847 an irate Bethlehem Township man wrote of the "itinerating peddlers, . . . nearly as thick in our country as the frogs were in Egypt . . . and nearly as great a nuisance." He declared that all of the "insufferable pack of swindlers" had "a tincture of yankeeism." [5] Others were equally unkind

[5] *Hunt. Gaz.*, Aug. 9, 1843.

in their comments; nevertheless, peddlers filled a place in the economic structure. One of them, known to us only as David, exceptional in that he was generally well liked and trusted, had a regular route through the western part of the county in the 1850's, along which he dispensed articles of clothing, cloth, cutlery, books, brushes, thimbles, pins, needles, and a multitude of small articles needed in the household. Linen peddlers were usually Irish and were considered somewhat aggressive. The "plaster of Paris salesman" carried a small table on his head. When he came to a home, he would set it down and proceed to make small plaster animals and other toys to sell to the children. Peddlers were often welcomed because they were bearers of news and because they gave provincial people a glimpse of a different world.

After the Civil War many new types of itinerants appeared. Vendors of lemons, coffee, patent medicines, clothes lines, sewing machines, silver polish, tin goods, cheap jewelry, timepieces, tooth paste, corn eradicators, wart removers, freckle cures, cigars, spoons, spectacles, and soap walked up and down the country roads. Someone in 1880 wrote, "It is estimated that one-sixth of the earth's surface is covered with the tracks of lightning rod men and book agents." [6] The twentieth century added the Larkin soap agent, the sewing machine "repairer," the fruit tree "specialist," and the salesmen and saleswomen for Fuller brushes, tinted photographs, silk stockings, and magazines. During the whole period a great deal was said concerning the rascality and tricks of these traveling salesmen, and harsh remedies involving tar and feathers, dogs, and shotguns were often suggested. Occasionally a peddler was bitten by a dog, was mobbed by indignant customers, or was fined or jailed; but people continued to buy, and the agents of various types apparently thrived. The way in which fake Indians and other patent medicine vendors could delude both villagers and rural people year after year seems almost unbelievable, but even at that their concoctions were probably as effective as the patent medicines sold by druggists and grocers.

One type of huckstering was merely an extension of the commercial activities of local stores and taverns, and therefore not subject to the same obloquy as was the trade of ordinary peddlers. In the early eighteenth century the wagons from which pies, cakes, beer, apple-jack, and other delicacies were vended to trainees and spectators were important adjuncts of military training days. Hucksters of shad and other fish had regular routes through the countryside until the beginning of the twentieth century. The oyster soup men of country vendues have

[6] *Hunt. Co. Dem.*, Nov. 16, 1880.

been replaced in modern times by dispensers of sandwiches, candy, and soft drinks. Of greater importance than any of these were the covered huckster wagons sent out from village stores on regular routes of trade. During the last several decades of the nineteenth century and the first two decades of the twentieth, housewives, without leaving their doorsteps, could trade surplus eggs, poultry, and butter for groceries and other merchandise.

The tavern as a merchandising medium has not yet been discussed. Because of the opprobrium now attached to the name, it is difficult to understand the important place of the tavern in the early community. It was much more than a vending place for liquid refreshments; it was a hotel with facilities for boarding and lodging travelers, a common meeting place for public and private business, and a veritable club room for the men of the community. The need for taverns or inns in early days was universally recognized, and they were encouraged and regulated by law. The preamble of an act of 1738 states that "the true and original Design of Taverns, Inns and Ordinaries, was for the accomodating Strangers, Travellers and other Persons, for the Benefit of Men's Meeting together for the dispatch of Business, and for the entertaining and refreshing Mankind in a reasonable Manner." Early court records of Hunterdon County mention the establishment of taverns, but are not specific as to location. Few communities seem to have been destitute on this score. In 1764 there were ten licensed taverns in Bethlehem Township; in Kingwood Township there were only four, perhaps because of the influence of the Quakers in the eastern part of the township.

A number of the county's taverns in Colonial days, most of them on main lines of travel, were locally famous. Among these were the Kingwood Tavern in Kingwood Township, Rittenhouse Tavern at present-day Rosemont, Van Sickle's Tavern near present-day High Bridge, Hickory Tavern in Bethlehem Township, Bonnell's Tavern at Hunt's Mills (Clinton), Fleming House at Flemington, the tavern at Everittstown, and "the White House" from which the villages of Whitehouse and Whitehouse Station later derived their names. On the Old York Road were Coryell's Tavern at Georgetown (Lambertville), Ringo's Tavern at present-day Ringoes, and taverns at Greenville (Reaville), Three Bridges, and Centreville. Traditions have it that Washington visited several taverns of the county during the Revolution, but in no case has positive proof been found. The Holcombe house, still standing near Lambertville, was used as his headquarters for a fortnight, but it was probably a private home and not a tavern.

As population and trade increased, taverns on main routes of travel became more commodious. In 1800 Moore Furman wrote as follows concerning the new tavern which he was soon to build at Pittstown:

The first story level with the road is to accommodate the common people that travel, such as carters, &c.—The second Story is for more respectable travellers, and public meetings, and for that purpose the end of the house next the Kitchen may be left for one room, and will be about 16 by 28, sufficient to dine upwards of 40 people and the other must have a swinging partition so as to be thrown into one room when necessary, then there could be 70 or 80 people accomodated at once,—the 3rd Story must be cut up into Bed rooms for traveling customers. If I build, it shall not be a Henroost.[7]

Taverns on the stage routes benefited from the increase in travel. Those at Flemington profited from the spectators who came to see court sessions, and others from public gatherings held within their doors. Travelers recorded their progress as being from tavern to tavern, and "dry spots" along the way were soon eliminated by the building of new establishments. In 1826 there were about seventy licensed taverns in the county as then constituted, not counting those of Trenton. Competition caused innkeepers to sponsor horse races, cock fighting, puppet shows, dancing, and other attractions in order to lure trade their way. Innkeepers along the Delaware rowed out in boats to serve their wares to raftsmen who were floating by.

The temperance movement apparently did not seriously hurt the liquor trade for several decades. In 1857 there were still over fifty licensed taverns in the county, and the number increased somewhat during the next decade; but by 1870 the temperance people were making their weight felt, and a sudden decrease resulted. In many cases licenses were refused to taverns long regarded as ancient institutions. The one at Rosemont was discontinued in 1870 after 113 years of steady business. "Temperance saloons" were tried, but they seldom lasted long. The coming of rail travel and the decline of the stage coach, by cutting off a remunerative part of the trade, were additional factors in bringing on the decline of taverns. Those which were left gradually became hotels in the modern sense. Since the repeal of the prohibition amendment in 1933 there has been a return of the tavern, but its atmosphere is much more that of the city saloon than that of the tavern of the old days.

Trade necessitates the working out of methods of exchange. In primitive societies a form of barter which is simply the trading of goods may

[7] Leigh, *op. cit.*

be practiced. A method more conducive to the flow of commerce, however, is that by which one or more common articles are used as standards of value and mediums of exchange. The next stage in commercial progress is the use of a money economy. In Colonial America barter of one type or another was usual in pioneer regions. As communities developed, however, it became a standard practice to use money valuations even though there was little monetary exchange. The small amount of specie in circulation was of Spanish and other foreign coinage, and this silver money and the bills of credit issued by the various Colonial governments constituted the chief currency in circulation. Few English coins were used here, though English terminology was ordinarily employed in stating prices and keeping accounts.

Because of the scarcity of money most commercial transactions were done through use of credit. Reckonings between persons were often delayed for years or until the death of one party concerned. Stores gave book credit and accepted farm produce, furs, forest products, and the labor of artisans and carters as payment. Tavern keepers found it more difficult to accept produce, but did give their customers credit. When John Coates of Kingwood Tavern died in 1751, nearly ninety per cent of his estate was on paper. Surveyors, weavers, and tanners did their work on time or accepted "country pay," and a doctor at Frenchtown in the 1750's accepted hay, rye, buckwheat, and cider as fees from his patients. A miller in Readington Township ground grain for a neighboring store, and accepted in return liquor, groceries, and articles of clothing. Such a system of credit put a considerable strain on human nature; many accounts ran on for so long a time that they were classed as "desperate." Moving out of a neighborhood without paying up one's debts was looked upon as a serious offense.

During the Revolution and after, the large issues of paper money by the Federal and state governments caused considerable disturbance in the economic structure. An entry in the vestry record of the Zion Church at New Germantown in 1781 reads as follows:

> There never was a time in any age or any country [when there was] such a perplexity and uncertainty of money affairs and settling estates as we have seen in our present unhappy war since the year 1776. The Continental or circulating money had lost already in '77 half his [its] value, . . . in '78 it became five for one, in '79 twenty for one, in '80 sixty for one, in '81 one hundred and fifty for one, whereupon the money came entirely to [havc] no value at all.[8]

[8] R. R. Honeyman, "More Local History—New Germantown."

Complicated arithmetical computations were necessary to compute money values. An entry of August, 1779, in the justice of peace docket of Ezekiel Cole reads as follows: "£309–10–0 Contentantel [*sic*] is 825 30/90 Dollars in hard by the scale." Because of the rapid depreciation of paper money both barter and the use of "commodity money" became common. In the latter case bushels of wheat were often used as a standard of value. Salaries were paid, promissory notes redeemed, and even land purchased on a commodity basis. Due bills and promissory notes were much used in place of money. Paper money circulated so rapidly that the depreciation of a given note was in most cases divided among many people, and those who were forced to accept notes at par or at a ratio set by the government were decidedly unlucky. The several trunkfuls of Continental money which Abraham Coryell finally received for ferrying Washington's army and supplies across the Delaware in 1777 and 1779 had little value and soon became entirely worthless. The family preserved a trunkful of the notes as souvenirs for many years.

The formation of the Federal government, the coming of better times, and the stabilizing of state finances brought a change after 1790. Bank notes, certificates of governmental indebtedness, and the considerable specie brought in by foreign trade constituted the circulating media. Difference in value of "hard" and "soft" money was a nuisance for some years, but the two gradually came nearer in value to one another because of commercial prosperity and public confidence. Occasional notes of the Bank of North America also circulated in this area. The use of pounds, shillings, and pence as standards of value persisted until late. Ezekiel Cole began using dollars and cents in his justice of peace accounts as early as July, 1799, but others used the old notation, in whole or in part, for two decades longer.

The increase in the amount of circulating media did not immediately have any marked effect on the time-favored credit structure of rural areas. The methods of general stores in Hunterdon County long remained the same as they had been in an earlier day. Their customers bought and sold over long periods of time without making a settlement. The same type of arrangements were carried out wherever there were exchanges of goods and services. A teacher in Flemington in 1804 accepted in lieu of cash for tuition such diverse articles as furniture, clothing, cakes, and beer. Dr. Bowne, who was a farmer as well as a physician, was well situated to accept farm produce and labor in return for his services. Over a period of years he gave credit on his books for flax, grain, seed, fish, cheese, vegetables, poultry, meat, livestock, fruit, cider, whiskey, honey, leather, rails, furniture, tools, a windmill, ground feed, service by a stallion, weaving, spinning, tailoring, shoemaking, blacksmithing, wheelwrighting, sawing, milling and common labor, both slave and free. Hiram Deats accepted goods and services of all kinds in return for the things made at his machine shop. The editors of the local papers allowed subscribers to pay their bills with cordwood, fruit, and vegetables, hired men on farms were paid partly in produce, and blacksmiths and tailors exchanged their labor for meat and grain. The following due bill in the Case records well illustrates a system of exchange not very far removed from barter: "August 6th 1824. Due Paul H. M. Prevost Esqr. one dollar and Fiftey Cents for one hundred C. wt. of rye Flour to pay the same in leather. Philip Case, Tanner."

The working of this elaborate system of credit had its tragic side. A man who stretched his credit too far or made a change in his trading habits often found himself involved in a lawsuit before the nearest justice of the peace, who in short order changed his book debit into a judgment. When a creditor died or moved away or a partnership broke up or a business ceased to function, a great effort was made to collect debts by the same means. Debtors who did not flee the country and were unable to borrow elsewhere might, at the pleasure of their creditors, be jailed. The jail at Flemington was usually full of insolvent debtors. In 1838 there were twenty-two of them there. A man often languished in jail for months, at his own expense, until someone advanced him sufficient money to pay his debts or until the courts were satisfied that the debt could not be paid. In 1842 imprisonment for debt was abolished as the result of a bankruptcy law, and the jail was rapidly emptied. This, of course, did not help in the least the position of creditors, who on their part tried from time to time to break the vicious circle. During the depression of the late 1830's and early 1840's many stores changed to a "cash and produce" system, and the idea was tried again in the 1850's and early 1860's by various merchants, but in no case was the experiment a success. In 1869 the plan was revived by the merchants of Flemington; the agreement was kept by all for six weeks, then completely abandoned. The *Hunterdon County Democrat* of October 20 commented that "during the existence of the cash system, Flemington was as dead as a mackeral."

Although the use of credit remained a necessity throughout the nineteenth century and into the twentieth, the proportion of cash transactions increased and the necessity for long credits diminished. An important factor was the increase in the amount of money in circulation, and that was naturally tied up with the growth of banking. The chief function of early nineteenth century banks was the emission of bank notes, often on a scale detrimental to both themselves and the community. Two new banks at Trenton were used by people of Flemington as early as 1809 and 1813. Checks of these banks were used as negotiable instruments, but the notes which they issued were of far greater importance in business transactions. By 1825 the notes of a dozen banks in New Jersey and several in Pennsylvania were circulating in Hunterdon County. Such paper money was easily counterfeited, and everyone had to be constantly on guard against spurious issues. Worse yet, very few banks had sufficient capital, and the number of failures, with consequent losses to those using their money, was very large.

Hunterdon County was not without its example of poor banking practices. The charters given to the New-Hope Delaware Bridge Company by New Jersey and Pennsylvania allowed this corporation to act as a bank. This right was used so extensively for some years at the New Hope end of the bridge that in 1823 the company could no longer redeem its notes. However, by manipulation of some kind a new issue was put out in 1825, and by 1828 it, too, was almost worthless. The New Yorkers who owned the bridge successfully prevented a bank from being started in Lambertville, and when they themselves resumed operations as bankers in 1831, it was at the Lambertville end of the bridge. During the boom days of the 1830's the company was apparently doing well; then came the Panic of 1837 and the hard times which followed, and for a time all of the banks in the state suspended specie payments. The bridge company might have survived this crisis, but early in 1841 its chief asset, the bridge, was carried away by a flood. The bridge was rebuilt, but the financial strain was too great for its owners, who were unable to redeem the $150,000 or so of notes which they had in circulation. New issues were tried but were not welcomed by the public. By 1848 the notes of this bank were worth only 30¢ on the dollar. The company ultimately forfeited its banking rights, and in 1853 lost the bridge also. Four years later the new owners managed to secure banking privileges in Pennsylvania, but apparently they made no great use of their right to issue notes.

Except for this one unhappy experience Hunterdon County was without banks until the early 1850's. The Hunterdon Bank of Flemington, organized by Isaac G. Farlee and other prominent citizens of the town, was in operation by 1854. Banks were also started at Frenchtown, Lambertville, and Clinton in short order. The panic of 1859–60 caused all of them to limit specie payments on their notes, but none of them failed. During the Civil War the note issues of the banks were supplemented by the legal tender notes issued by the Federal government. The greenbacks, as they were popularly called, were fiat money, and because of popular distrust soon depreciated in value as compared with gold and silver. Therefore, gold and silver coins were hoarded, and soon there was no money of small denomination in circulation. Token money issued by private firms, postage stamps, and United States Treasury notes in amounts as small as five cents were soon in common use. "Shinplasters" remained the chief fractional currency until about 1877, when small coins came out of hiding because of a positive governmental policy as to redemption of greenbacks and a marked decline in the market value of silver. The contraction of greenbacks led to the rise of the Greenback party, and in Hunterdon County several hundred Greenbackers under Dr. George H. Larison were active politically for several years. The free-silver movement had few followers, but in 1896 the county, always preponderently Democratic until recent times, voted for Bryan and Sewall.

With the closer regulation of banking and of paper money by the Federal government during and after the Civil War, the amount of currency in circulation in a given locality ceased to be dependent on the action of local banks. In spite of the political activity of Dr. Larison and his Greenback followers, there are no indications that Hunterdon County ever suffered from a currency stringency after this time. More important than an increase in currency have been the development of the use of negotiable instruments, especially the bank check, and the growth of bank credit. As the need of banking service increased, the number of local banks has grown. Their growth has coincided with the increased commercialization of agriculture and with the shift in trade from a credit to a cash basis.

FIG. XIV. An octagonal schoolhouse and its ultramodern successor. This photograph won a prize at the Buffalo Exposition of 1901.

FIG. XV. Green Sergeant's Bridge on Wichecheoke Creek.

SWIFTSURE REVIVED

BETWEEN

Philadelphia and New York.

Splendid Troy Coaches,

In connection with the

Steamboat at Elizabethtown Point

FOR NEW YORK.

THE public are respectfully informed that this line has undergone a thorough change, and will leave **GEORGE VOGEL'S, Second Street House, between Market & Arch Streets,** Philadelphia, every **Monday, Wednesday, and** Friday, at **9** o'clock A. M. via **Jenkintown, Willow Grove, Buckingham, New Hope, Lambertsville, Ringoes, lodge at Flemington—thence to Centreville, Somerville, Boundbrook, Plainfield, Scotch Plains, Westfield, Elizabethtown Point,** then take the steamboat **JOHN MARSHALL, and arrive in New York early the second afternoon.**— Fare through, **$4 25**

☞ The Coaches leave **FLEMINGTON** for Philadelphia and New York at **5** o'clock in the morning of Tuesday, Thursday, and Saturday.

The public may rest assured of every accommodation. Packages, boxes, and parcels of every description will be taken in this line at moderate fare. This line is well calculated to accommodate persons wishing to travel at their leisure, as it passes through a delightful, fertile, and pleasant country.

N. B. All Baggage at the owners' risk.

George Carter & Co.
Philadelphia.

John A. Weart,
Trenton.

Anderson & M'Cutcheon.
New York.

[PROPRIETORS.

STOVES! STOVES!

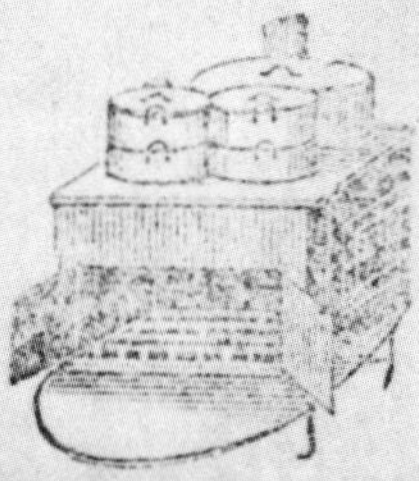

THE Subscriber would again inform his friends and the PUBLIC generally, that he is well prepared to accommodate them with STOVES of any description that can be desired, both for coal or wood for cooking and heating, of the best and latest patterns that can be had from any part within our reach. He being well acquainted with the business, feels himself capable of judging quite correctly of the different kinds, would say that there are more or less alterations in this business every year: a part of which is more for speculation than for merit or value; and he will be bold enough to say that there are no better kinds made than he has on hand —especially Cooking Stoves— he would say that there is a great difference in the value of the tin and copper work, of which considerably the largest part are made of inferior kinds—especially those coming from the city. He having his fixtures made at home, under his own direction, will recommend them to be a good heavy article; and being so largely in the business enables him to sell very low for CASH or OLD METAL, and if the community have not yet learned that the City is the wrong place to purchase Stoves, if they will call on him or at any of his stands they can easily be convinced of that fact. Those not having Cooking Stoves of these modern kinds are not aware of the advantages to be received in using them, or they will not be without them.

A very handsome metal and Sheet Iron Stoves for Parlor use.

He would return his sincere thanks to a generous public, and solicits a continuance of their custom, believing that he can give them good satisfaction.

These Stoves will be kept at the following places: G. W. Risler, Flemington; S. W. Dilts, Lambertville; Charles Vorhis, Milford; A. Stiger & Son, Clinton; and Hoffman, Foster & Co., Clinton; in the brick store of Mr. Doughty, in care of David Hoff, Somerville, David Neighbor, German Valley.

HIRAM DEATS

Quakertown, Sept. 27, 1843.

FIG. XVI. LEFT: A stage coach advertisement from the *Hunterdon Gazette*. April 27, 1832. RIGHT: A typical advertisement by Hiram Deats which appeared in the *Hunterdon Democrat,* October 11, 1843. Plows and other farm machinery manufactured or assembled by him were advertised regularly.

CHAPTER XI

Industry

THE NEEDS AND DESIRES of a people engaged in agriculture naturally caused the development of many subsidiary occupations. Pioneer farmers were jacks-of-all trades, and by their own efforts supplied most of their wants. Some of their home industries, for one reason or another, were retained until a comparatively late date; but as communities developed, specialization of labor took one type of work after another out of farm homes and put them into the homes or shops of individuals who made part or all of their living by doing custom work. Eventually artisans and mechanics of various kinds made the clothing, the home furnishings, and the agricultural equipment which farmers needed, and processed the farm products which the farmers brought to them. Artisans of early days usually practiced more than one craft, but as custom increased there was further specialization. With the coming of modern manufacturing the making of goods and the processing of farm products have been for the most part removed from the rural areas. There are still many activities which supplement agriculture, however. The artisans of the countryside have gradually been replaced by men engaged in the many service occupations necessary to our modern economy. Specialization of work created the artisan in the first place, but further specialization and the minute division of labor of modern industry have made him largely unnecessary. The "mechanic" of a century ago was a creator, the one of today is a repairman.

Sawmills were among the first industrial establishments of pioneer communities in Hunterdon County. Even tiny streams were harnessed to operate small saws during part of the year, thus making it possible for nearby farmers to convert logs from their clearings into lumber for frame buildings. The old "up-and-down saws," many of which were used far into the nineteenth century, were not very speedy, and the effect of a growing population was to increase the number of sawmills. In the early nineteenth century every stream of any size had one or more in operation, and waterpower was insufficient in many cases except after a rain. Jacob Kugler, who had a sawmill at the mouth of a tiny tributary of the Delaware at Tumble in the early nineteenth century, would get up in the night to put his machinery in motion after a shower. His establishment was locally known as "Thundergust Sawmill."

Sawmills along the Delaware, using in some cases the water of the river but usually that of tributaries, sawed the logs of their own localities and some of those which were floated down as rafts from the Lehigh Valley. One of the busiest was the sawmill at Durham Falls in Holland Township, which operated until about 1900. As mentioned previously, the feeder canal, which paralleled the river for twenty-three miles, was used to bring log rafts to sawmills on its banks. The late E. T. Bush once wrote that these establishments were in his boyhood "nearly as common as gas stations today and much sweeter smelling." [1] In 1840 there were sawmills in every township and a total of sixty in Hunterdon County. At about this time portable steam outfits began to replace the old water-driven ones. The result, so far as the county's timber resources were concerned, was about the worst possible. So long as logs were transported to stationary mills, there was little incentive to cut other than the largest trees, but as soon as portable sawmills were taken into the woods, all trees of a usable size were cut. As lumber prices increased, timber operators slashed through more and more of the forests of the county.

A mill for processing locally raised grain was as necessary to a community as a sawmill, and often the two were combined in the same establishment. The "tub mills" of early days did not take a great flow of water and were not very expensive to build. Therefore, communities with sufficient grain production did not in most cases have to wait long for mills. William Dawlis is known to have built one at Amwell near present-day Ringoes in 1727 and another in 1735. Other millers soon established themselves at Alexauken and Phillips' Mills, and Edward Rockhill, a Quaker, built a mill at Pittstown at about the same time. John Opdycke began operations at Headquarters soon after 1738; and traditions, often supported by better evidence, point to early mills at two or three points on the South Branch, at Idell, at Spring Mills, at Lamington (Pottersville), on the Rockaway Creek, and at several other places. Until local mills were built, the farmers of a community took grain by wagon or pack horse to distant ones or else used hand mills to make flour and meal for home use. According to a tradition at New Hampton pioneers in that locality used such devices when the roads were bad rather than make the long trip to Pittstown. The community's

[1] *Hunt. Co. Dem.*, Oct. 17, 1929.

need is said to have been filled by the building of a local mill in the 1740's.[2]

As the population increased, more mills were constructed. Not all were equipped for the making of flour and meal, and a distinction grew up between "flouring mills" and "grist mills." The latter term was locally used then, as now, to designate only those mills which converted grain into feed for livestock, and will be used in that sense in this discussion. Except on the largest streams these establishments were necessarily small, and maximum use of the power of a stream was gained by construction of a series of mills. On the Wickecheoke Creek at various times were built four grist and flouring mills, six sawmills, an oil mill, a fulling mill, and a number of clover hulling mills. Capoolong Creek had in less than four miles in the Kingtown area five grist and flouring mills, three sawmills, and a fulling mill. On the Nishisahawick Creek there were, in addition to the ones located on tributaries, important mills above Everittstown, in Everittstown, and in Frenchtown. Mill Brook, a small stream near Ringoes, had a regular succession of mill dams and mills of various kinds, which gave employment to the residents of the village of Amwell, which has since disappeared.

On the South Branch there were mills at High Bridge, Clinton, Hamden, Rowland's Mills, Mettler's Mills, and two or three other points. Lambertville suddenly became a milling center in the 1830's as a result of the power furnished by the feeder canal. Nearly every village had at least a grist mill, and many mills were located in the open country. The larger ones offered a variety of services, as shown by a description of Saxton's Mill on Rockaway Creek in 1845:

> The mill . . . has about 24 head and fall on a never failing stream, it has two pair excellent burr millstones and one pair of country stones, with cracker, etc., for grinding plaister. There are in the mill two bolts, one pair hulling stones, fan screen, elevators, etc., sufficient for country and merchant work.[3]

During a long period Hunterdon County was well served by its many small milling centers. Their products, such as various grades of wheat, rye, and buckwheat flour, corn meal, shorts, middlings, and bran, were regular items of local commerce. In 1840 there were seven flouring mills and fifty-seven grist mills in operation in the county.

Many mills had facilities for the extraction of oil from flaxseed and for the grinding of the residue into oil meal, which was sold to farmers

2 *Hunt. Rep.*, Mar. 31, 1870.
3 *Hunt. Gaz.*, Jan. 8, 1845.

for cattle feed. There are said to have been oil mills at Amwell and Alexauken at a very early period. In 1840 there were ten such mills in operation in Hunterdon County. Soon after the close of the Civil War the production of flax declined, and one after another of the oil mills was discontinued. The last one in the county was at Frenchtown and operated until the 1870's.

Steam-driven mills began to displace those operated by water soon after the Civil War, and towns with no water power, such as Flemington, soon acquired mills. Those in the country began a slow decline, though some of them were operated well into the twentieth century. In the 1880's the roller process began to displace the old methods of flour manufacture, and a few of the newer local mills established trade names and built up markets. A mill at Flemington manufactured Pride of Flemington, White Wonder, and Imperial brands of wheat flour; the one at High Bridge made White Rose, White Lily, and White Silk; the one at Milford, Crown Sheaf and Silver Spray; and the one at Three Bridges, Golden Rod and Daisy. In 1926 the Flemington mill was making seven kinds of flour and numerous kinds of dairy and poultry feeds. Since that time flour milling has declined in Hunterdon County, and only one or two mills still continue. On the other hand, increased dairying and poultry raising have led to a greater need for grain feeds. Grist mills, driven by steam or by motor, continue to supply part of the demand, though much of it is met by the purchase of processed feeds brought in from the outside. Many farmers today mix their own feeds, either using their own hammer mills or hiring the services of portable ones driven from farm to farm by motor truck.

A much-needed artisan in every community and one of the first to appear was the blacksmith. As early as 1710 one of the squatters on the Bryerly tract was a smith named William Titfoot. Inventories of early days show that these workmen usually made an adequate living, though, like most artisans, they often had small farms and supplementary trades to give them work when custom was lacking. Their raw materials, such as bar iron and "country steel," were regular items of trade at the general stores. A store at Clinton in 1831 carried tire iron of Swedish, American, and English manufacture, round and square iron of various sizes, nail rods, hoop iron, rivets, sheet iron, and cast, blister, and common steel. Early blacksmiths did all sorts of work in metal and wood. One who was located in Kingwood Township in 1797 shoed horses, mended vehicles, made spikes by hand, put steel cutting edges on axes, repaired tools, and made fireplace irons. Thomas Carhart, located at

Flemington in 1817–18, made nails, staples, rivets, burrs, corn knives, and spades, mended axes, hoes, shovels, and neck yokes, put iron hoops on pails, and shoed horses. In 1829 a Mechanicsville (Whitehouse) smith advertised that he did blacksmithing, wheelwrighting, cabinet making, and painting, and in addition manufactured plows. The making of wrought nails and, later, of cut nails remained a job usually done by the local smith until the day of the factory-made product. Many blacksmiths burned charcoal as a side line, and sold to others the surplus above their own needs.

Tendencies toward specialization appeared fairly early, and wheelwrighting as such became a separate occupation for many artisans during the eighteenth century. For example, a wheelwright shop was one of the early establishments at Amwell. Flemington, which was becoming an important trade center, had two or three wheelwrighting shops in the 1790's. In the nineteenth century this trade was nearly always divorced from blacksmithing, though the two supplemented each other and were often carried on side by side. There were eighty-one men engaged in the making of wheeled vehicles in Hunterdon County in 1840.

Blacksmiths and wheelwrights remained important artisans through most of the nineteenth century. The 237 farmers of Raritan township in 1850 were served by twenty smiths and fifteen wheelwrights. In 1860 there were in the county thirty smithies employing sixty-seven men, seventeen carriage shops with fifty-nine men, and twelve shops with twenty-two employees making wagons and carts. The master of one of these small shops often employed one or two apprentices and sometimes a journeyman. His investment in every case was small. By 1890 the concentration of the manufacture of wheeled vehicles in larger urban centers had drained many wheelwrights from rural regions, but the industry persisted in Flemington, Lambertville, Frenchtown, Clinton, Califon, Lebanon, Whitehouse Station, and Ringoes. By 1920, however, wheelwright's shops had entirely disappeared from Hunterdon County. The number of smithies also decreased markedly during the late nineteenth century and the early twentieth. A few, by concentrating on the types of work remaining for them or by combining their trade with other kinds of repair work, still function in the larger villages.

Agricultural tools in early days were usually made by blacksmiths or carpenters, who sold their products directly to farmers or through stores. There are no records of toolmaking specialists in Hunterdon County until the nineteenth century. The census of 1820, however, reported an establishment which specialized in the making of sieves and riddles. A decade or so later a man by the name of Gary was making various tools and implements in his shop north of Croton. His two sons learned the trade from him, and each later set up a shop on his own farm in the "Gary neighborhood." Among other things they made grain cradles, horse and hand rakes, oat forks, harrow frames, hoops for covered wagons, ox yokes, and wooden handles for tools. The Gary pride in workmanship is shown by an advertisement in 1860:

New Counterfeit! A lawless character, by the name of Joe Marshall, is manufacturing an inferior quality of rake and stamping them with the name of J. Gary. We would say that there are no genuine "Gary Rakes" made but those that have John S. Gary or H. Gary's stamp on. Croton, Apr. 20, 1860, Hiram Gary.[4]

The three-story factory which Hiram Gary built in 1855 used steam, wind, and horse power. This plant, operated by a fourth generation of toolmakers, was still turning out some work in the early twentieth century despite the decreasing use of wooden tools.

Another interesting maker of farm implements was Robert L. Williams, who settled at Frenchtown in 1836. He made grain cradles, brooms, and other articles at first, but soon specialized in the making of fanning mills. With the help of eight men he turned out four hundred of these in 1849. His factory was still operating as late as 1881. Occasional other toolmakers received mention in the records, and doubtless some blacksmiths and carpenters continued to make tools during slack seasons until large scale factory methods destroyed the industry.

Hunterdon County was not blessed with large mineral deposits, but there were iron veins at a number of points and small pockets of copper in two or three places. Except for possible use as pigments the Indians did not use iron ore, but the whites were quick to utilize this resource. Close to the iron deposits were plenty of wood for making charcoal and lime to use as a flux. A vein near present-day High Bridge was already opened and a bloomery in operation prior to 1742. The tract of land which Allen and Turner leased from the West Jersey Society in that year included the mine but not the bloomery. The new owners at once built an iron furnace on Spruce Run, where there was sufficient water power, and during the next few years installed a slitting and rolling mill. Despite the British prohibition of slitting and rolling iron in the colonies after 1750, this mill was in operation, at least for local markets, in the 1760's. In 1752 Allen and Turner enlarged their holdings, and soon opened a mine and erected a new furnace or forge between High

[4] *Hunt. Rep.*, May 4, 1860.

Bridge and Cokesbury. This purchase included the ancient bloomery on the South Branch.

A Swedish traveler in 1759 visited the Allen and Turner establishment and wrote as follows concerning it:

They have two mines distant one and a half miles from each other. The iron is of good quality in both. . . . Here are two furnaces and two forges, each with two stacks. Also a trip hammer and a flatting hammer.[5]

Evidently the Spruce Run furnace, destroyed in the land riots in 1748, had been rebuilt. The second furnace may have been the Norton furnace, which used ore from a very early mine at Van Syckles. Tradition says that both packhorses and wagons were used for the carrying of ore to the furnaces. The superintendent of the Spruce Run furnace in 1784 has left the following record of his operations: "June 6, put fire in the furnace, put ore on Thursday the 14th at 6 o'clock in the morning. Began to blow at 9 o'clock on Friday the 15th and made the first casting on Saturday the 16th at 6 o'clock in the morning." [6] The memorandum goes on to list all iron produced during the summer and autumn, something over two hundred tons. About twenty tons per week were produced when the furnace was in full blast. Apparently the pig iron was carted to New Brunswick and sent by boat to New York for transshipment abroad. Allen and Turner, who also owned Andover Furnace in Morris County, were doing a prosperous business in the early 1760's, but further British legislation hurt their foreign trade. In 1769 John Allen wrote, "The wrong headed schemes on your side of the water have fully taken up our attention. . . . The enumerating of the Commodity has put most of our works to the stand, and Mr. Turner and I intend to let ours . . . lie till better times." [7] During the course of the Revolution Allen and Turner lost their iron works, though not all of their land. Union Furnace, as the iron works were now called, was acquired by Robert Taylor, who had been superintendent for over a decade.

Smaller ventures in iron making were made at a number of other places in Hunterdon County during Colonial days. On Lackatong Creek in Kingwood Township there were, prior to and during the Revolution, two forges for making bar iron from pig iron, but it is not clear as to where the pig came from. Another forge on Capoolong Creek near Pittstown probably hauled its pig from the Bloomsbury region by way of the "king's road" which ran along the top of the Barrens. Hacklebarney,

5 Israel Acrelius, *A History of New Sweden*, p. 166.
6 Quoted in O. M. Voorhees, "Iron Industry at High Bridge," p. 108.
7 *Ibid.*, pp. 114–15.

or Budd's Forge, at Potterstown secured its ore from a mine in Morris County. It was apparently not built until 1763. A forge known as Hughes' Forge was located in the north part of Alexandria Township, and used ore from a pit nearby. Several forges were situated on the Musconetcong. Changewater Forge, seventeen miles above its mouth, and Bloomsbury Forge, ten miles downstream, were both on the Warren County side, though using Hunterdon County ore, and the same was probably true of the Greenwich and Chelsea Forges farther down. Tradition places a forge at Finesville, on the Hunterdon County side of the river, and there may have been others. The famous Durham Iron Works were across the Delaware in Pennsylvania, and there was a forge at New Hope, opposite Lambertville, which used iron from Durham.

The furnaces of this area were already on the decline by the time of the Revolution. The Union Furnace made cannon balls and other supplies for the American army, and in 1780 was still producing nail rods and other iron for the local market. There is no evidence except tradition that it operated after that date, and it certainly had closed before the turn of the century. Hacklebarney Furnace ran sporadically until nearly 1800. The Durham Furnace closed definitely in 1789, and the other furnaces of the region ceased operations during this same period. Rutherford in 1786 stated that the iron industry was being hurt by the scarcity of wood for charcoal, the high cost of labor, the loss of the British market, and the importation of iron from the Baltic. The first-named is the traditional reason given for the closing of the Union Furnace. An iron furnace in the charcoal era required large quantities of timber, and the encroachment of agriculture may well have been a serious factor. At any rate, the furnaces fell into decay, and for over a half-century the mines, too, were abandoned.

Just before the Civil War came a sudden revival of iron mining in this region. By use of the Bessemer process ores of a high sulphur content, which included those of Hunterdon County, could be used. One mine was already operating in 1860, and within a few years there was a marked boom, centering chiefly about the High Bridge-Cokesbury and Asbury Station areas. The High Bridge deposits were developed by three different companies and yielded large quantities of ore. Most of it was sent to the anthracite furnaces of the Lehigh Valley in Pennsylvania, some to the revived Durham Furnace. Among those who benefited were the descendants of Robert Taylor, who opened a new and very productive vein of ore on their lands. Soon after 1870 another iron region was developed near Jugtown (West Portal) by a Pennsylvania company which opened two productive mines, the Turkey Hill and the Swayze. During the next few years new mines were started near Pottersville, at Mt. Lebanon, near Glen Gardner, west of Clinton, and in Holland Township; and some of the old shafts, as the one near Van Syckle's, were reopened.

Not all of the new ventures were successful, and none paid as well as the High Bridge and Jugtown mines. The latter employed 380 men in 1881. Unfortunately for the success of the new enterprises, the development of Lake Superior ores brought prices below the cost of production in New Jersey, and the mines were abandoned one after another. The ones at Jugtown put in labor-saving equipment and carried on spasmodically until 1888, when the promoters went bankrupt. During the last decade of their operations they are said to have produced over a million tons of ore. Abortive attempts were made to open the mines in 1897 and again in 1917. Due to a demand for ore of a low phosphorus content for armor plate, some of the old mines have been put into operation again during World War II.

Iron working, as distinguished from iron smelting, was until comparatively late a function of local blacksmiths in this area. In the nineteenth century, however, the operations of some individuals expanded to factory proportions. The best example is that of Hiram Deats, who began making plows, stoves, sled shoes, school desks, kettles, and other articles at his shop between Quakertown and Cherryville in 1831. For over seventy years he and his nephew of the same name operated foundries and machine shops making the above-named articles and many others, but specializing more and more as time went on in the making of agricultural machinery and stoves. In the 1840's part of Deats' operations shifted to Pittstown, and in 1850 he was making plows, corn shellers, threshers, sheet iron products, and stoves at four different establishments in Franklin Township. A branch factory at Brookville, now part of Stockton, was opened in 1852 and operated for nearly half a century. During at least a portion of that period it was used as a stove foundry. Hiram Deats ended a very active life in 1887, but his work was continued by his nephew until 1904, when the business was sold. A little later the foundry and machine shops at Pittstown burned, bringing to an end an interesting local development.

Hiram Deats was only one, though the most successful, of a number of iron manufacturers. A foundry and machine shop was established in Lambertville before 1840, but the Lambertville Iron Works, which operated during most of the second half of the nineteenth century, was not started until 1859. The local census schedules of 1850 show a small foundry and machine shop in Tewksbury Township, and this may have been the same one which was functioning at Pottersville four decades later, while the census of 1860 shows that three foundries were in operation in Hunterdon County in 1860, probably the three mentioned above. Another was started at Frenchtown in 1870 but led a very checkered existence. Hiram Deats eventually purchased and held it until his death, when it was sold. A foundry which specialized in the making of iron bridges was operated at Flemington by J. W. Scott during the latter part of the nineteenth century. The operations of these and other foundries and machine shops of the county were never large, and with the concentration of industry in bigger factories they were put at such a disadvantage that most of them eventually failed.

A successful foundry in recent times has been the Foran Foundry at Flemington. In 1893 John Foran acquired a small plant which had been built at that place a few years earlier. He started with fifteen workmen, and within a few years was hiring over one hundred. In 1901 this foundry was supplying supports for the new subway in New York City and steel girders for the Gimbel Building in Philadelphia. Governmental orders for materials used at Panama and in Hawaii, shipments of lighting standards and other equipment to several cities, and armament orders during World War I gave employment through the years. Since that time the plant has weathered the boom period of the 1920's and the depression of the 1930's. In 1938 it was employing 135 men. During World War II the plant has been extremely active.

The largest iron and steel plant of the county, that at High Bridge, claims as its progenitor the old Union Furnace, though no iron making was done in this area from before 1800 until after the Civil War. The Taylor family during this time continued its ownership of ore lands, and with the revival of mining after 1860 began selling ore to Pennsylvania furnaces. The revival of iron working at High Bridge had little or no connection with this development, however, but stemmed rather from a small bloomery forge and manufactory for wagon axles started by Lewis H. Taylor about 1851, after the coming of the New Jersey Central Railroad. The plant gradually shifted over to the production of railway car axles; and after the Civil War Taylor formed the Taylor Iron and Steel Company, dammed the South Branch to create Lake Solitude as a source of water power for trip hammers, and went into the business of making car wheels and other railway equipment. Pig iron was brought in by railroad, and if there had been any serious thought of smelting local ores, the idea was eventually dropped. By 1872 the Taylor concern was making 175 wheels and seventy axles per day.

About 1892 Taylor pioneered in the use of manganese steel, and within a few years was making manganese steel castings such as rails, safes, and mining and excavating machinery parts for shipment to all parts of the world. The teeth of many of the big power shovels which dug the Panama Canal were made at High Bridge. During the Spanish-American War this plant made shell cases, during the World War it did a large business in armaments, and in World War II it again became very busy with war orders. Through an amalgamation the company became the Taylor-Wharton Iron and Steel Company in 1930, with a second plant at Phillipsburg in Warren County.

The development of copper mining, though very spectacular, was far less important than that of iron. Old mining shafts and tools found a mile south of Flemington point to early operations, which traditions say occurred before the Revolution. A copper plate for the issuing of stock in an abortive company bears the date 1817 and shows that the presence of copper was known at that date. A number of unsuccessful promotions were tried during the next two decades. Finally, in 1836, local persons, including Hugh and John H. Capner and their cousin Samuel Hill, organized the Neshanic Mining Company. The old shafts were opened, but the venture was dropped when it proved unprofitable. In 1848 a new company, the Central Mining Company, started operations on a larger scale, and for a time turned out small quantities of good ore; but its interests were eventually acquired by the Acorn Copper Company, which in turn sold out to the Hunterdon Mining Company. The latter, a stock-jobbing outfit capitalized at a half million dollars, used the mine as a base of operations for a huge swindle. The only lasting result of the presence of copper at this place was the naming of the village of Copper Hill.

The other scene of actual copper mining in Hunterdon County was one fourth of a mile west of Flemington on the Capner Farm. One promoter, the Flemington Mining Company, failed in 1838, apparently without doing any mining. Interest was aroused again, however, when workmen digging a cellar uncovered a surface vein of copper. Hugh Capner chose a safe course this time, and sold his farm at a fancy price to the Flemington Copper Company, of which he was one of the original incorporators. Probably he was the only man who made any honest money from the copper craze in Hunterdon County. The new groups soon began work in earnest. At first their ore was taken by wagon to

Stockton, where it was loaded on canal boats and started toward its ultimate destination, a smelter in Baltimore. A geologist hired to investigate the mine in 1846 praised the quality of the ore, but advised against overexpansion. Nevertheless, the company hired a large crew of Irish miners, opened a number of experimental shafts, and put in an anthracite smelter. Almost at once it was in financial difficulties and forced to close down. New capital permitted resumption of operations in the fall of 1848, and at least part of the labor force was kept until 1849. After another slump some digging was done again in 1853 and 1854, but it soon ceased, and the company eventually sold its holdings to the Hunterdon Mining Company. The latter carried on extensive operations for a time, and in January, 1860, was shipping out sixty tons of forty per cent ore each week.

Despite early successes the new company soon failed as dismally as had its predecessor. The Irish and English miners were suddenly dismissed in 1861, and a riot of some violence was precipitated. The mine was acquired by the Kent Copper Company in 1864 and operated for a few months, but was closed permanently in 1865. From time to time in the present century the old shafts have caved in, thus reminding people that copper was once mined here. It is doubted that ore was ever obtained on a profitable basis; stock manipulation was more important to the various promoters than was actual mining, and many persons lost their money as a result of it. There were through the years a number of other ventures based on supposed copper deposits in Hunterdon County. These were purely wildcat schemes, and only enough digging was done to uncover ore which had been "planted" to fool a gullible public. Hunterdon County had some good copper ore in small veins, but operations were never geared to the size of the deposits.

The copper deposits of Hunterdon County were of less lasting importance than were those of stone. The men who built stone houses in early days carted their building material from small quarries near at hand, and occasionally owners of the best quarries sold stone to others for houses or public buildings. The demand was enlarged by the bridge building of the nineteenth century and increasing amounts were needed for farm and village structures. Local quarries throve, sometimes selling stone for construction quite a distance away, as when granite from Rocktown was used for buildings in Flemington. Certain types of stone had particular uses. For example, argillite from west of Flemington furnished building materials for foundations, cellar walls, floors, and sidewalks.

The greatest development of quarrying of building stone was in the Raven Rock-Stockton-Lambertville region, from which came the excellent brown and gray stone used for dwellings at an early period. The first intensive development occurred, however, when the feeder canal was constructed, and later the Delaware and Belvidere Railroad used great amounts of the same stone. The best of it was transported by canal boat and used for fine churches and bridges in Trenton, Camden, Philadelphia, and other places. The demand declined with the increasing use of concrete, and the old quarries gradually fell into disuse. The presence of another type of stone, trap rock, greatly in demand for stone roads and for mixing in concrete, has led to the creation of new quarries near Lambertville.

Something has already been said concerning the quarrying of limestone for agricultural uses. As the use of calcined lime increased, limestone quarries were extensively developed. The limestone belt which extends from southeast of Clinton to High Bridge and west to Pattenburg had numerous quarries, as did the smaller deposits between Little York and Spring Mills, in the Califon area, in Holland Township, and at a number of places in the Musconetcong Valley. The ruins of limekilns and the evidences of quarrying are still to be seen in many places. Lime for plaster, and other building purposes was also in steady demand, and with the coming of concrete some of the quarries furnished lime for making cement. In 1918 the state purchased the quarries near High Bridge as sources of agricultural lime for the state farms at Clinton and Glen Gardner. Otherwise the quarries of this area have not been worked much in recent decades, probably because of the competition from areas where limestone is found in larger and more workable deposits.

The making of clay products on a small scale began at an early date. In the Colonial period occasional farmers built houses and other buildings from bricks which they manufactured from clay dug on the spot. It is said that there was a brickyard at Amwell during or soon after pioneer days. Bricks apparently made locally were sold at Flemington in the 1790's, and by 1825 sizable brickyards were in operation both there and at Lambertville. Smaller kilns were constructed at Croton, at Clinton, and perhaps at other places not long after. Yet by 1860 there were only two brickyards in the whole county, and a small one built at Baptisttown a few years after did not long survive. The brick makers of this area had great difficulty in competing with mass production methods elsewhere. A fairly large concern at Flemington lasted for several decades, and was making 50,000 bricks daily in 1907; but it, too, was eventually discontinued.

There is no clear evidence as to when pottery was first made in this region. Dishes and utensils of wood, pewter, and other materials were used in pioneer days, and it is probable that pottery making had no great incentive. This changed, however, before long. There is an old story that Potterstown received its name from an early pottery at that place, and another story that there was one at the village of Amwell near Ringoes. What were probably remains of a pottery were found at the latter place in 1913. There may have been others, as pottery was very popular, at least by the 1790's. The first important potter, however, was Samuel Hill, a nephew of Joseph and Thomas Capner. Having learned his trade in New Brunswick, he came to Flemington in 1814 and, helped by Thomas Capner and others, launched upon a successful and significant career. Until his death in 1850 his earthenware manufactory was one of the most important businesses in Hunterdon County. Some of his pottery was sold locally or traded for firewood, but much of it was packed in straw and hauled to New Brunswick, Newark, and other places. In December, 1827, he advertised in the *Hunterdon Gazette* that he had wagon loads ready for the carters. Census records show that in 1820 four potters were employed in his plant, in 1840, nine, and in 1850, six.

At his death his manufactory was purchased by Abraham Fulper, one of his workmen, who for a time put all of his efforts into the making of drain tile, but eventually reopened the "pot shop" and made both earthenware and stoneware. Labor-saving devices and plant expansion brought production by 1897 up to 1500 pieces daily and further improvements in the plant were made in 1901. In 1915 the Fulper Pottery won several awards for fine glazes at the Panama-Pacific Exposition in San Francisco. Pottery production in Flemington came to an end in 1929 with the partial destruction of the main plant by fire, and the Fulper pottery which has been sold in Flemington since then has been made elsewhere.

Many other potteries have been started at one time or another in Hunterdon County. A small one at Lambertville hired three men in 1840 and one in 1850, but had disappeared by 1860. Another in Bethlehem Township in 1872 did not last very long. A plant established near the Lehigh Valley Station in Flemington in 1889 did well for a time, but failed in 1892. A pottery started in Lambertville in 1908 lasted over twenty years, but was eventually discontinued. A very successful plant which uses clay, not of local origin, is the Frenchtown Porcelain Company, which started on a modest scale in 1910 and now employs several hundred workers. Its specialties are spark plugs and certain types of electrical and plumbing supplies.

In the working of wood skilled men had little difficulty finding employment in the day when so many articles of everyday use were made from that material. Carpenters or "joiners," as they were commonly called, were among the earliest settlers in Hunterdon County. The carpenter of that period had to be proficient in the squaring of timbers and in fastening them in place with wooden pegs, and had to be able to make lath and shingles by riving them out with a frow. In early days the carpenter's trade was often combined with farming or with other trades which might be practiced during slack periods, but as population increased, and especially as villages grew up, carpentry became a trade apart. In 1850 there were forty-six carpenters in Raritan Township. Lebanon Township, much more rural, had but fifteen. There have been no statistics on a county level since that time, but the carpenter has remained an important artisan. Machine production of building materials has changed his trade to some extent in recent decades, but has not displaced him.

Other workers in wood for a century and a half were the coopers who made the barrels, kegs, firkins, wooden pails, and other containers in which many types of produce were carried to their ultimate markets. As the need of barrels increased, the number of coopers was augmented. In the nature of the case they were dependent on local woodlot owners for seasoned staves, hoop poles, and heading materials. White oak was used chiefly for the rugged containers for pork, linseed oil, butter, applejack, and vinegar, while red and black oak were used for making flour and meal barrels. Coopering declined before 1850, probably because of the competition of larger concerns outside the county, and by 1860 the number of establishments had declined to one. The industry did not die out entirely, however, until 1914, when the stave factory at Bloomsbury moved to Warren County.

Cabinet making, which included the manufacture of all kinds of furniture, branched off from carpentry at a fairly early period, at least in the larger towns. The cabinet maker at Flemington in 1793 combined the making of sleighs with that of furniture. In the early nineteenth century these artisans were established in all the larger villages and were producing from mahogany and pine as well as local woods such furniture as bureaus, secretaries, tables, chests, workstands, bedsteads, cupboards, desks, candlestands, and coffins. The making of chairs was a special type of furniture manufacture, usually carried on in separate establishments. A shop specializing in this business is said to have existed at Alexauken at a fairly early date. The most interesting and most famous of Hunterdon's chair makers was John Volk, who started business in 1830 at Flemington, where he made rush bottom, table, children's, rocking, and common chairs. After a few years he moved his establishment to the vicinity of Croton. In an advertisement of 1840 he stated that he had "procured a steam power by which he intend [ed] to execute all kinds of turning in wood." [8] In 1850 he was employing four professional chair makers. The census of that year shows that there were a dozen or more cabinetmaking and chairmaking establishments in the county, but by 1860 the number had declined to four of the first type and two of the second. Outside competition in furniture manufacture further decreased the number during subsequent years, and eventually all of them disappeared. A chair factory at Frenchtown lasted from 1861 until nearly 1900.

After the Civil War another type of woodworking, that of making spokes, reached factory proportions at Frenchtown, Lambertville, and Stockton. The products of shops in these towns soon gained world-wide markets. During two or three decades these establishments used up most of the easily available hickory on both sides of the Delaware, and eventually discontinued for lack of raw materials. The making of frames for pictures and mirrors was another interesting type of industry utilizing the local wood supply. The census of 1860 shows that there was already one such plant in the county, employing four workmen. The most important factory of this kind, however, was the one started at Clarksville (Glen Gardner) by Dr. T. E. Hunt about 1866. At first water power was used, but in 1870 steam power was added and the plant was expanded to make bookcases and desks. During the two decades following, the factory worked intermittently and was still making several kinds of woodenware in the 1890's. Another "frame factory" operated for a time at Changewater in the 1870's and 1880's, and another at Milford in the 1880's. This type of industry was never very successful financially, and succumbed rather easily to outside competition.

The manufacture of leather in early days was usually a home industry. Pioneers built simple vats and used the bark from trees in the nearby woods to tan the hides of animals which were killed for meat. Sooner or later some individuals enlarged their operations, accepted custom

[8] *Hunt. Co. Dem.*, Mar. 18, 1840.

work, and bought surplus hides from their neighbors. Of interest is the partnership agreement of two tanners at Quakertown in 1736, in which they agreed to share equally the "costs and charge of purchasing skins, pelts, and hides and of getting bark for tanning and also of making tan pits, vats, and mills for grinding bark." [9] During the following half century occasional tanners are mentioned in the records, and it is likely that most of the hides of the county were tanned within its borders.

The records of the Case tannery near Flemington provide an excellent source of information for the tanning industry from 1783 to 1851. The Cases did custom work and bought hides not only in the neighborhood but from as far away as High Bridge, Readington, and Frenchtown. Hides and skins of cattle, calves, horses, sheep, hogs, and even dogs were brought in. The tanning industry also provided a market for tan bark. The bark of red, white, black, pin, Spanish, and chestnut oak, beech, birch, and hemlock was regularly purchased to supplement the supply which the Cases secured from their own woodlots in the Great Swamp. The bark of willow was sometimes used to tan sheepskins. Lime for the removal of hair was hauled in regularly by wagon, at first from New Brunswick, later from local kilns. Soon after 1800 the local supply of hides proved inadequate and new sources had to be found. Through agents in the city of New York the Cases bought both "country slaughter" hides and ones imported from New Orleans, Caracas, Pernambuco, Buenos Aires, Maracaibo, and other places. The tanned leather, either in finished or unfinished form, was graded before sale. Although John Hall reported in 1795 that the Case leather was not up to the grade of English leather, it found a ready local sale and surpluses were sold in New York by agents. A by-product easily sold was the hair, which was used in preparing plaster mortar.

Whatever may have been the case in the eighteenth century, there were certainly many tanneries scattered about the county in the early part of the nineteenth. Quite typical was the one at Whitehouse, described as follows in the *Hunterdon Gazette* of January 24, 1827:

> Near 30 vats, and water running through the yard, a good currying shop, four vats below in the beam house, and a marble table above, etc.; a bark mill, with patent mill and stone wheel, two bark houses, with 40 or 50 cords of bark. A dwelling house near the yard, a good spring house and spring, a bark kiln, etc.

[9] Vail, *op. cit.*, p. 9.

The census of 1840 reported twenty-three tanneries employing eighty-two men in operation, and an occasional farmer had a few vats of his own. The number of tanneries had not changed greatly by 1850, but by 1860 had declined to seven, with fourteen workers, for the whole county. A few, like those at Sergeantsville, Fairmount, and New Germantown (Oldwick), were still in operation in 1882, but the day of the tannery in Hunterdon County was past, and in the end all were discontinued. One factor was the decline of the bark supply, but the chief cause was the competition of larger tanneries elsewhere.

The largest users of leather have always been the shoemakers and harness makers. Pioneers from necessity tried their hand at all types of work, but it may be surmised that many of them did not do too well at the job of making shoes. Thomas Budd, writing from West Jersey in its early days, pointed out the advantages of the new land for professional shoemakers. The number of cordwainers, as makers of shoes were long called, was not at first great, but increased with the population. The shoemaker usually did part of his work in a shop or in his home, but often "whipped the cat," that is, traveled from home to home making and repairing shoes on the spot. The leather in such cases was furnished by the farmer, and the same was often true for "bespoke work" contracted for in advance and to be done in the shop. Ann Capner wrote that the shoemaker at Flemington in 1790 could make "court heels" and "French heels," but that his work was "clumsy" by English standards. Dr. Bowne during the first third of the nineteenth century regularly hired shoemakers among his patients to make shoes for his family. The price during most of the period was 3*s.* 9*d* per pair, "leather and fixings found." The various needs of the trade, including tools, were kept at country and village stores.

Shoemaking was sometimes a part-time job of millers, farmers, and others. In larger villages, however, it became a full-time occupation. When not busy with bespoke work, the shoemaker and his helpers fashioned ready-made shoes for direct sale to customers or to stores. The average shoe shop in 1850 was a small, owner-operated concern, at most employing one or two apprentices and journeymen. An exception was a manufactory in Lambertville which employed ten male and three female workers. The twenty-four shoemaking establishments of the county in 1860 employed thirty-four persons. After 1840 the evolution of the shoemaking industry was markedly noticeable. A shop in Flemington in that year sold not only its own product but also shoes which were made in "an extensive establishment" somewhere else. Shoemaking shops by degrees became either stores or repair shops, and shoemakers became either retailers or shoe repairers. After the Civil War shoes made by machines rapidly displaced those made by hand. Shoe repairing, which itself has made an increasing use of power machinery, remains a local occupation in most villages. Short-lived shoe factories have been established at various times within the county, as ones at Lambertville in the 1870's and in 1891, and at Clinton in 1890.

The manufacture of harness and saddles has had a history similar to that of shoemaking. A Ringoes tradition reports that Henry Landis, a leather worker from New Germantown, Pennsylvania, came to that locality in 1737, and that he and his sons and grandsons made saddles and harness there for over a century.[10] Saddlers, though not numerous in early days, probably took care of most of the local requirements. Oftentimes they combined their trade with some other occupation. For example, the owner of a tavern in Kingwood Township was a saddler by trade. Saddle and harness makers, like shoemakers, often "whipped the cat," and many of the customers of the Case tannery were farmers buying leather which they would later have made into harness, saddles, or shoes in their own homes. One of the Cases paid a saddler $17 for making a woman's saddle in 1805, and at about the same time one of Dr. Bowne's patients paid his bill by making the doctor a set of driving harness. With greater use of the horse as a draft animal the need for harness makers increased, and in the early nineteenth century nearly every village had its harness shop. Artisans who worked in leather vied with one another as to "elegance of workmanship," and often showed their work at county fairs. The average shop at the middle of the century was small, seldom employing over two men, and the ten harness-making establishments of the county in 1860 employed only twenty-seven men. Partial dependence on machine-made harness and parts of harness was noticeable as early as the 1850's, and harness shops soon began to put in new equipment for doing simple operations formerly done painstakingly by hand. By combining repair work with the sale of "patent harness" made elsewhere, many shops survived until the coming of the automobile and the tractor, and since that time occasional ones have carried on a losing uphill fight.

Alcoholic beverages in a new country are often made from surplus farm products. In Hunterdon County an important drink for two centuries was fermented cider, which could be made in the home, though custom cider pressing was often done by local mills. In 1786 Ruther-

[10] *Hunt. Rep.*, Jan. 13, 1870.

ford wrote, "Cyder is made on almost every Farm, which beside a pleasant Beverage, gives a plentiful Supply of Vinegar much used by all Ranks in Vegetable and Salt Diet." The distilling of fermented cider to make apple brandy—variously called applejack, apple whiskey, and "Jersey lightning,"—was a common practice. There are said to have been distilleries at Headquarters and at Philip's Mills on the Neshanic at a very early date, at New Germantown in the 1750's, and on the Nishisakawick during the Revolution; and without doubt many other communities had establishments of their own. Cazenove, speaking particularly of Morris County, mentioned the "many distilleries where spirits are extracted from cider." Cider presses were often run in connection with the distilleries, and farmers who brought their apples to the press received in return part of the cider derived therefrom or else a proportionate amount of "apple whiskey." The distiller, who in most cases was either a miller or a farmer, sold his share of the brandy to the local taverns, and the tavern keepers completed the circuit by serving it over the bar to farmer customers.

The production of apple brandy flourished in Hunterdon County and adjacent areas until after the middle of the nineteenth century. A store at or near present-day Oldwick in the 1820's purchased or accepted in trade large quantities of this beverage each year and carted it to the New York market. In 1825 a local distillery was buying from its patrons thousands of bushels of apples each year. Nearly every farmer who patronized it took part of his pay in brandy at the standard price of 75¢ per gallon. An observer of the time reported that almost every village had a similar plant, and the census of 1820 showed that over 170,000 bushels of apples were used for brandy making in Hunterdon County in the previous year. Large orchard owners often had their own cider presses and distilling equipment, and usually did a certain amount of custom work. For over three decades Dr. Bowne had the apples of his small orchard converted into brandy, and accepted besides large quantities of the beverage from his patients in return for services. It is not clear from the accounts whether he used it all for home consumption or not.

Other types of liquor have also been manufactured in this area at one time or another. The early Dutch settlers are said to have made beer, both in their own homes and at local breweries. The grain distilleries at Amwell, at Headquarters, at Philip's Mills, and at other mill sites used large quantities of local grain in the latter half of the eighteenth century. Moore Furman had a similar establishment at Pittstown soon after the Revolution, and there was another a little later at Calvin's

Mills (Pattenburg). A side line for grain distillers was the fattening of cattle and hogs, to which were fed the slop left behind after the extraction of the liquor. A field south of Ringoes, more productive than other land nearby, is still pointed to as the feed lot for the Dawlis distillery. Some grain distilleries kept up operations into the nineteenth century, but the growth of outside markets for grain hurt their position and eventually caused their discontinuance. The only beer-making establishment with any considerable success was the one at Lambertville which operated from before 1850 until at least 1869.

The decline of liquor manufacture was related to the temperance movement, but since it began sooner than the decline of local taverns, we must conclude that other factors, such as outside competition and a change in tastes, were more important. Between 1840 and 1860 the number of distilleries in the county fell from thirty-eight to twelve, and during the next twenty years most of these were discontinued. In Delaware Township, where there had been eight distilleries in 1838, only two remained in 1885. A very few carried on until the turn of the century, and one, at Fairmount, survived until the war measure of 1917 stopped all liquor manufacture. The decline of apple growing and cider making in recent decades has resulted in less use of cider as a home beverage. After the repeal of the prohibition amendment an attempt was made to manufacture applejack at Flemington on a large scale, but it proved unsuccessful.

A more utilitarian product made from apple cider was vinegar. Because of the ease of home manufacture vinegar seldom appeared as an item of local trade in the eighteenth century. In 1805 half a barrel of "winegar" was worth one pound York money or about $2.50. By 1825, however, it was a regular item of commerce at six cents per quart retail. The surplus cider of orchardists and cider makers, when not made into brandy, was converted into vinegar. There were no vinegar manufacturing plants as such reported even in the census of 1860, but this may have been because vinegar making was an industry subsidiary to distilling. At any rate, there was a "cider and vinegar works" in Lambertville in 1861. As the consumption of brandy decreased, distilleries probably turned more to the production of vinegar. One vinegar works which started from humble beginnings lasted for over half a century and became a large concern. It began as a side venture on the farm of John P. Case near Clover Hill in 1878 or soon after, and Case and his sons carried on an increasingly large business until 1907. In that year the business was sold and moved to Flemington, where it prospered for three more decades. In 1937 the Distilled Liquors Corporation, to which it belonged, failed as a result of the mistaken venture in applejack mentioned in the last paragraph, and the plant fell into disuse.

From early days the settlers of West Jersey made linen and woolen goods from flax and wool of their own production. Spinning wheels, often made locally, shears, wool cards, needles, thread, knitting needles, redwood, alum, indigo, copperas, and other needs of cloth makers were commonly offered for sale at general stores during the whole period of home manufactures. Spinning wheels, looms, dye tubs, and other paraphernalia were often listed in the inventories of estates. Native bark, leaves, and roots were used as dyes in early days, but as time went on they were in large part replaced by imported vegetable dyes. On November 28, 1787, Ann Capner wrote from Flemington, "They . . . sew their own flax, swingl [*sic*], etchel and spin it and some weave it, then they whiten it till it is as white and as thin as Irish cloth." Both she and her mother praised American spinning and knitting. The spinning of flax, tow, and wool and the twisting of yarn to make rope and twine occupied many a winter evening for farm families. Not all cloth was made in the homes, for even in Colonial times stores had English weaves for sale, and in 1786 Rutherford complained that imported cloth was hurting home manufactures. In the nineteenth century American factory-made cloth also became common. Nevertheless, home industries persisted for some time, and sixty-two households of Tewksbury Township reported home manufactures in 1850. The county fairs had some entries for spun yarn and home-woven cloth in the 1850's, but almost none by the 1880's. The making of knit goods, such as gloves and stockings, was common for a time even after home spinning had nearly disappeared.

In pioneer days the entire process of manufacture from raw material to woven cloth was in many cases completed in the home, but before long certain tasks were taken over by persons specializing in those operations. For instance, the fulling for a whole neighborhood might be done by a single family, and farmers like Derrick Hogeland in the 1740's and Abraham Hageman in the 1780's often did custom weaving in their spare time. In the 1770's a store at or near present-day Whitehouse wove cloth for its customers, and special shops for textile manufacture were sometimes established in the villages, as those at Amwell and Alexauken at an early date. The accounts of a small carding, weaving, and fulling shop at Flemington in 1810 portray well the difficulties of the trade, worst of which was that of securing trustworthy labor. The shop was evidently one of the many ventures of Thomas Capner, probably in partnership with his cousin, Richard Choyce. As the carding of wool and the fulling of woolen cloth became mechanized, local mills often took over these operations and sometimes that of weaving. Weaving was often done at or near taverns; and in neighborhoods where there were no weavers taverns, stores, and mills were used as collecting and distributing points. The weaving shops usually limited their operations to the making of a few kinds of cloth, commonest of which were worsted, flannel, broadcloth, linsey-woolsey, blankets, coverlets, and spreads, and to the weaving of rag carpets. Custom knitting and spinning were often done by women in their own homes under agreements with their neighbors or with local stores. The Case tannery accepted labor of both types in payment for leather, and Dr. Bowne customarily had some of his patients spin and weave the wool which he produced on his farm or which he received in payment from other patients.

As cloth making became a factory operation elsewhere, Hunterdon County textile makers tried to keep pace. To the carding and fulling machines were added those for spinning and weaving, also run by water power. A complete woolen factory near Ringoes was started in 1811, and there were in the county in 1820 at least one hundred fifty spindles, three other spinning machines, nine looms, and five carding machines, but because of the influx of British goods and the effect of hard times not all were in operation. Cloth factories established at Clinton, Lambertville, and Frenchtown in the 1820's and 1830's were not reported as operating in 1840. Though times were not prosperous at the latter date, the factory near Ringoes was in operation, as was a linsey woolsey factory at or near Flemington. In 1850 a woolen manufactory in Lebanon Township was using water power to operate three spinning jennies, three looms, a carding machine, and fulling machinery. Probably this was Wise's Mills, said to have been in operation already in the 1820's. Young's Mills, five miles north of Flemington, made many kinds of cloth. The owners, Shepherd Brothers, acquired the plant near Ringoes also, and operated both until about 1880. Eventually, however, these establishments gave way to the same economic conditions which killed all woolen manufacturing in this area. The building of the Ringoes factory is still standing, part of it in ruins, the rest used as a poultry house.

The manufacture of other kinds of cloth was never large. A linen manufactory which started at Lambertville in 1838 and was said to "make flax cloth look like cotton" employed sixteen persons in 1840 and six in 1850, but had failed by 1860. The sudden revival of flax production and manufacture under the stimulus of bounties in the early 1880's and the sudden decline thereafter has been mentioned previously. A cotton manufactory which started in Lambertville soon after the Civil War failed in 1877, a total loss to the stockholders. In the twentieth century there have been a number of short-lived ventures in silk making in several towns of the county.

The manufacture of cloth into garments was long a household occupation, and even today some clothing is made in the home with the aid of the sewing machine. There have always been persons, however, who needed or wished to have this work done by others, and in early days they often paid the wives of their neighbors to do their sewing for them. When there was sufficient custom, tailors appeared to make men's and sometimes women's clothing. The Capners reported three tailor shops in Flemington in 1793, and two years later encouraged an English tailor to settle there also. Every village of any size had one or more master tailors, who set themselves up as exponents of good taste and were always in possession of knowledge as to the latest English and French styles. Sometimes a tailor took his lapboard and goose and journeyed from house to house, doing his work in the homes of his patrons. A man wanting a suit first bought the cloth, then hired the tailor to fashion it. In 1849 J. B. Thompson, a Readington boy later famous as a preacher, came home from college for vacation and went to his favorite tailor for a coat. In his diary he recorded, "The cloth I got from Uncle Aaron and the trimmings from H. Hageman." Larger general stores often hired tailors to make clothes for their customers. In 1850 these artisans were represented in nearly every village, and there were several in every township of the county. Most tailors moved about, continually looking for new and greener pastures. An exception was Asher B. Stout, who for fifty years after 1867 plied his trade in Flemington.

Tailors often made women's clothing in early days, but the ladies came to prefer the judgment and services of members of their own sex. The larger villages eventually supplied a sufficient trade to support mantua-makers or dressmakers, and Flemington acquired a second one in 1795. Dressmaking and millinery advertisements cleverly drew feminine attention to the bonnets and hat, the ribbons, artificial flowers, and other trimmings, the "French worked" collars, the imported scarves and handkerchieves, the newest weaves of cloth, the artificial curls, the shell combs, and other feminine needs which were for sale at the moment. After about 1850 the use of ready-made clothing caused a gradual decrease in custom for both tailors and dressmakers. There were local attempts for a time to enlarge certain shops to factory proportions, but conditions were not right for success. There are not many tailors in the county today, and their work is largely that of alteration and repair, combined with the sale of ready-made clothing. Dressmakers and milliners in like manner became retailers of ready-made clothes for women. In 1881 Flemington had seven milliners to advise its women how to dress. The dressmaking shop as such has now disappeared, being replaced by the "women's shop."

The manufacture of other types of clothing has engaged workers of the county at various times. An establishment for making men's hats existed at Lambertville prior to 1825, and during the following years the industry was continued spasmodically. The 1850 census reported three men and two women making 2500 fur and silk hats yearly. Other small shops, located at Quakertown, Flemington, and Clinton, were engaged in hat making in 1831 but little is known of their history. In 1840 the makers of hats, caps, and bonnets in Kingwood, Amwell, and Alexandria townships employed twenty-two persons; but by 1860 there were no hat-makers left in the county.

The making of hosiery on a factory basis was unsuccessfully tried in Flemington in 1891, and a shirt factory worked spasmodically there in 1896 and 1897 before moving to Philadelphia. An overall and work pants factory was started at Milford in 1897, a glove factory at Flemington in 1909, a braid factory at Dart's Mills in 1913, a woman's hat factory and a fur coat factory at Flemington in 1919, and a pants factory at the last named place in 1925. Though some ventures lasted for several years, few proved successful over a long period. The factors which are necessary for the manufacture of clothing were not present here. The longest-lived concern of this type has been the shirt factory at Clinton, which by skillful management was kept going continuously for over thirty years, beginning in 1898. In 1931 it ceased operations as a result of the business depression, but has since resumed.

The manufacture of rubber products has been a development of some consequence in the county. The Lambertville Manufacturing Company was making "India rubber goods" already in 1860, and after recovering from a fire of 1868 was employing fifty-six men in 1870. Under various names the company lasted for sixty years. Though it made many kinds of rubber goods, its best known product was rubber footwear, sold under the trade name of "Snag-Proof." The company was eventually absorbed by the Goodyear Rubber Company, and the factory was discontinued in 1928, throwing four hundred men and women out of work and giving the city of Lambertville a serious economic blow. For a time before and after 1910, there was also a small factory making rubber goods at Stockton. A reclamation plant for scrap rubber operated for a time in Lambertville in the 1890's. In 1917 Flemington entered the rubber goods business with the coming of the Dural Rubber Corporation. This concern made tires and tubes at first, but was soon busy with government orders for airplane parts. The postwar slump brought financial disaster, but a new company was organized and soon put the plant on its feet. During the depression of the 1930's the factory was in part-time operation only, and changed hands again in 1939. At the present writing it is again busy with war orders.

Paper making has been another important industry of Hunterdon County. A small paper factory north of Lambertville was already in operation in 1831, but lasted only a few years. Another opened in that town in 1862, and during the next two decades three separate concerns built up a considerable industry, which carried on despite ups and downs until 1928. Their discontinuance in that year was another blow to the prosperity of Lambertville. A small paper manufactory at Stockton in the 1870's and 1880's specialized in the making of paper receptacles. The most successful development in paper making in this region was that of the Warren Paper Company, which expanded to Hunterdon County in 1910 after nearly five decades of successful operation across the line in the southern tip of Warren County, at Riegelsville, Warren Glen, and Hughesville. The plant built at Milford in that year was used for the making of glassine paper, copied from a German development. It became the largest manufactory of glassine paper in the world and is today the largest of the plants of the Riegel Paper Corporation, as the company has been called since 1930. In 1938 the four Hunterdon and Warren county plants of this company were employing 1150 men, of whom half were working at Milford.

Several types of industry were mentioned in earlier chapters—the making of cordage, corn husk articles, iron bridges, calcined lime, peach baskets, brooms, linseed oil, and railroad rolling stock. Besides these and the ones mentioned thus far in this chapter, there have been many other types of manufacturing of one kind or another. At Flemington a brushmaking factory using locally produced bristles operated for two decades beginning in the 1840's. Other brush factories at Clinton and Glen Gardner were started in 1896 and 1898, but lasted only a few years. The manufacture of tobacco products has been carried on at times. Chewing tobacco, snuff, and cigars were made at a shop in Lambertville from the early 1840's until at least 1860, and cigars were still made in the town in 1891. Cigar manufactories also operated in Clinton and Frenchtown during the 1880's. The largest venture of this kind was that of the Changewater snuff factory, a family-owned concern which operated for many years. In 1893 it produced $200,000 worth of snuff. When it was "gobbled up by the snuff trust" in 1900, the village was seriously hurt by the loss.

A few other industries of the county should receive mention. A factory for making pianos at Bloomsbury lasted from 1894 to 1897, then moved from the county. An especially interesting development has been that of glass cutting at Flemington. In 1902 the Empire Glass Company was started, and in 1908 the Flemington Cut Glass Company. The former moved away after a time, but the latter expanded until it was using seventy hands in 1913. Since then it has operated most of the time, though sometimes at only partial capacity. Other small glass-cutting establishments have also started up in recent years. A modern development has been the manufacture of electricity in Holland Township above Milford. The huge plant which was built there by the New Jersey Power and Light Company cost over six million dollars. It was started in 1922 and put into partial operation in 1930. It now furnishes current for most of northwestern New Jersey.

Manufacturing in the modern sense has never employed a very large portion of the labor force of the county, but it has played an important part in the economic life of the people. A marked change in recent decades has been the decline in the use of raw materials produced locally. The following table gives a few statistics taken from census records. The figures are for the year previous in each case.

Census	*Establishments*	*Capital (dollars)*	*Labor Force*	*Raw Materials (dollars)*	*Value of Production*
1860	267	899,190	998	1,152,374	1,881,345
1870	614	2,136,681	2,273	3,025,765	4,754,685
1880	374	1,973,625	1,260	2,040,428	3,104,070
1890	242	2,281,058	1,710	1,758,738	2,956,637
1900	298	4,230,462	2,444	2,918,949	5,074,736
1920	145	2,832,766	2,800	7,516,653	14,077,001
1930	51	(not given)	2,385	5,511,736	12,991,140
1940	40	(not given)	2,519	6,700,210	13,951,946

A trend toward centralization and expansion of operations is noticeable, as well as the employment of a greater number of men and women. The value of final products has not changed greatly during the past three decades.

CHAPTER XII

Labor

IN BOTH AGRICULTURE AND MANUFACTURING individual farmers and artisans soon expanded their operations to a point where they desired supplemental labor. This was even more the case as regards mining and other extractive industries. Labor in addition to that of the family was secured in a variety of ways—by indenturing and apprenticing, by acquiring slaves, and by employing wage hands. Unfortunately, each plan had its drawbacks and limitations: indentured servants and apprentices served for definite terms only; slaves, except in certain staple crop areas, were often a questionable investment because of the necessity of supporting them for life; wage employees in new lands usually received high wages and too often were attracted away by the possibility of setting up in farming or business for themselves. The shortage of labor was seldom acute, but both skilled and unskilled labor were usually in sufficient demand that free laborers could afford to be highly independent. In Hunterdon County, which was a general farming area typical of the Middle Atlantic Colonies generally, all of the above methods of getting laborers were used. However, in the final analysis here as elsewhere a man's family was usually his greatest asset.

The large families of the eighteenth and early nineteenth centuries may or may not have been due to economic factors. At any rate the shortage of other sources of cheap labor led to a maximum utilization of the family labor supply. Family records show that each generation in turn followed the Biblical injunction to multiply. Albert Opdycke of Hopewell Township (1685–1752) had five sons and three daughters. Of the four sons for whom there is a record, three settled in present Hunterdon County. John, the eldest, had four sons and five daughters. The three sons who grew to manhood all followed the trade learned at their father's mill. One of them, at Milltown (Idell), had two sons and eight daughters; another, at Sergeantsville, had one son and six daughters; the third, at Locktown, had one son and four daughters. John Opdycke's four grandsons had seven, twelve, eight, and ten children respectively.

Joshua, brother of John, had eight children, two sons and six daughters. The two sons had families of twelve and seven children and grandchildren almost innumerable. Benjamin, the third of the immigrants into Hunterdon County, reared two sons and four daughters. The two sons had six children each, but westward immigration carried away all of the twelve except two. These two, however, replenished the loss by bearing eleven and three children respectively. It will be noticed that the Opdycke clan produced more daughters than sons. Nevertheless, the name became one of the most common in this section and was borne westward to new lands by each succeeding generation. The family labor force, male and female, was in nearly every instance large. The nine six-foot sons of Richard Opdycke of Kingwood Township, a son of Joshua, could without doubt take care of a considerable amount of farm work.

Other families could serve as well as examples. John Peter Rockefeller, a German immigrant in Amwell Township, had two sons and seven daughters. The two sons perpetuated the name by giving him twelve grandsons, who in turn passed the name on to fifty-one great-grandsons. Some of the descendants of Peter Rockefeller are now famous, but those in humbler walks of life number into the thousands. For many generations they were farmers and artisans and tradesmen, and the size of their families comported with the economic situation. Henry Landis, an Amwell saddlemaker, is said to have had twenty-four children by two wives, and one of his daughters was the mother of fourteen children. Samuel Fleming, an early settler after whom Flemington was named, had ten children; and one son-in-law, Thomas Lowrey, had eleven. The numerous Apgar family of this and neighboring counties are all said to have descended from one German immigrant who had ten sons.

The conditions which encouraged large families in pioneer days obtained throughout the general farming period. Mrs. Mary Capner, who in her capacity as midwife traveled about the countryside a great deal, was amazed at the fecundity of Americans. In a letter of February 7, 1791, she mentioned a woman who at death left 17 children, 130 grandchildren, 100 great-grandchildren and one great-great-grandchild. In the same letter she wrote concerning a Yorkshireman who had just arrived:

> He expects his wife and seven children in the spring. I tell him this is a fruitt-full country, so he may expect to be father of Seventeen (Aew, Nae, he said) and that his daughter will be married and have children before her mother has done.

A woman once apologetically told Mrs. Capner that she had "but eight." Johannes Schmidt (John Smith) of Stanton when he died in 1790 left behind nine sons and eight daughters, all with families of their own. Daniel Moore, a farmer north of Croton, was survived by fourteen children when he died in 1808. Between 1816 and his death in 1850 Wilson Bray of Kingwood Township had fifteen children. The mother of this brood lived to the age of seventy-six. In 1850 a Bethlehem Township farmer aged forty-one and his wife aged thirty-six reported children aged five months, two, four, five, six, seven, nine, eleven, twelve, fourteen, and sixteen years, the eight eldest of whom were attending school during part of the year. In 1869 members of five generations of one family were living under the same roof in Centreville. It is little wonder that the many "family associations" of the county have had and in cases still have a large membership.

Not all families were large, of course, but the above listing, which could be carried to great length, shows something concerning the tendency of the times. The average family was large and often spanned a considerable age variation. Widows and widowers usually remarried, in many cases uniting their respective families. The size of families may have been due to lack of knowledge of birth control measures and to religious and other factors, but the fact that children could pay their way in the world was something of an incentive. Farmers, of course, could and did give boys and girls regular tasks from an early age. Artisans and mechanics put their whole families to work. Ministers, doctors, lawyers, and teachers did not have this same opportunity; but their sons could hire out and their daughters could spin and sew and cook until marriage solved the problem of their disposal. As time went on, and especially in the twentieth century, there has been a change in the situation. The costs of raising a family increased, the use of machines has lessened the need of having as large a home labor force, a change in the attitude toward child labor has developed, and a knowledge of birth control has been disseminated and utilized to a greater degree. The size of rural families has as a result decreased markedly, though it is still much larger than that of urban families. Hunterdon County, predominantly rural, still has relatively large families.

The use of indentured servants, *i. e.*, servants legally bound to their masters for a period of time, was not so common here as in plantation areas, but was nevertheless of some consequence. The first settlers in West Jersey provided that, where no agreements had been made, adult servants should work four years to pay their passage and minors a term set by judicial action. When John Reading died at Stockton in 1717, the remaining time of a servant boy and two servant girls was considered as part of the estate, and the probate papers of other estates show similar instances. Kalm in 1748 reported that the number of servants in West Jersey was very large. The use of indentured servants in Hunterdon County apparently increased after the middle of the eighteenth century, but declined rapidly after the Revolution. Most servants were employed on farms or in homes, but occasionally an artisan or professional man sold his services for a term in order to get passage to America. In 1757 Casper Berger, a German stone mason, was "bought" by Cornelius Van Horn, who lived near present-day Whitehouse. In lieu of a stated term Berger agreed to build three stone houses. That done, he settled down in the neighborhood, and eventually became a prominent citizen. He was one of the signers in 1775 of the following interesting advertisement in the *New York Journal:*

Runaway from the Subscribers, living in Reading Town, Hunterdon County, West New Jersey, on Saturday night, 15th instant, an indented servant man named Hugh Smith, about 5 feet 6 inches high. Whoever takes up and secures said servant in any of his Majesty's Gaols, or delivers him to any of the subscribers, shall have five dollars reward and all reasonable charges paid by us subscribers who bought him to teach a school. Thomas Stout, James Van Der Veer, Casper Berger, James Van Sickle, John Van Sickle.[1]

Cazenove in 1794 wrote that for poor Scotch, Germans, and Irish to indenture themselves to secure passage was a "custom of that class." The Capners were interested in the fact that their neighbors had bound servants, and determined to try them out for themselves. They and their relatives brought over both parish apprentices and "free willers," as those who came of their own free will were called in contrast to paupers and felons who were forcibly transported. In order to keep their servants contented, the Capners paid them a small wage besides paying their passage. John Hall thought indentured servants superior to native labor. However, the indenturing of servants for the payment of ship's passage or other debts gradually became less common. One of the last servants of whom we know in this area was a German "bought" by Hugh Exton for $70 in 1804. About this same time Paul H. M. Prevost at Frenchtown paid passage for a German and his wife. After serving their term of four years, they settled down on land near Frenchtown which was given or sold to them by Prevost.

Apprenticeship was similar to indenture in its legal aspects, but ap-

1 Reprinted in *Hunt. Co. Dem.*, Feb. 14, 1924.

plied only to children and had as its reason or excuse the teaching of a trade. The parent or guardian made a contract which bound the child until it was of age or for a definite term of years. Poor children might be bound out by the overseer of the poor with the consent of a justice of the peace. The form used came by custom to include certain terms. The apprentice was to serve his or her master faithfully and to watch over his interests, was to be of good moral behavior, and was not to frequent taverns or play at cards and dice or be absent without consent. The master on his part was to teach the apprentice a trade or business, in many cases that of farming or housewifery. He must furnish food, clothing, and lodging, must teach the apprentice to read and write, and at the end of the period must furnish a complete outfit of working clothes and another for the Sabbath. Often there were certain other specified privileges and emoluments; apprenticing had definite advantages for both parties concerned in most cases. The apprentice was in a minor position, however, and might be mistreated by a harsh master. In such cases a justice of the peace might free him from bondage. In 1799 the parents of a boy who had been apprenticed to a blacksmith brought suit before Ezekiel Cole in Readington Township, and as a result the boy was freed because of "cruelty and ill usage" by his master. Children of immigrants were often apprenticed, and in some cases parents who remained in Europe sent their children to America so that they might learn trades in this manner.

Advertisements of runaway apprentices show that relationships between masters and servants were not always happy. Apprentices with a grievance were likely to leave in the night, taking with them whatever of value they could lay their hands on. At such times owners often advertised more than nominal rewards. Usually, however, a master offered a few cents plus a reasonable amount for costs involved, his chief concern being the avoidance of debts contracted by the runaway. Since the use of apprentices was a definite part of the economic system, there were always persons on the lookout for boys interested in learning a trade. Tailors, tanners, shoemakers, harness makers, blacksmiths, wheelwrights, carpenters, cabinet makers, chair makers, newspaper editors, masons, and painters advertised for boys "of good morals and industrious habits." The local census schedules in 1850 show that the household of many artisans included one or more apprenticed lads. The indenturing of Negro children, usually for farming or housework, will be discussed later. With the decline of the small workshop and the encroachment of the factory has come a gradual breakdown of the apprentice system. Paralleling this change there has developed a much greater emphasis on

FIG. XVII. Breaking and spinning flax.

FIG. XVIII. Copper Hill blacksmith and wheelwright shop of Lambert B. Lake. One of the last of its kind, late 1930's.

FIG. XIX. Spoke works along the Delaware in the 1880's. These plants were dependent upon upriver stands of hickory, and failed after the hickory disappeared.

education and a proportionately greater attendance of the public schools.

Negro slavery was tried throughout the English colonies as a means of coping with the labor shortage, but only in certain areas, notably the South, did it prove successful enough to remain over a long period. Slavery in Hunterdon County in Colonial days seems to have been quite widespread, though the ratio of Negroes to whites was never very large. Slaveholding on a large scale was unusual, which probably explains why later traditions stressed the few slave gangs, such as the one employed at the iron works of Allen and Turner. The ferry at present-day Lambertville used eight blacks, and there were other examples of men owning five or six slaves. Most masters, however, had one, two, or, at most, three slaves. Often a substantial part of the personal property of an estate was the one or two slaves which the deceased had owned.

At the time of the first national census in 1790 there were in Hunterdon County, which at that time included two townships of present-day Mercer, 1301 slaves in a total population of 20,153. This means that approximately one person in fifteen was a Negro slave. There were at that time only 191 free blacks in the county. By 1800 the number of slaves had fallen to 1220, while the total population had climbed to 21,261, of whom 520 were free blacks. The slave ratio had fallen slightly, but the colored population had increased faster than the white. By this time only seventy per cent of all Negroes were slaves. In 1810 there were in the county 1119 slaves and 687 free Negroes in a total population of 24,556, or 4.5 slaves and 2.8 free blacks in every one hundred persons. The slave population declined rapidly after that time, and by 1820 there were only 716 slaves in a total population of 28,604. On the other hand, free Negroes now numbered 1444, or two for every slave. By 1830 the number of slaves had fallen to 171, of whom 131 lived in Hunterdon County as now constituted. There were 1770 free blacks, of whom 855 were within the present bounds of the county or, in other words, thirteen freedman for every two slaves.

Before the census of 1840 Hunterdon had been cut down to its present boundaries. By then the number of slaves had fallen to thirty-five, of whom thirteen lived in Raritan Township, in which Flemington was located. The free black population had dwindled to 778. By 1850 there were only four slaves left in Raritan Township and one each in five other townships, though there were at this time 808 free colored persons in the county. For all practical purposes slavery was dead. The nine remaining slaves ranged in age from fifty-eight to ninety years. Six townships had no slaves at all, and even the free Negroes were disappearing from the rural areas by this time. Lebanon Township had only sixteen Negroes, Kingwood but eight, and already the sight of a Negro was a novelty in some neighborhoods. A small boy near Frenchtown had never heard of such a thing as a black man. When he saw his first Negro, he ran excitedly to his mother and proclaimed that he had "seen the devil himself down in the meadow by the goose's nest." [2] By 1860 there were only four slaves left in Hunterdon County and only eighteen in the whole state. None of the slaves could have been considered, at their age, economic assets by their masters. The county's free Negroes numbered 796.

The provincial slave code in New Jersey was harsh, the fundamental law being that of 1713. Slaves wandering from home were whipped, and their masters were required to pay the informers and the constables who applied the lash. The personal liberty of slaves was restricted in many ways. A law of 1751 specifically prohibited them "from meeting in large Companies, from running about at Nights, and from hunting or carrying a Gun on the Lord's Day." These colonial laws became a part of the state law during the Revolution, and were reaffirmed with modifications in the act of 1798 which summarized all legislation regarding slavery to date. There are indications that these laws were winked at in practice during the later period. The attitude of New Jersey colonial law toward manumission is well-expressed by a statement in the 1713 act, which read: "It is found by Experience, that Free Negroes are an idle, slothful people, and prove very often a charge to the place where they are." It was required that persons of property must sign a "security" for the support of any freed slave. At the same time it handicapped freedmen by not allowing them to own real estate. A law of 1769 reaffirmed the stipulation of security and named the specific sum of two hundred pounds. It is little wonder that few slaves were freed during nearly a century of Hunterdon County's history.

Beginning at about the time of the Revolution, however, there was a strong reaction against slavery in New Jersey. This was shown by the provisions of the law of 1786, which in its preamble contained a severe indictment of the institution, and by those of a supplementary act of 1788. Slave trade across state lines was rigidly prohibited. Cruelty against slaves was made an indictable offense, punishable by a fine of five pounds for the first offense and ten for the second. A slave between the ages of twenty-one and thirty-five might be freed, provided that two township overseers of the poor would attest to his soundness of mind and body.

2 Vanderbilt, *Hunt. Co. Dem.*, Feb. 20, 1912.

The essentials of this legislation were repeated in the law of 1798, and the age limit for manumission was increased to forty years. Nevertheless some handicaps to the wholesale freeing of slaves were retained, and two owners of property must sign a bond for one hundred dollars that a newly freed slave would not become a public charge. Despite these hampering restrictions the number of free Negroes, as previously noted, was increasing in Hunterdon County by 1800. An interesting feature of the law of 1786 was the requirement that slaves must be taught to read but it is doubted that this provision was seriously enforced.

Slavery in New Jersey received its death blow in 1804 with the legislation entitled, "An Act for the Gradual Abolition of Slavery." A child born of slave parents after July 4, 1804, was theoretically free, but was to remain a servant of his mother's master "in the same manner as if such child had been bound to service by the trustees or overseers of the poor." This servitude was to end at the age of twenty-one for females and at the age of twenty-five for males. The child of a slave must be registered with the county clerk, but at any time before it was a year old the master might signify his desire to abandon his rights, in which case the overseers of the poor might bind out the child to someone else. If he did not so signify, he was under obligation to support the child for the regular period. In practice these children were considered slaves and so counted by the census takers. In Hunterdon County only some fifty or sixty slave children were ever registered. Slavery in Hunterdon County evidently was allowed to die out even faster than had been intended by those who drafted the law of 1804.

The laws of 1798 and 1804 and subsequent legislation of 1812 and 1818 were codified in the act of 1821. The act of 1818 had made even more stringent the restriction against exporting slaves or colored servants from the state. Later legislation was not too important in changing a trend already under way. The state constitution of 1844 stated that "all men are by nature free and independent," but the courts immediately upheld the right to property in slaves. The law of 1846 theoretically abolished slavery in New Jersey. By this act remaining slaves became "apprentices" for life under the same restrictions as if they still were slaves, and children of slaves were free at birth.[3] After this time there was no further legislation regarding slavery in New Jersey. Negro children were bound out, even after the Civil War, but were, legally at least, on the same plane as white apprentices. Available evidence indicates that

3 *N.J. Session Laws,* 1846, pp. 382–90.

slaves in Hunterdon County were freed at a rate which outstripped the provisions of the law. The county book of manumissions has not been found, but the study of 115 certificates of manumission which have been preserved indicates that a wave of reform struck the county soon after 1810. These certificates are from every part of the county. They were originally brought to the county clerk for recording, but the newly liberated slaves never bothered to redeem them by paying the required fee. The laws of 1798 and 1821 required that freedmen in order to be employed or to leave their home counties must have such certificates available. These provisions of the law were evidently disregarded by whites and blacks alike.

The chief occupation of male slaves in Hunterdon County was evidently general farm work. Most of the owners of slaves mentioned in probate records were farmers. The records of the Case tannery show numerous examples of slaves on farms, and the accounts of Dr. Bowne show dozens of cases during the first two decades after 1800 in which the good doctor bled the slaves of his farmer neighbors and administered to them the cathartics and emetics which constituted the principal remedies of the time. The owners of runaway slaves mentioned in the *Hunterdon Gazette* were usually farmers, and advertisements of slaves for sale sometimes extolled a slave's agricultural abilities. Notice in a Trenton paper in 1781 read as follows:

> A Negro Man and his Wife, with two male Children. The man is a complete farmer. The wench is a good dairy woman, and can wash, iron, and cook very well. They are honest and sober. Enquire of the Printer.[4]

The usual female slave seems to have been a maid of all work and general helper in the household, and Negro men, too, were often partially employed in the innumerable odd jobs and chores which went with the maintenance of a home in the days before our modern labor-saving devices. Negro farm hands were apparently allowed that freedom of action which it is necessary to give to helpers on a general farm, where the variety of tasks makes any other arrangement difficult. The account books mentioned above give numerous instances in which slaves ran errands and made trips with teams for their masters. In many cases the transaction involved an exchange of money, thus showing the master's faith in the honesty of his servant.

Men of other occupations sometimes found it advantageous to keep a slave labor force, especially in early days when hired labor was very scarce.

[4] *N.J. Archives*, 2d ser., V, 205.

Emmanuel Coryell used slaves to operate his ferry in the 1740's, and when John Opdycke died in 1777, he left each of his three sons not only a mill but a slave helper. The Case tannery generally used one or more slaves, and a widow in Bethlehem Township ran her deceased husband's still with slave labor. The fact that occasional slaves were trained in various trades makes it seem likely that artisans, too, sometimes employed this kind of labor. The largest slave gangs were used by the iron furnaces of the north part of the county prior to the Revolution. When Hacklebarney Forge was sold at sheriff's sale in 1764, the advertisement mentioned "sundry Negroes, Women and Children." [5] The ironworks of Allen and Turner had a labor force of eighteen slaves in 1747, and is said to have employed forty just before the Revolution. When war came, the Tory owners took the slaves out of the state in order to prevent their confiscation. According to tradition one of the Negroes escaped to the woods, another was retained as a personal servant, and the other thirty-eight were chained together and driven to Virginia.[6]

It was not unusual for a man to hire his slaves out to others. A typical case was shown by an entry in John Lambert's memorandum book in 1774, which read, "Ambrose Barcroft Allowed 6/ for black Jacks hawling logs to Snyders Saw mill 1/2 day." In 1813 one of Dr. Bowne's patients paid his bill by sending the doctor a Negro wench to serve as a maid at five shillings per week. Sometimes slaves were hired out for a term of years. The commonplace book of the Hankinson family mentions two pickaninnies whose master had died and who were bound out until twenty-one years of age. One, born in 1805, would then be legally free; the other, born in the first half of the previous year, would still be legally a slave and would revert to the estate. The practice of hiring out Negroes reflected in all probability the difficulty of always finding work for them to do. Unlike a hired laborer a slave could not be dismissed when idle, and the cost of keeping him continued. As has been pointed out, many owners released their slaves when they might have retained them for life or until a given age. It is probable that economic motives often flavored their acts of kindness, for once a slave was over a specific age he must be supported for life. Since a laggard slave had no fear of dismissal, slave labor was often inefficient. The long-time disadvantages of such labor constituted the underlying cause for the disappearance of slavery here as elsewhere in the North.

The lot of the slave in Hunterdon County was seldom a hard one.

5 C. S. Boyer, *Early Forges & Furnaces in New Jersey*, p. 82.

6 *Hunt. Rep.*, Jan. 27, 1870.

Personal relationships between the slave and his master's family were naturally very close. Only a few stories of extreme cruelty have come down to us, and many traditions bear out the supposition that the average slave was treated with kindness. Corporal punishment was administered on occasion, and a public whipping of disobedient slaves was sure to draw a crowd. But personal influence and lighter punishments seem to have been the rule. Perhaps from lack of incentive slaves in many cases did not work as hard as their masters. The kinds of work which they liked best were such pleasant tasks as going to market. They were very much in attendance at the many community frolics, where they were allowed to mingle and drink grog with their masters. No neighborhood dance was complete without two or three Negro fiddlers and a circle of Negro onlookers. The noisiest, most excited spectators on general training days and at the crossroad tavern horse races were the blacks. In many cases the servant was treated as a respected member of the family, and at death received a decent obituary notice in the local paper. Often these family retainers took the surnames of their masters, and persons of color bore such distinctive names as Ten Eyck, Van Derveer, Van Horn, Hoagland, and Apgar. The black slave in this area was not considered the equal of his white master, but he did hold an assured place in the social structure, and by his own standards had many of the things which go to make for happiness.

The feeling against Negroes as a race was not serious until the 1850's, and then largely as a result of the slavery issue in national politics. During the period of slavery the terms "wench," "Negro man," and "person of color," were used with none of the malice later attached to the word "nigger." In general, it seems that so long as blacks "knew their place" the relationships between the races were harmonious. No opprobrium was attached to Mrs. Mary Capner acting in her professional capacity as midwife for a slave woman in 1787, and Dr. Bowne treated black and white patients alike. Old stories report that Negroes often were made the butts of tavern jokes, but they evidently liked the attention and came back for more. However, the bitterness and strife of the 1850's resulted in a change of feeling on the part of the whites. Blackwell had been an indulgent and paternal master, and as late as 1856 reemployed a Negro couple who had been lured away a year previous by the pleasures of small-town life in Clinton. But his tolerance changed to disgust for all "niggers," especially after the start of the Civil War, which he regarded as purely and simply the result of abolition agitation.

Even during the period of general tolerance, however, Negro children, slave or free, who attended schools were exceptional. In 1817 when the Presbyterian minister and the local teacher jointly started a Sunday afternoon school for Negroes at Flemington, the local opposition was so strong and a minority faction of the school board so vociferous that they decided not to use the school building for colored scholars. When a Sunday school was organized the following year, a separate class was provided for Negro children. Census returns in 1850 showed that few blacks attended the schools. In Raritan Township, for example, with its Negro population of 117, not a single colored child went to school. The friction of the following years did not improve the situation.

Negro participation in religious services, on the other hand, was encouraged from an early day. The various New Jersey laws which forbade Negroes to congregate together specifically stated that there was no intention of preventing church attendance. Perhaps it was thought that religion made slaves more tractable. At any rate, no church group barred them from membership. At the first recorded Lutheran baptismal service, in 1714, one of the three children baptised was a Negro; and the sponsors were a freedman and his wife, both good Lutherans. The Quaker group at Quakertown was particularly friendly to blacks, and all of the Dutch Reformed churches welcomed Negro members. Incomplete records of the oldest one, that of Readington, give between 1720 and 1894 the names of twelve slaves and seventeen free Negroes who were members. The blacks at this church apparently of their own accord chose to sit together at the far end of the gallery. The Presbyterians, the most numerous sect of the county, also welcomed colored members. Converse, the historian of the United First Presbyterian Church of Amwell, reports that the revival of 1843 brought in twelve Negro converts. He writes, "The seats in the gallery to the right and near the pulpit were long appropriated by the colored people." Such segregation seems to have been usual, but was probably as satisfactory to the blacks as to the whites. After 1853 the clerk of the last named church stopped designating Negroes as such in the church records. The records of the earlier Baptist congregations are rather scanty, and no history of a Baptist church now within the county mentions Negroes. However, Hopewell Church, just outside the present county, had a large number of colored members. A diary entry of Peter Haward in 1811 casually remarks, "Heard a black man preach at the Baptist Church." The Baptists were doubtless as free in allowing Negro members as were other sects, and the same thing must be supposed for the Methodists. The largest Negro group of the county, that of Flemington, organized its own African Methodist Episcopal Church after the Civil War.

Perhaps the hardest feature of slavery was the fact that the slave was subject to sale. In 1773 at a vendue in Amwell Township "a likely Negro boy" was sold along with the stock, farm equipment, and household goods; and a will of 1777 provided that a Negro girl was to be sold and the proceeds given to the grandchildren of the testator.[7] In a single letter from Flemington in 1787 Mrs. Capner mentioned that a Mr. Williams had lately sold a twelve-year-old girl, that "Lawyer Smith" had paid one hundred pounds currency for a Negro man, and that a neighboring farmer had purchased a Negro farm hand. A scrapbook made for the anniversary exercises of the Readington Reformed Church in 1894 contains seven bills of sale for slaves sold between 1805 and 1825. The "slaves for life," all young, brought prices up to $300 each. A seven-year-old girl, who would be free at twenty-one, brought only $17, but a boy of fourteen, who would be free at twenty-five, brought $130. The new master of the boy was bound by contract to teach him to read and write. In the issue of the *Hunterdon Gazette* for September 1, 1825, two slaves were offered for sale. One, who was described as "a healthy young black woman . . . principally employed in cooking and housework," had three and one-half years yet to serve. The other, also a house worker, was a slave for life. On January 19, 1831, a man in Clinton advertised as follows: "For Sale the time of a Black Man Who has four Years to Serve. He has been accustomed to Farming." These examples are typical of slave traffic in its later days here. The last sale for which we have a record was one in 1836. The young Negro who was sold did not like his new master and soon ran away.

The sale of Negroes would have been far less tragic had family ties been preserved. Mention has been made of one case in which a family was offered for sale as a unit, but the evidence supplied by probate records indicates that a master seldom owned a whole family. In one case, for example, a man owned four Negro women and two children, but no menservants. At any rate the conditions were hardly conducive to morality. In the minutes of the Hopewell church was written the following terse notice of excommunication: "Nelly, a Black, for bearing a bastard child." The fact that so many mulattoes are mentioned in various records points to considerable miscegenation, both before and for two or three generations after the gaining of freedom. In 1831 a runaway

[7] *N.J. Arch.*, 1st ser., *XXXIX*, 37, *XXXIV*, 417.

Negro girl carried with her her thirteen-months-old mulatto child. In 1850 Blackwell recorded in his diary without comment that his Irish farm hand had left hurriedly for New York and that his colored servant girl Tamer had given birth to a child which was promptly named Florance Donavan. The local census takers mentioned many mulatto children in 1850. Between 1795 and 1848 only twenty-three marriages were recorded by the county clerk in which the parties were labeled as Negroes. Though the list is undoubtedly incomplete, the smallness of the number would seem to indicate that during that period the marriage ceremony was not considered very important for blacks, whether slave or free.

A slave who did not like his lot had the somewhat dubious choice of running away. His most serious problem was finding where he could go for safety. This fact probably explains why the advertisements for runaway slaves were fewer in number than those for indentured servants, particularly during the early period. The runaway mentioned in a 1749 advertisement was thought to have headed for Maryland. Another master remarked that his slave, a cooper by trade, might well pass as a free Negro. In one instance a slave named Brunce had been practicing up on the use of a new name, Tom, before leaving. Hopes of acquiring permanent freedom were naturally alluring. If worst came to worst, the fugitive could submit to capture and return to take his medicine. By the time of the first newspaper in Flemington, the *Hunterdon Gazette* in 1825, runaway slaves were fairly numerous. Each advertisement was headed with a neat little illustration of a Negro trudging along, all of his worldly goods wrapped in a handkerchief and carried at the end of a stick over his shoulder. It is not always clear whether the runaway was a slave for life, a slave till the age of majority by the law of 1804, or an apprentice bound out by the overseers of the poor. It is doubtful that in actual practice there was much distinction. In 1852 Blackwell spoke of "manumitting" his Negro bond servant. An example of a runaway "colored apprentice" occurred as late as 1867.

The slave advertisements throw some interesting sidelights on the period of slavery. In early days there were occasional "new Negroes" who could not speak English, and in one case a slave spoke "low Dutch," but only broken English. Many of the slaves bore the indelible marks left by smallpox. Most of them wore the usual linsey and tow clothing of the working classes, but an occasional runaway was decked out in fancy frock coat, swansdown vest, or other castoff clothing which he had been given. The advertisements seem to indicate that runaway slaves were not as light fingered as were fugitive white servants. Masters were not usually very concerned about their loss and sometimes were very tardy about advertising. Rewards offered were never very large in comparison to the market values of slaves. In early days the usual offer was twenty shillings and reasonable charges, though an occasional master offered more. One irate owner in 1765, who figured that his slave had been enticed away, offered three pounds for the return of the slave and an extra seven pounds for the conviction of the thief. In later days it was common to offer a nominal reward of a few cents. By such an advertisement the owner charged officers of the law with the task of apprehending the fugitive and avoided obligations which the runaway might incur in his name. Slaves were evidently not considered very valuable. A Lambertville story tells of the runaway who was surprised that his master had offered a reward for his return, and came home under his own power after escaping from his captors.

In many cases the owner deliberately ran down the character of the slave who had run away. This probably indicates that the basic cause of the effort to escape was a lack of adjustment in personal relations, though in some cases it may have been designed to discourage anyone from harboring the runaway. A Hunterdon County master in 1754 described his Negro as "about 30 years of age, near five feet high, has a flat nose, much pock-marked, a lover of white women, and a great smoker." [8] In the *Hunterdon Gazette* of January 19, 1826, an owner advertised for a young black as follows: "He has a down cast look, avoids looking a person in the face—and is very fond of liquor." The best example found of such defamation is an advertisement of 1838: "Said Catherine is a noted liar, and will steal everything she can lay her hands on; and to say the least, this is the fair side of her character." Those who emphasize the mutual benefits of slavery forget the cases where there was only antagonism between possessor and possessed.

On the other hand, it is to be doubted that the average Negro of Hunterdon County gained much by becoming a free man. His freedom did not bring equality, and seldom were his relations with the whites as happy as before. A striking example to the contrary was that of the Aree van Guinea or Aray family of Readington Township in early days. "Harry from Guinea" may have been an Arab or a Moor, as some of his neighbors thought, but more probably was "a very respected Negro," as has been said since. Provincial law to the contrary, he held considerable land. The first German Lutheran services in New Jersey are said to have

[8] *N.J. Arch.*, 1st ser., XIX, 383.

been conducted in his home. His daughter belonged to the Dutch Reformed Church at Readington, but the usual appellation, "colored," was not placed after her name. Tradition has it that the members of this family were on a social level with their neighbors, and, if this is true, the Arays had surmounted a very high barrier. Unfortunately for the social experiment, they moved from the county some years before the American Revolution. There are a few stories of free Negroes at a later date working out for themselves a not unhappy place in society and in the economic structure. A pleasing tale is that of the Negro lad who was freed at nineteen, "married a wench of unuzual [*sic*] respectability," became a farmer and a "worthy citizen," and "raised a family of respectable children." [9] The census taker in Lambertville in 1850 reported that one Negro had real estate valued at $400, but it is noticeable that none of his eight children attended school.

Unfortunately, few freedmen seemed able to hoist themselves by their bootstraps. A lucky group were those who had paternal guidance during their period of transition, and especially those who worked for their old masters or lived near them. Dr. Bowne retained his freedman, Daniel Williams, as a general handy man, and kept scrupulous accounts with him, even charging him for castoff clothing. Blackwell helped his servant girl and her husband to start up housekeeping and gave both employment on occasion. At Potterstown a colored woman named Mat lived with the Wyckoff family for ninety-nine years as slave and freedwoman. The following report probably shows a not unusual situation:

> The last slave in Hunterdon County is a mulatto, familiarly known as "Bob." . . . A home has been found for him during the past sixteen years in the family of Ingham Waterhouse, near Sandy Ridge, where he has been kindly treated and cared for.[10]

The extent of such paternalism is not known, and at any rate it was not a permanent solution. The drift of the free colored population of the county toward the villages has already been pointed out. The census of 1850 showed relatively few Negroes in truly rural districts, and those who remained were almost invariably farm hands. In the villages the blacks worked as day laborers in hotels, livery stables, taverns, and stores. Blackwell, impatiently waiting for his garden to be spaded on May 7, 1860, wrote, "All negroes are humbugs. No dependence to be placed upon them—." In Lambertville in 1850 all adult Negroes with occupations were classed as laborers, and most of them lived in a cluster of houses

[9] Larison, *Fisher Family,* pp. 60–61.
[10] *New-Brunswick Daily Fredonian,* Apr. 1, 1868.

in their own part of the village. The colored population was becoming a segregated group, seemingly doomed to a position of inferiority. Better times might be ahead for them, but the first generations out of bondage were not to be envied.

The first group in the county to oppose slavery wholeheartedly were the Friends of the Bethlehem Meeting at Quakertown, and even they on occasion had to recognize the institution, as when one of the members acted as executor for an estate which included three slaves in 1744. There are indications that these Quakers, either as a group or as individuals, helped several slaves to gain their freedom. A Negro by the name of Mingo Whano by some means had obtained his freedom, and in 1797 borrowed or was given by the Quakers £75 with which to free his wife and two children. Copies of the manumission certificate and the receipt given by the master were incorporated into the records of the Meeting. Six years later the manumission of a young Negro girl was recorded in like manner.

Although other groups were slow to take a stand, many individuals abhorred slavery at a fairly early date. When Moore Furman, the foremost citizen of the county, freed a Negro in 1784, he stated that he was "convinced of the iniquity and inhumanity of slavery and desirous of discouraging the same." [11] One of the Capners copied an antislavery poem or song into the family commonplace book, the first stanza of which read as follows:

> I [was] once posse[ss]or of my native plain
> Content I triumph'd indigent and free
> But wretched now I live a life of pain
> For white man's land is naught but misery.

Joseph Capner put his philosophy into practice by manumitting his mother's maid, Hannah, in 1799; but his nephew, Joseph Exton, was in 1850 the last slaveholder in Bethlehem Township. Another antislavery poem has been found in a commonplace book of the family of John Blackwell, long the county clerk of Hunterdon County. The attitude of such leading families reflected the feeling against slavery which brought the law of 1804. It must be understood, however, that the movement was local in its application. The slavery problem was settled sensibly in New Jersey by compromise between reformers and slave owners. There is no evidence in Hunterdon County, at any rate, that the wave of reform was tied up with the idea of forcible abolition elsewhere than in the state.

11 *Letters of Moore Furman,* pp. 9–10.

The movement for manumission was connected closely with Negro colonization, and especially so after the free Negro problem became pressing. Charles George, the able editor of the *Hunterdon Gazette* from 1825 to 1838, was a warm supporter of the New-Jersey Colonization Society and gave it a great deal of free publicity at all times. A typical article in one of his early issues portrayed the need of funds for use in sending to Liberia a shipload of free Negroes. He pointed out that the already numerous free blacks could never hope for "equal privilege" here, and that due to the lack of education and economic opportunity a very high proportion of them were in penal institutions. There was sufficient interest in the matter to warrant public lectures, and Rev. Charles Bartolette, the Baptist minister at Flemington and Sandy Ridge, was long one of the leading proponents of colonization for free Negroes. Rev. Peter Studdiford of Readington freed his own manservant and probably helped him get to Africa. At any rate, this Negro, John L. Stryker by name, became a prominent merchant in Liberia. Although the State Legislature gave money to the Colonization Society as late as 1852, the colonization movement lost its momentum during the 1840's. An interesting revival in new form occurred just before the Civil War. A group which included Thomas Bartow, editor of the *Hunterdon Republican,* favored purchase of lands in Central America and colonization of all Negroes there. Most people of the county probably recognized early that colonization on a large scale was unfeasible. As the number of Negroes decreased, the problem of free Negroes became of less consequence locally.

During most of the history of Hunterdon County hiring has been the most important method of securing labor other than the family. In early days, however, it was very difficult to find efficient hired help at a reasonable cost, which explains the use of supplementary types of labor. Farmers used indentured servants and slaves for the most part because of the free labor shortage. Artisans preferred journeymen when they could get good ones, but sometimes trained slaves as their helpers because of necessity. Apprentices, because of the cheapness of their services, were in demand, but too often were not efficient workers. Early industrial enterprises, such as the iron furnaces, hired free workmen for skilled work, but found it very difficult to hold them. For unskilled work they used slaves or indentured servants to a considerable extent. As the extreme scarcity of labor of pioneer days lessened and as the use of indentured servants and slaves for one reason or another declined, the use of hired labor increased.

The hired man and the housekeeper, working by the month or the year, became the most common types of employees. John Lambert, a miller and farmer, was in 1774 and 1775 regularly hiring three or four persons. His housekeeper was paid by the quarter. His hired men usually came and went, but a certain Jacob Butterfoss worked for him continuously during the period, at first contracting by the month, later by the year. Part of Lambert's record for him is quoted:

Jacob Butterfoss Begun his year April the 24th 1775.—
Jacob butterfoss lost 3 days 1/2
May the 27th 1725 1 day June 13th
July the 20th 1 day & 1/2
Augst. 21st lost 7 days
September the 4th lost 1 day
September the 4th to Cash paid Butterfoss 3:0:0
Octbr. the 7th to Cash paid Butterfoss 0:15:0
Octbr. the 7th lost 1 day. Octbr. the 30th lost 1 day and 1/2—
Octbr. the 30th 1775 Jacob Butterfoss to pay his tax £1:10:6
January the 16th to one pair of breeches 1:2:6
 also lent at the Weding 2/6 2:6
February the 15th Lost 8 days and 1/2 day a training
 to mending shoes 2/4 to mending shoes 2/ to makeing
  one pair of shoes 7/
 at Natl. morer for Butterfoss 3/6

Besides his regular hands, Lambert hired neighbors for occasional work. In 1775 he was paying harvest hands five shillings per day.

To both foreign travelers and local observers labor seemed scarce and expensive during the latter quarter of the eighteenth century. John Rutherford wrote in 1786 concerning West Jersey, "Labour is very high, as Day labour is from one-third to two thirds of a [Spanish] Dollar a Day, and from Sixty to Eighty Dollars by the Year." In 1787 Mrs. Capner wrote that her son had been paying a master carpenter 4*s*. 6*d*. per day and his helpers in proportion. This carpenter stayed "until the building was reared," but thereafter only rode by each morning on his way to another job, stopping long enough to give his men instructions. Laborers on the courthouse in 1791 were paid 4*s*. 3*d*. and 4*s*. 9*d*. and painters received 6*s*. In 1792 the Case tannery paid a laborer £36 per year in "hard" money. Beginning in 1793 Joseph Capner paid Mary Choyce 12 guineas per year and keep as a dairymaid. After 1795 the notation was changed and her salary, practically unchanged in value, was $60. In 1795 Thomas Capner wrote, "A labourer in this country gets from 3/ to 4/6 per day, a carpenter from 5 to 7/6, a millwright from 6/6 to 10/, a mason in the country 6/6 to 7/6." [12] Immigrants were often hired immediately on arrival, but it proved hard to keep them unless they were paid in accordance with existing wage scales.

After the turn of the century the labor situation became more stabilized. A more or less standard wage for many years was 3*s.* 6*d.* for ordinary work and 7*s.* 6*d.* for work in harvest, sawmill work, and other hard labor involving long days. Prices between these two levels were sometimes agreed upon for certain kinds of work. A first-class laborer received $120 per year and keep, and others were paid somewhat in proportion. There were slight changes from time to time, corresponding in general with the situation elsewhere. These fluctuations were especially marked during the second war with England. After 1815 it became customary to use American notation, and $1.00 per day in harvest and from 50¢ to 60¢ for winter work were more or less standard for a time. These wages tended to rise slightly in 1817 and 1818, then fell off for several years during the hard times after 1819. Wages for harvest hands and even for carpenters and masons fell to 75¢, while ordinary labor brought about 40¢. Wages by the year fell to $60 for top hands.

By about 1825, however, the wage scale had climbed back to its former levels, and maintained itself rather persistently for some time thereafter. There was a fluctuation upward during the prosperous mid-1830's and a downward one for some years after the Panic of 1837. In 1850 the local census takers reported $9.00 and $10.00 per month as average wages with board, 50¢ to 75¢ for day labor with board, and 81¢ to $1.00 for day labor without board. During the 1850's it became increasingly difficult to find laborers. Many native hired men had gone west and their place had been taken by Irish workers, untrained in American ways and too inclined to leave without notice. Wages rose rapidly during this period. Blackwell was sarcastic about the "modesty" of laborers who asked $2.00 per day for harvest in 1860.

Wages, hours, and working conditions of a neighborhood were usually fairly uniform. If serious variations developed the farmers might by group action make an effort to even things up. Usually an employer kept an account with each man over a period of time, debiting in addition to money advanced such items as house rent, produce, groceries bought, and the grazing of stock owned by the laborer. At the end of the contract or when either party chose to terminate the arrangement, there was a settlement. All time lost was charged against the hired man, usually on a

12 Letter to Mrs. Hugh Exton, Jan. 4, 1795, Capner Papers.

per diem basis. One hired man in 1827 was docked for time lost in bad weather, for time making garden and mowing hay for himself, for sickness, for attending his sister's funeral, for staying home with a sick child, for going hunting, for attending a frolic, for attending a special church meeting, and for a day's work for a neighbor. The following quotation of a notation scribbled on the flyleaf of a farm account book among the Case records illustrates how an employee who did not work "from sun to sun" might have his wages reduced: "By parts of a Day, Munday 23, Half past 7 in the morning Came, had Breakfast, and begent to mow Half past 8. [A]bout 5 in the afternoon Began to Grind Sith, had one of my boys to turn the Stone." Harvest hands could get in full days by hoeing corn or doing other work in the morning before the dew dried off. Certain kinds of labor were sometimes paid for on a piecework basis, as mowing hay and grain, cutting wood, and splitting rails. In 1817 Dr. Bowne paid 75¢ per cord for cutting wood and the same amount per hundred for splitting rails.

Some hired men stayed with employers for years or alternated among the farmers of a neighborhood. Apparently many of them were satisfied with their occupation, and passed it on to their sons. But others were wanderers, and when their contracts were up—or before—hunted for new surroundings. On April 12, 1853, Blackwell hired a man for a year at $10 per month. Later he wrote in red ink, "Is an unprincipled villen [*sic*], and went off without fulfilling his engagement." Some of his workers drank heavily, others "sloaped," and still others "fooled away their time" so that he "had to give them the sack." One temperamental individual became "miffed" and left. Like others, Blackwell tried Irish immigrants, and though more patient than many employers was often annoyed by their lack of knowledge of American ways of doing things. Once he wrote, "Scant of help—hands disappoint—bad—trying to learn an Irishman to mow—pretty hard work." Later he inserted, "He *came* it." One Irishman stayed for two years and was adjudged "a pretty hard hand." But several proved untrustworthy, and one hard-working Irishman left hurriedly, as noted above, after seducing Blackwell's Negro maid. A German couple, on the other hand, were very satisfactory, but left when their year was up. In harvest time it became increasingly difficult to hire hands, though wages were greatly increased. It was chiefly because of this situation that Blackwell bought labor-saving machinery.

Growing children and women who hired out were not in as advantageous a bargaining position as were adult agricultural laborers, but even their condition improved with time. As a general rule boys received about one half what men got for similar work. The Case tannery in 1829 paid the lads who worked there 25¢ and 31¢ per day. A standard wage at corn-dropping time was 25¢, though a boy who could "keep his heels out of the way" for two coverers might be paid a little more. Hired girls and housekeepers were seldom very well paid in early days either. In 1815 Dr. Bowne gave a girl 5*s*. 6*d*. per week and keep, or 10½¢ per day. In 1821 he paid 50¢ per week, and in 1829 raised his wage scale to 62½¢ for a woman who had been with him for several years. By 1850 wages had advanced so that the average girl received 75¢ or $1.00 per week. In the 1850's immigrant Irish and German girls were serving as housekeepers, chambermaids, and tavern helpers. Blackwell employed one or two on occasion but preferred native girls. In 1850 he paid $1.50 per week; in 1854, $3.00; in 1855, $5.00.

The hired helpers of artisans were known as journeymen. Generally speaking, their labor was considered worth less than that of their employers. In 1791 a painter employed on the new county courthouse was paid six shillings per day and his journeyman helper half that amount. The accounts of Richard Choyce's small wool carding shop at Flemington about 1810 show something of the difficulty of hiring and keeping good journeymen. After various experiences the employer was able to secure a man who was an excellent workman—when sober—and managed to keep him for several months. As was usual, journeymen were expected to do other work, farm labor in this case, when not kept busy at their trade. It was very important that an artisan secure skilled helpers. In 1829 Bray and Taylor of Clinton advertised in the *Hunterdon Gazette* that in their spinning, weaving, and fulling factory they had "men of good character and long experience." A Flemington blacksmith in that year hoped to employ a journeyman who understood "working Lehigh Coal." Coopers, wheelwrights, cabinet makers, tailors, bricklayers, and others advertised so often that it would seem that there was a perennial shortage of good workmen. When wages were mentioned, it was usual to promise "customary wages." A journeyman's agreement in 1850 specified that the employer was "to find a house and garden and pay the Sum of one hundred and Twelve dollars for one year's work at the Tanning and Curring [*sic*] business and work at the farming business if wanted." [13] The usual artisan's shop of that time had one, two, or three journeymen. The tailor who worked for William E. Anderson and Com-

13 Case tannery, day book, entry of Apr. 1, 1850.

pany, operators of the largest general store in Flemington, used a half dozen or more workers. A journeyman received from $14 to $30 per month in 1850 and a little more by 1860.

After 1860 the increasing industrialization of the eastern part of the United States had a marked effect on the farm labor situation. Even after the return of the soldiers farm labor was not very plentiful. Farm wages in Hunterdon County in 1868 ranged from $2.00 per day in greenbacks for harvest work to $1.00 for winter tasks such as butchering and sawing wood. Greenbacks at that time were worth about 72¢ in gold. By 1873, when they were worth about 90¢, West Amwell Township was paying its road workers $1.50 per day, and in 1876 the county paid $2.00. The scale of wages near Readington in 1876 was as follows: $125 per year and keep for a hired man; $1.00 per day for ordinary work, with more for harvest work; $1.50 per day for a mason or carpenter; 50¢ per day for a woman worker; $1.50 per week and keep for a hired girl; and from 50¢ to 60¢ per day for a boy dropping corn. Despite the rise in greenback values to par by 1879, money wages during the next decade did not change greatly, day laborers being paid $1.50 per day in 1883 and mechanics up to $2.25.

In the 1880's money wages tended to rise, though probably no faster than the cost of living. Elmer Bonnell paid varying wages to a number of persons in his employ at this time, the lowest amount paid to a man being $120 per year and keep and the smallest sum for a boy $4.00 per month. The following quotation of a diary entry of November 27, 1888, concerns his agreement with his best-paid man: "We are to pay him $250 in money, find him house and keep a cow. Pen for hog. 6 Bus. winter apples and peaches for family use. . . . Commence work April 1st, 1889." Though wages were greater than a few decades earlier, there had been no diminution in hours. During hay and harvest season, hired hands were roused at 5:30 for breakfast. They worked at making fence, hoeing corn, and other tasks until the dew was off, then in the hay fields or harvest fields until night. A short period was taken out for the noon meal and a shorter one for an afternoon lunch served on the job. When it was too dark to work any longer, they ate the heavy meal of the day and after smoking and conversing for a time were glad to retire for the night.

From figures given above it will be noticed that wages stayed up at a time when farm valuations were rapidly declining. However, as the situation became progressively worse during the 1890's, wage levels fell somewhat. In 1891 Bonnell was paying hands from $120 to $200 per year and $15 per month for extra labor during peach harvest. During the following winter he paid a man $13 per month. Thomas Lauderdale of West Amwell Township in 1893 paid $10 per month and keep for regular hands, 75¢ per day for day labor in winter, and $1.50 per day during harvest. During the winter of 1898 he was able to secure a man for a time at 65¢ per day. After the turn of the century the farm situation began to mend, and at the same time, because of the lure of industrial cities, labor became increasingly scarce. As a result wages began a slow but steady advance. A good farm hand in 1911 could expect $25 or $30 per month and board. In the period from 1912 to 1915 ordinary work paid $1.00 to $1.50 per day and harvest work $2.00 or more. Road labor in West Amwell Township in 1914 was paid 20¢ per hour.

One answer which farmers made to the labor shortage of the period of World War I was the purchase of machines. Nevertheless, as war industries and the army pulled more and more men away, the situation became critical at a time when the national government was urging an increase in production. Efforts to ease the situation by bringing in city workers and college students did not help much. Farmers complained that the former did not know enough and the latter too much. In 1916 laborers were paid $3.00 per day at silo filling time, and men hired by the month, $35. In 1918 monthly wages were about $50, and in 1920, $70. Day laborers during this period received from $4.00 to $5.00 per day. During the postwar slump of the early 1920's wages sagged for a time, then slowly climbed again. In the period of so-called prosperity which followed, war levels were nearly regained. In 1931 a survey of 116 farms in the Flemington area showed that one fifth of all farm costs were for labor. During the depression of the 1930's wages on the farms fell in general, but the trained farm hand, who today is something of a mechanic and skilled workman, did not suffer greatly. The unemployed of cities and towns who flooded the countryside were untrained and in a poor bargaining situation. As a result the range of wages paid to farm labor was probably greater than at any other time in the county's history. During World War II farm labor of all types again became scarce and expensive, as the armed forces and war industries depleted the usual sources of supply.

Other types of labor than farm labor have been of some consequence in this area. The roistering Irishmen of an early day and the Negro, Italian, and Hungarian workers of a later time dug the canals, laid the railroads, constructed the improved roads, and put in the pipes for water, oil, and sewage. Usually they have lived in shanties erected on the spot, stayed until the particular job was completed, then wandered away to other regions. The labor gangs who worked the stone quarries and iron mines stayed longer, though they, too, were generally itinerant laborers. Their wage levels fluctuated with the times and were never very high. Prallsville quarries paid $1.30 per day in 1873 and the iron mines less than $1.00 in 1884. During depressions wages were suddenly decreased, or, worse yet, ceased altogether as operations were suspended. For example, over one hundred stonecutters were thrown out of work when the Prallsville quarries closed down during the 1870's. When iron mining ceased in this area, several hundred men were forced to look elsewhere for jobs. During the two decades just past the largest construction gangs have been those which laid the concrete pavements. Wages of these workmen have varied with the fluctuations of the wider labor market.

The industrial development after the Civil War caused an increasing employment of laborers and skilled workers in factories. Wages were high during and immediately after the Civil War, but declined greatly during the Panic of 1873. The wage earners were in a poor bargaining position, and for several decades wages remained low. The factory which made picture frames and mirrors at Glen Gardner paid an average of less than a dollar per day, even after raising wages in 1880. The Taylor Iron and Steel Company at High Bridge paid an average of about $10.00 per week in the late 1890's, and somewhat less during the next decade. The Fulper Pottery paid a carpenter 25¢ per hour in 1904. Employers sometimes made up in part for lowness of wages by benevolence. In 1904 the Lambertville Rubber Company gave its 210 married workers a turkey apiece at Christmas, and other presents to its 240 unmarried workers. Yet two years later the officials of this corporation, when faced with a strike involving hours and wages, closed up their plant because of the "unwarranted interference." In 1914 the Fulper Pottery shut its doors in order to prevent unionization of its labor force, and other concerns in similar fashion usually fought off the demands of labor. The union activity of this period undoubtedly reflected a general dissatisfaction with wage scales, hours of labor, and working conditions.

During World War I scarcity of labor and the prosperity of business in general brought industrial wage levels up rapidly, though probably no faster than the scale of living costs. Constant wage increases and bonuses were given until the labor situation eased in 1918. Wages in factories were so high that contractors in Flemington had to pay carpenters, masons, and painters $4.00 per day in order to hold them. In the 1920's there was usually enough work for everyone, but during the depression of the 1930's many factories were closed down. Lambertville lost its rubber industry before the depression was really under way, and was badly hurt by that event and by the loss of other factories. Those which were able to keep going resisted New Deal legislation and avoided unionization. Many factories had their first taste of union activity at this time. As times improved in the late 1930's, wages increased, but at a slower pace than living costs. The workers were now in a better bargaining situation, and carried out a number of strikes, some of which were partially successful. During World War II wages have risen somewhat despite efforts to stabilize them, but probably have not kept pace with living costs in most cases.

Women workers except in home occupations were rather rare before 1850. Occasionally they helped in taverns and stores. A tobacconist's helper in Lambertville in 1850 received the stipend of $10 per month. As already pointed out, wages for household workers increased somewhat prior to 1860, but both farmers' daughters and immigrant girls were available in sufficient numbers during the remainder of the nineteenth century to prevent any large further increase. The attraction of the large cities in the twentieth century has taken away many of the young women, and those remaining behind have been in sufficient demand to command somewhat better salaries and working conditions, especially during times of general labor shortage. An increasing number of women have entered industry as time has passed. In 1860 a few hundred females were laboring in small factories making men's shirts and other men's wear, in millinery establishments, and at the cordage plant at Lambertville. The shirt factories, silk mills, textile plants, and rubber factories have used a large proportion of female workers since that time, and tomato canneries during their brief annual period of operation also used women. Wages were always low. The twenty-eight workers at the shirt factory at Clinton in 1898 averaged less than $4.00 per week. World War I increased the demand for women workers and brought up their wages somewhat. For instance, the Crescent Fur Company was paying $10.00 per week in 1919, though hours were very long. The depression beginning in 1929 naturally brought a decrease in wages, but the ground lost was more than regained in the latter part of the 1930's. Certain types of work in recent decades, as telephone service, beauty treatments, some kinds of clerical work, and stenography have been completely feminized.

Boy workers, too, have been usually at a disadvantage. Many continued to work on their fathers' farms, and some occasionally hired out to neighbors. In the 1880's Elmer Bonnell usually paid boys who worked in his peach orchards from $4.00 to $5.00 per month. Dropping corn and other kinds of light work remained boys' work until the coming of machines to replace them. The rubber goods, textile, and other factories also employed boys to some extent. Just before the panic of 1873 wages of $2.50 per week were paid in Lambertville. In general during this period boys have been at more of a disadvantage than they were under the apprentice system. Except during the war periods their wages have always been low. State legislation and federal laws until declared unconstitutional have at one time and another interfered with the employment of children, and the expanding of primary and secondary education has of course taken more and more young people out of both home and factory industry.

In both agriculture and industry in this region hiring became the chief method of securing labor. Nation-wide economic forces constantly operated to change wages and working conditions. Special factors were the pull of free fertile lands in the West, the attraction of the shorter hours and greater creature comforts of the cities, the coming of labor-saving machinery, and the operation of the economic cycle. The number of agricultural laborers in the county has fluctuated, and their wages have been influenced by the times. A comparative scarcity of cheap labor was one of the chief causes for the mechanization of agriculture. The farm laborer of today is a skilled worker, much in demand, and his real wages are probably higher than at any time in the county's history. The local industrial development was never great enough to create a demand for factory labor larger than the supply. As elsewhere where this was true, laborers were at a disadvantage as regarded wages and working conditions, and labor unions to fight the laborers' battle were slow in coming. In recent times, however, industrial laborers have been able to better their condition considerably.

CHAPTER XIII

Ways of Life

THUS FAR we have been much concerned as to the means by which the inhabitants of Hunterdon secured a livelihood. What that livelihood included is also of some interest. The standard of living of pioneer families in Hunterdon County was, as elsewhere, naturally low. The amount of household equipment which could be brought from older settlements by pack horse or by wagon was definitely limited. A log hut or cabin with earth or puncheon floors, gave shelter from the elements but was not a comfortable dwelling. A fireplace built at one end of the building must furnish warmth and a means of cooking. The light of day came dimly through a small window or the open door, and at night the only illumination besides that from the fireplace was given by a lard lamp or tallow candle. Tables and benches were constructed from rough hewn or sawed boards. The whole family often lived in one room, which gave privacy to no one; at night the parents slept on a platform bed in one corner of the room and the children on trundle beds or on cots in the attic. Bedding consisted of straw or feather ticks, linen sheets, woolen blankets, and down-filled comforters, all of home manufacture. Extra bedding and clothing and prized possessions were kept in a homemade chest. Dominant pieces of furniture were the spinning wheels used for the spinning of flax and wool and the loom for making cloth. The few iron tools and utensils were supplemented by rude wooden ones, "gums," gourd dippers, reed baskets, and other homemade equipment. The food was simple. Corn and other grains, wild fruits, and game were cooked in the fireplace by use of skillet, spit, and other devices. Clothing brought from older settlements had to last until flax and wool could be produced.

Since the settlements farther south in West Jersey were not far distant, primitive frontier ways were left behind fairly rapidly in Hunterdon County. Log cabins were improved or replaced by homes of frame, stone, or brick. Better furniture was made or purchased, tools brought in from outside, home manufactures built up, and a greater variety of food crops grown. As equipment improved, the standard of living advanced to the stage of that of older communities. John Reading, who died in 1717, possessed iron and brass utensils, good cutlery, plentiful woodenware, and many tools. Probate inventories show that his neighbors owned similar equipment, though not always in such quantities. Mortars and pestles lessened the work of preparing grain, then fell into disuse as mills were built. Churns, dough troughs, and fireplace ovens simplified the task of preparing food. Pewter, glass, and earthenware dishes gradually replaced those of wood. Bedsteads, chairs, tables of various types, cupboards, pictures, mirrors, and books began to appear among household furnishings.

As time went on, families of relative prosperity acquired possessions in harmony with those of the "better classes" elsewhere. Even the homes of ordinary citizens were furnished with an eye for show as well as utility. In 1788 Mrs. Capner thus described a neighbor's house near Flemington:

There is a bed, Mohogany [*sic*] Bedsteads, printed cotton curtains, mohogany chairs, Harvateen seats, mohogany chest of Drawers and a mohogany Dining table. . . . From this room you descend some Steps into a piazza, from thence you may descend to the ground either to the right or to the left. In this room hang to [*sic*] or three pieces framed with glass before them. Another of cut paper, another of Silk to represent artificial flowers are of Needle work. . . . In the window was a harp placed organically [*sic*] under the Sash for the wind or air to blow thro. As it blew Stronger or weaker the Sounds were so too.[1]

Not all homes were so carefully kept or so tastefully furnished. Cazenove wrote in 1794 of the "lack of neatness and furniture in the farmhouses" of nearby Morris County. According to tradition there were no carpets in the Milford region until the 1790's, when Thomas Lowrey used them in furnishing his new house. An absconding debtor in Readington Township in 1803 had only the following household goods: a table, six chairs, a chest, a desk, two beds and bedding, some iron pots, two trammels, a tub, and two pails. Utility was in most cases considered of primary importance. The following is a description of a typical home in the northern part of the county:

Single story house, big cellar, chimney in the middle, wide fire-place in the kitchen and stove room at the other end. A long pine table was fixed in one corner of the stove room, with benches on either side. On the upper floors, a portion of the grain was kept. Beds were generally filled with straw

[1] Mrs. Mary Capner to Mrs. Hugh Exton, May 18, 1788, Capner Papers. The harp referred to was an Aeolian harp, common at the time and later.

or chaff, with a fine feather bed for covering during winter. An indispensible appenage [*sic*] of the German's house was the piazza, in which the saddles, bridles, and sometimes plough and wagon harness were hung up.[2]

The advertisements of household equipment in the *Hunterdon Gazette,* which began circulation in 1825, show a multiplicity of "goods and gear" which would have been undreamed of a century earlier. Houses were larger and held more furniture, and the cabinet makers of the countryside were kept busy satisfying the demand. Bureaus, dressers, cupboards, stands, tables, fancy bedsteads, sofas, settees, washstands, desks, bookcases, clothespresses, patent churns, rocking chairs, stoves, mirrors, brass clocks, and carpets were among the contents of the ordinary home, in addition to the innumerable tools, receptacles, and containers so necessary in that day. The multiplication of the worldly possessions of the ordinary family and an increasing ornateness of design were national trends which were modified only toward the end of the nineteenth century and not very rapidly even then. An interesting development was the creation of the parlor as a sort of *sanctum sanctorum.* This room remained a very important part of every home until the twentieth century. A good description of the Hunterdon County parlor in the period just after the Civil War has been given by Stillwell:

> Those who could afford it had an ingrain or brussels carpet on the parlor floor. This room was usually furnished with a marble top table in the center, a haircloth settee and a few chairs covered with the same material. In one corner you would find a whatnot and on the mantel some candelabra with prisms. This room was seldom used except for weddings and funerals. If there were a young lady in the home it was used for the reception of her gentlemen callers. If there were several young ladies in the family they would either arrange for separate nights or each lady would reserve a corner for her particular use.

A number of developments in answer to household needs are worthy of mention. In the eighteenth century the use of whitewash on interior walls was common except in the houses of the wealthy. This practice continued until after the Civil War, but the custom of papering the walls and ceilings gradually became more and more common and, in the end, almost universal. Another change was that from wooden and earthenware receptacles and utensils to those of "Yankee" tinware, sold by peddlers and by village stores. Leaky pans and tubs were kept in repair by traveling tinkers. Later in the century the development of "agateware" or enamelware led to a partial replacement of tinware, and in

2 Traditions of Our Ancestors," *Hunt. Rep.,* July 7, 1870.

the twentieth century aluminum and steel utensils entered the scene. The prosaic but still much-used oil cloth came into general use in the 1840's. Venetian blinds were introduced before the Civil War, but never had a great vogue in this area. Various inventions during the past century have added to or brought changes in home furnishings. Washing machines, sewing machines, pianos, victrolas, refrigerators, and radios have been among the most revolutionary of these innovations.

One of the greatest changes has been in methods of heating. The cheery fireplace around which the whole family gathered on winter evenings had its advantages, but it was not efficient enough to heat a whole house and was a great consumer of fuel. It devoured "backlogs" eighteen inches or more in diameter, smaller "foresticks," and additional "longwood" with great rapidity. Small charcoal heaters were used as foot warmers in early days, but their effect was definitely only local, and charcoal pits were used rather ineffectively in public buildings. Because of the limitations of the fireplace and other devices mentioned above, the German method of heating by stoves was adopted. Already in the 1740's "ten-plate" stoves, so-called because composed of ten castings or plates, were being manufactured at the Durham Furnace and other ironmaking establishments. An inventory of the goods of an Amwell Township farmer in 1765 mentions "the Stove Room" of his house. A little later Franklin and other stoves developed in America came into common use. On November 17, 1787, Mrs. Capner wrote, "Bro. has bought a stove for his room, its not a ten plate Stove, neither is it a dutch Stove nor a Franklin Stove, it is a Rittenhouse Stove." The new stoves were usually supplementary to fireplaces until fairly late, but they were sometimes used alone, as at the Van Fleet School near Three Bridges, where a ten-plate stove was the only means of heating in 1813. Teacher and pupils often roasted potatoes and apples on its broad top.

The earlier stoves were huge and cumbersome and were eventually replaced by smaller models, which were more efficient for cooking. The old ten-platers gave way to stoves made from seven or nine plates. Hiram Deats often accepted the ten-platers in trade when selling his smaller models. For heating purposes stoves of the Franklin type were gradually replaced by "airtight" and other wood burners. The coming of cheap anthracite coal about 1840, however, led not only to the nearly complete abandonment of the fireplace but cut deeply into the use of all kinds of wood stoves. Coal-burning "parlor stoves" rapidly became the fashion. In 1843 Deats was selling a "very handsome metal and Sheet Iron Stove," and other makes were sold by stores under various fancy names.

The Stanley stove in 1847 was pictured in the *Hunterdon Gazette* as a box-shaped affair mounted on four stubby legs. A good parlor stove cost Blackwell $10 in 1853. Since that day stoves have changed in shape, but except for the modern development of the circulator have changed not at all in principle.

Anthracite coal from the Lehigh Valley could be bought in Delaware River towns or at stations on the Delaware Division Canal at from three to four dollars per ton in the 1840's. For many years Blackwell hauled his coal from Center Bridge and Frenchtown; but after the coming of the railroad to Flemington he usually bought it there, even though it cost half again as much. Residents of villages not on railroads often hired carters to bring their coal to them. Other types of fuel for heating purposes were also tried out at various times. For a while after gas came to the villages of the county, gas heaters were used, but they proved too expensive in operation to suit most people. A more important development was that of the coal-burning central heating plant. Steam heating became of importance soon after 1885 and hot water heating not much later. At first, steam and hot water systems were used only in public buildings, but gradually made their way into private homes. Hot air furnaces also came to have something of a vogue in this area. All three types are in use today, and the stove still holds an important place.

The open fireplace of early days was important for cooking as well as for heating purposes. Open fire cooking, an art not easily learned, has been described thus by a resident of a neighboring county:

> At this vital spot where the fire was kept burning nearly all the time, and where, because cooking food is shadowed by pot or pan the lard lamp hung on its prong from a crack in the lintel beam, the old cook learned how to manage a small fire rather than a large one, how to reduce it with water or increase it with bellows, how to keep two or three beds of cooking coals going at once, how to escape smoke, save ashes, bury and revive embers, rekindle from coals carried about in iron boxes or shovels, or light up from a tinder box.[3]

At the back of the fireplace was a lug pole or back bar which supported pothooks on chains of varying length known as trammels. The early lug poles were of green wood and slowly charred away so that they became dangerous if not replaced. At a later time iron was substituted. Another hanging device, replacing or supplementing the lug pole, was the crane. A movable, three-legged device for holding utensils in the fire was called a trivet. An old expression, "As right as a trivet," has come down to us

[3] H. C. Mercer, "Open Fire Cooking in Bucks County."

today. The Dutch oven, a deep kettle with three legs and a lid, was used for baking bread.

Other implements were the gridiron, the waffle iron, the toaster, and the coffee roaster. Skillets for frying had either legs or long handles. An iron device for making buckwheat cakes hung from a trammel and was known as a "lazy-back bake-iron." Brass and iron kettles of various sizes were among the family's most prized possessions. They were usually found hanging from the trammels in the fireplace. Meat was roasted by use of a twisted cord hung from the ceiling, a spit turned by hand or by a dog on a treadmill, or a "jack" operated either by the draft of the chimney or by a spring. "Tin ovens" or "tin kitchens," operated on a hand-turned spit, were sometimes used for baking. A portion of the cooking was transferred to the ten-plate stove when it came into use, but some open-fire cooking was retained into the nineteenth century. As cook stoves became lighter, cheaper, and more efficient, open-fire cooking, except for the making of apple butter, became a lost art; and fireplace utensils became valueless or nearly so. At Blackwell's vendue in 1858 a tin bake oven brought only six cents, a pot brought seven cents, a skillet two cents, and a miscellany including a gridiron eight cents.

For the large-scale baking of early days special stone bake ovens were usually constructed. Sometimes they were in one end of the kitchen, sometimes out of doors in a separate building. The making of bread was no light undertaking. Dough was mixed in the evening in the dough chest or dough tray, and left overnight to respond to the effects of home-made hop yeast. By morning the "sponge" had risen sufficiently to mold into loaves. The cook built a fire in the stone oven, preferably using hickory or chestnut wood to give the bread a flavor. Short pieces of wood such as broken rails are sometimes called "oven wood" to this day in rural areas. After the oven was sufficiently heated, the fire was removed and the bread put in by means of a special shovel. Pies, cakes, and other delicacies were also baked at the same time. In season fruit was dried in the oven after the baked goods were taken out. Although proponents of the old method argued long as to the excellence of this kind of baking, most of the ovens were in disuse by the time of the Civil War. Those within doors long remained as part of the houses into which they were built, but have for the most part been eliminated by remodeling during the present century.

The old ten-plate stove was used for cooking as well as for heating. From it evolved both the wood-burning and the coal-burning cook stoves. The latter came into use in the 1840's, but a popular prejudice in favor of cooking and especially of baking with wood held back their adoption. The wood-burning cook stove or "range" had a tremendous vogue, however, and particularly the excellent type made locally by Hiram Deats. In an advertisement in the *Hunterdon Democrat* in 1843 he stated that there were "more or less alterations in this business every year, part of which is more for speculation than for merit or value." The stove which he advertised at that time was a square box-like affair with pot lids at the top and with an oven underneath and to one side of the firebox. The Deats stoves had numerous rivals within the county. J. P. Bodine, a hardware dealer in Flemington in 1858, advertised as follows:

> If you want a Stove of all Stoves the best,
> Go and get "The Pride of the West,"
> Of all "Cooking Stoves" this is the Queen,
> Cheaper, more beautiful, there never was seen,
> Which to bake, and burn, is a regular "brick!"
> The fire box is large, takes a two foot stick,
> Buy one, and try it, if it don't suit you to a hair,
> Just take it back and your money is there.[4]

As time went on coal-burning cook stoves gained new users, and Blackwell in 1853 paid $32 for his first one. Bodine's store in 1858 was selling the William Penn, a coal-burning cook stove, and Stewart's Air Tight, a cook stove using either coal or wood. "Combination ranges" like the latter eventually became most common, and a great many are still being used. Some of those manufactured today are streamlined in order to look well in modern kitchens. With the coming of gas many people in villages turned to its use for cooking. A rather recent development has been the use of bottled gas for cooking purposes on farms and in villages where there are no gas mains. Kerosene has been used for cooking purposes almost since its discovery in 1859, and oil stoves were commonly advertised for sale in the 1870's. The harnessing of electricity brought electric cooking, and in recent times rural electrification has given country people of Hunterdon County an opportunity to use electric stoves.

The development of methods of kindling fires has its own history. The most common fire-making equipment until just over a century ago was the flint and steel. The tinder box in which these were kept also contained sulphur-tipped "spunks" and a linen cloth or other tinder. A fire was started by striking flint to steel and transferring the spark by

[4] *Hunt. Rep.*, Dec. 1, 1858.

means of tinder to the spunk, which in turn was used to ignite kindling. In an emergency a fire could be kindled by use of the flint lock gun. A little powder was put into the pan and ignited from the flint; the flame was then transferred to tow or other combustible material. By carefully covering a fire at night, it was usually possible to avoid trouble in rekindling. Neighbors living close together depended on one another in emergencies, and often did not even keep tinder boxes. Long after the custom of borrowing fire had disappeared, there persisted a local country saying, "Don't be in a hurry; you didn't come for fire, did you?" The invention of the friction match, which was in common use by 1835, brought at last an effective answer to the problem of fire making. These matches, at first called "lucifer," then "locofoco" matches, were very large. Eventually they were made smaller and put into boxes, and packaged matches became common items of trade. The unfortunate liking of mice for sulphur has resulted in the destruction of many homes by fire, and though the coming of the safety match provided a remedy, the old-style match has persisted.

Although we sometimes sigh for the simplicity of the past, most of us would hesitate about going back to old methods of illumination. The chief lighting device in Hunterdon County for over one half of its history was the tallow candle, made either by dipping wicks repeatedly into hot tallow or by use of molds made for the purpose. The former method required less equipment but was somewhat more laborious. Candlesticks were invariably part of the equipment of the Colonial home. Although most people made their own, tallow candles became an item of retail trade. Their price in Flemington in 1791 was a shilling per pound, exactly the same price as in 1858 in the same town. Since many farmers kept bees and beeswax was a common item of trade at country stores, it is probable that wax candles were used also. These were made by pressing wax around a wick by hand. The decidedly superior spermaceti candle was too expensive for general use until fairly late, but the factory-made paraffin candle largely displaced other types soon after the Civil War. Although candles were the chief illuminants until at least 1830, there was always a certain amount of competition from lamps. During the Colonial period the lard-burning betty lamp was most common. Later, brass, pewter, tinware, and glass lamps of various shapes were made, usually burning either lard or whale oil. A lard lamp manufactured by an artisan of Lambertville in 1843 was advertised as giving "a brilliant mellow light."

In the 1830's a new illuminant was developed. It was made by mixing turpentine and alcohol, and was sometimes known as "lamp oil" or "burning fluid," but more commonly called by its trade name of "camphene." Never very cheap, camphene sold for 44¢ per quart in 1849, and for 33¢ in 1855. The "fluid lamp" was never entirely trustworthy, as its contents sometimes exploded and caused serious fires. About 1860 kerosene made its appearance under the name of "coal oil." This highly volatile fluid, with its comparatively brighter light and cheaper cost, soon drove camphene from the field. But because of the high gasoline content of kerosene there were many accidents, often fatal, because of its use. A dangerous practice was that of carrying lamps from place to place. Nevertheless the new fluid rapidly became the chief means of lighting. Kerosene, lamps, lamp wicks, and chimneys or "globes" became important articles of everyday commerce. In 1864 the trustees of the Cherryville Baptist Church authorized the alteration of their camphene lamps so that they could use kerosene. Street lights of some of the villages of the county used kerosene, and the same was true of railway cars. In 1879 a newspaper correspondent wrote, "What a change would now be felt, were kerosene done away with, and we had to come back to tallow candles for our light." [5]

Since gas lighting came into common use at about the same time as kerosene, many people living in villages changed from candles to gas lights instead of to kerosene. A local group, of which John C. Hopewell was a leader, built a gas plant in Flemington in 1859. At once "gasaliers" and fittings were advertised, and people were soon trying out the results of "Hopewell's Folly." When the gas works were destroyed by fire in 1866 and again in 1877, they were loath to turn to "dingy, troublesome lamps" and glad in each case to have the gas plant rebuilt. Lambertville had gas as early as 1868, and other towns followed suit sooner or later. In the twentieth century the various plants have been purchased by the New Jersey Northern Gas Company. Use of gas was the most satisfactory way of lighting streets until the arc light, though both Flemington and Lambertville tried naphtha and gasoline lights at times because they were cheaper; and Bloomsbury was using kerosene lights yet in 1891. Ringoes had gas street lights as late as 1913; and even after the coming of electricity, some gas lights were used in houses and stores, chiefly because of the success of the von Welsbach mantle. This method of lighting gradually declined, however, and gas was eventually relegated to the field of cooking.

With the invention of electric lights came immediate agitation for

[5] *Hunt, Ind.*, clipping, 1879.

their adoption. Lambertville was talking of trying the new lights as early as 1881, but put up with gas lights until 1893. In that year arc lights were installed on its streets and incandescent lights in its homes. Clinton three years earlier had put in the first light plant in the county, but some of the citizens were dissatisfied with the incandescent lights on the streets, and in 1896 the town copied its neighbors by putting in arc lights, which in turn were discarded for new type incandescent lights in 1914. Flemington in 1894 put in forty-two arc lights of 1200 candle power each, besides some incandescent lights, to replace its ninety-four gas lights, which had been of only twelve candle power. The new lights were first used on January 1, 1895. There was great excitement, and citizens walked about for hours admiring the brilliance and proving to themselves that newspapers could be read on the streets. It was reported that one old lady shook her head and said, "Well, I don't know about this—turning night into day." [6] Frenchtown, after long agitation, secured electric lights in 1901. Changewater, by utilizing the dynamo installed in the snuff factory, already had electric lights in 1897.

During the next two decades the number of customers for electricity increased, but the small local power systems of that day had definite limits. Some additional plants were put in, as at Bloomsbury in 1917, Annandale and Three Bridges in 1920, and Quakertown in 1924. A plant installed at Milford was owned by the municipality, while that in the High Bridge area belonged to the Taylor Iron and Steel Company. Villagers and farmers not within reach of electricity sometimes put in acetylene lights, but the frequency of serious explosions acted as a deterring factor. One of the worst accidents killed ten people in the village of Lebanon in 1919. Others, both villagers and farmers, installed home electric plants, a procedure which called for a considerable outlay of capital.

The real extension of electric lighting came with the building of power lines and the consolidation of ownership under the New Jersey Power and Light Company. This company, which was incorporated in 1915, bought out the local concerns one at a time, and built lines to other villages. Most of its current now comes from the Gilbert Power Plant in Holland Township. In 1930 the Public Service Gas and Electric Company, which serves northeastern New Jersey, built steel towers and put in high tension wires from Philadelphia to their new switching station at Rosemont. The purpose of the line and station is to provide means for this company to lend current to or borrow current from the

[6] *Hunt. Co. Dem.*, Jan. 8, 1895.

CLOTHING!

All the latest and up-to-date Clothing at

LOVELL'S,

THE CLOTHIER.

Wear a smile and a Red Rover Hat.

Goods and prices all new.

FRED. E. LOVELL,

CLOTHIER.

The Spot to Buy. Opposite County Hotel.

NEVIUS BROTHERS!

SPRING CAPES and JACKETS!

EASTER

Is rapidly approaching in spite of snow drifts, and everybody will want

A NEW CAPE OR JACKET

for that occasion. ☞ We have all the

LATEST STYLES

in Capes and Jackets for Ladies, from $2.00 to $10.00.

☞ Misses' and Childrens'

REEFERS and JACKETS

in a great variety of styles from $1.25 to $7.50. You should see them before buying.

Respectfully,

Nevius Bros.

March 21, 1893-tf.

$20 Reward.

RAN AWAY from the subscriber on the 7th inst. a Negro man slave, named Harry, aged 25 years, 5 feet 3 inches high, very black, strong and well built: had on when he went away, a grey linsey sailor coat and pantaloons, wool hat and new shoes. He has a down cast look, and when addressed, avoids looking a person in the face—and is very fond of liquor. The above reward will be paid to any person who shall deliver said negro into the custody of the keeper of the jail at Flemington, New Jersey, sound and in health.

All persons are forbid harbouring said negro at their peril.

Tunis Quick.

January 19, 1826. 4

The Full-blooded

HORSE

POLYDORE,

WILL stand this season at the stable of the subscriber in Flemington, at SIX DOLLARS the season. Leaps and Insurances to be agreed for at the stand.

The pedigree and stock of Polydore are so well known in Hunterdon county, where he has stood for several seasons, that it is not deemed necessary to give them in detail.—His colts are more highly prized for Studs and Match horses than those of any other horse.

POLYDORE took the highest premium in 1822 at the New-Brunswick Cattle Show, and was supposed by the judges to be the best horse that had ever been exhibited on that ground; and is now in the prime of life. He was got by Old Nelson, and his dam by the Virginia horse Pantaloon.

P. C. REA.

Flemington, April 26, 1826. tf

FIG. XX. UPPER: The well-dressed young man and young lady in the Gay Nineties as shown in the advertisements in the *Democrat-Advertiser* and the *Hunterdon County Democrat*. LOWER: Common types of advertisements appearing in the *Hunterdon Gazette* in its early years.

Fig. XXI. Friends Meeting House, Quakertown, built in 1745.

Fig. XXII. Hunterdon County Court House, Flemington, in 1894.

New Jersey Power and Light Company, which serves all of northwestern New Jersey. There has been a rapid expansion of power lines to every village of the county and into the rural areas. In 1937 it was estimated that forty per cent of the farms of the county were electrified, and the New Jersey Power and Light Company now reports that the figure has risen to seventy-five per cent.

Another of life's problems, that of a water supply, was handled simply in early days. Pioneers used springs wherever possible, or dug shallow, stone-lined surface wells. Water was raised in heavy wooden iron-hooped buckets, often by means of a well sweep or a windlass. An occasional farmer had a windmill which operated a wooden pump made by laboriously boring lengthwise through oak or hickory timbers. Factory-made wooden pumps, known as "cucumber pumps," became common early in the nineteenth century, and by the 1830's were themselves being displaced by iron lift and force pumps. In the villages, too, the chief sources of water until a late date were individual wells. Citizens of Flemington tried an interesting plan in 1807, when they installed an aqueduct of bored logs to carry water from springs outside the town to the kitchens of the village. The system fell into disuse in the 1820's or 1830's, however, and the townspeople had to depend on wells and cisterns until 1859. In that year iron pipes were used to bring water into the town from springs two miles to the west. By 1880 it was necessary to increase the water supply, and pipes were laid to the South Branch, where a pumping station was installed.

Other towns of the county sooner or later acted in similar fashion to secure a water supply. Furthermore, the coming of electricity has made possible the installation of electric pumps in the homes of smaller villages and on farms. The custom of regular bathing, considered a fad by most people until a century ago, has been a factor in the situation. As people became more concerned with cleanliness, new equipment and a greater water supply were needed. A shower bath operated by a foot bellows was an invention of the 1840's. The editor of the *Hunterdon Gazette* soberly informed the public that he had improved his health and invigorated his mind and body by use of the device. As families brought running water into their homes, it became necessary to install bathroom and kitchen plumbing, and the county fairs gave prizes of various types of fixtures. As bathrooms became more common, the amount of plumbing supplies and fixtures handled by hardware stores increased greatly. The problem of sewage disposal for larger centers of population became a serious one, handled somewhat tardily. Flemington, not on a stream, was the first to solve the problem by installation of a sewage disposal works in 1898. Lambertville did nothing about the matter until a decade later.

A most interesting aspect of the life of a people is their food habits. In this matter the first settlers in West Jersey were forced to make many adjustments. Not many foodstuffs could be brought from the homeland, nor was there much exchange between settlements. In the 1730's the chief foods sold by Janeway and Broughton were salt, spices, molasses, and sugars. Peter Kalm listed in addition tea, coffee, and chocolate as items which were brought in from abroad in large quantities. By far the greater part of the food consumed, however, was produced locally. Kalm had little to say concerning the farm-grown grain and meat products which were staple foods, but he dwelt at some length, not always approvingly, on eating habits which seemed odd to him. The standard breakfast was buckwheat cakes or toasted bread, with butter and tea or coffee. Many foods were consumed raw, such as wild and tame fruits, nuts, cucumbers, and watermelons. Wild grapes were made into raisins, wine, and vinegar, and persimmons into brandy. Wild strawberries were eaten with either milk or wine and sugar. Fruits were also made into desserts of various kinds. Of apple dumpling Kalm reported, "It tastes quite good. You get as many dumplings as you have apples." The young leaves of many plants were used as greens, served with vinegar, and eaten with a knife. Pumpkins were eaten in many ways, sliced and roasted, boiled and served with meat, made into gruel and served with milk, mixed with meal and fried as pancakes, or made into puddings and pies. Dried pumpkin was a standby in the winter, as was sauerkraut, which was made from cabbage and preserved in large containers. The farm meat supply was supplemented by use of fish, squirrels, rabbits, whippoorwills, quail, and frogs' legs.

Although no records have been found which specifically describe the foods of early days in Hunterdon County, it is probable that Kalm's general description would have fitted that area. The ledger of a store, probably near Whitehouse in the early 1770's, mentions only a few foodstuffs: mutton, rice, ginger, allspice, tea, sugar, pork, veal, flour, butter, clams, wild pigeons, and molasses. With the coming of the Capners and their relatives our sources suddenly became abundant. These Englishmen were fond of food and were keen observers of American eating habits. John Hall once made the Capner's hired girl, who was serving the family with shad, weep by saying that "Americans were fit to live with the devil, they were so fond of pepper and salt." [7] On the whole, however, they liked American ways. Ann Capner, in a letter copied into her copy book, mentioned pickled pork, "hung" beef, bacon, and dried fruit, and observed, "The more we see of their ways the more we see it nessary [*sic*] to come into them." The following extracts from two of her mother's letters are particularly enlightening:

Buckwheat cakes are now in Season and are very good. [T]hey resemble pikelets. [N]ow they can be leavened in summer tho we eat them unleavened to breakfast all summer.—Since I have had workmen I see the Utility of pye or something of that sort to Breakfast. [W]hen the cake plate is empty of Buckwheat then they will take pye to their coffee. . . . [F]or supper we had Boiled beef, turnips, potatoes, pickles for sauce, fowls and Ham and rice pudding and apple pye.—I sat down with Mrs. White after and she eat pickles with fowl.—I saw our carpenters eat Bread and cheese with apple giblet pye.

Your letter dated September came to hand just as we was going to sett fourteen apple pyes into the oven with a batch of bread, two puddings and bit of veal. . . . [W]e as usual breakfasted on Buckwheat cakes and coffee. [W]e had three rockfish to dinner. Bread pudding with Stewed pares instead of raisings [raisins].[8]

In part of the first letter not quoted it was stated that a number of geese and four rabbits had been killed. Mrs. Capner figured that "these with a quarter of beef" would "make a pretty good larder for Christmas."

The quantity and type of food depended somewhat on who was to be served. The men at a "raising frolic" in 1787 ate a meal in which roast veal, beef, mutton, and potatoes were served with plenty of gravy. Five women who were waiting to usher a child into the world were served as follows: for breakfast, tea, bread and butter, cheese, pickled cucumbers, mince pie, and pumpkin pie; for dinner, roast beef, potatoes, cole slaw, and pickles; for supper, tea, very peppery hashed beef, cheese, honey, pickles, bread and butter, and two kinds of pie. Concerning slaw, Mrs. Capner wrote, "The weomen [*sic*] . . . seemed fond of the Slaw. I tasted it of Curiosity. It is raw cabbage cut as thin as possible, it looks like horse radish, then there is something warm poured upon it and peper [*sic*] is strewed on it." [9] For supper served to a visiting circuit judge in 1793, the Capners had tea, hot roast chicken, sliced cold beef, preserved quinces, bread, butter, and toast. They were amused that the

[7] John Hall diary entry, Apr. 17, 1787, quoted in *Hunt. Co. Dem.*, Jan. 21, 1913.
[8] Mrs. Mary Capner to Mrs. Hugh Exton, Dec. 22, 1787, Nov. 17, 1788, Capner Papers.
[9] *Ibid.*, Nov. 17, 1788.

visitor ate preserves with his fowl. Some of the above meals would have been considered too light for consumption by working men.

As a matter of course every family put up each fall huge quantities of pork and beef. On September 5, 1793, Mary Choyce wrote to her parents: "Our neighbor Case the tanner has just salted down upward of 17 hundred weight of pork and 8 hundred weight of beef for his family to live on this winter and next summer, which is the general custom here." Mutton and veal were also common foods, but generally eaten fresh; and poultry, prepared in a great variety of ways, was a common delicacy. Corn meal and wheat, rye, and buckwheat flour were used in a multitude of ways for cooking in both the Dutch oven and the large baking oven. Buckwheat and rye were used much more than today. Potatoes were by this time coming into considerable use and were stored for the winter along with apples, cabbage, and root crops. Apples and other fruits were consumed during the growing season and large quantities were dried for winter use. Mary Choyce once made the following observations:

> Apple pye and sauce on the table three times a day constantly. [T]hey eat apple sauce to almost everything here. [W]hen the green apples are gone we shall have dried apples to serve until cherrys are ripe and then we shall have Cherry pye till we are tired of it. Apples are [so] plentiful here that cousin Joseph says he has eat a waggon load this winter.[10]

To the staple fare secured from field, garden, orchard, and livestock were added various kinds of game—rabbits, squirrels, quail, pheasants, passenger pigeons, wild fowl, turtles, and freshwater fish of several kinds. Much use was made of wild fruits, plums, currants, mulberries, grapes, cherries, strawberries, raspberries, blackberries, huckleberries, bilberries, haws, persimmons, and such nuts as walnuts, butternuts, hickory nuts, and hazelnuts. Honey from both domesticated and wild bees was a tasty addition to any meal. Likewise there were a few imported items secured through local stores: tea, coffee, cake chocolate, salt, rice, molasses, brown sugar, lump sugar, raisins, and pepper. Clams, oysters, and fish were commonly bought from peddlers, but in Flemington there was a clam-buying club whose members took turns going to New Brunswick. The man making the trip received for his trouble a few hundred clams which he might sell to nonmembers. In July, 1790, the Capners purchased two hundred from a member of the club who had just carted twenty-nine hundred to Flemington.

Various descriptions of food during the next half century show no

[10] Mary Choyce to Mr. and Mrs. James Choyce, July 1, 1793, *ibid.*

fundamental changes in diet. Pork, beef, and mutton were still consumed in large quantities, winter and summer, usually with plentiful amounts of gravy. Rye bread remained a standby in summer, buckwheat cakes the usual breakfast in winter, and wheat bread something of a luxury at all times. Corn meal was used not only for corn bread, but also for mush, Indian cakes, Indian dumplings, and Johnny cake. Rye flour provided not only bread, but was mixed with wheat flour for making cakes, tarts, doughnuts, and other pastries which were eaten so plentifully by all. Fruit, pumpkin, and mince pies were always popular, and one of the greatest delicacies was rice pudding. Most breads and pastries were made at home, but bakeries began to appear in the larger towns by the 1830's. Crackers, used chiefly to eat with cheese, were introduced through the stores. Other innovations were ice cream, various confections, and peanuts. The same vegetables were retained and a few new ones added, most important of which was the tomato. Pumpkins were still dried for winter use, and cabbage, beets, carrots, potatoes, and apples were stored in pits. More potatoes were used than formerly, and many families now served them hot or cold three times a day. Home-grown beans were often baked in the Dutch oven.

The farm meat supply of this period was supplemented by a somewhat lessened use of wild game, and by purchases of fresh fish, clams, and oysters from hucksters and of salt mackerel, shad, halibut, and codfish from village stores. Salt fish had become a standard part of the family breakfast during summer and autumn. Fresh and dried fruits continued in use, and a welcome addition was the prune. Fruits were still used in making sweets and preserves, and apple butter made in the butchering kettle was an old standby. Oranges and lemons could be bought in the stores at times as early as the 1820's. The same sweetening elements were used as formerly, and a new one, white sugar, was gaining popularity. The consumption of spices—ginger, cinnamon, nutmeg, allspice, Cayenne pepper—remained large, as was the case with vinegar. Not all the dishes which were eaten sound appetizing today. Indian dumplings were made by boiling corn meal dough in the broth from cured pork, and were served with a dip made by mixing sour milk, sweet cream, and molasses. "Prenta Mella soup" was made by boiling fat pork and scorched rye, and was served with cold mush or cold buckwheat cakes. Leftover pork and potato stew was eaten cold with vinegar. Tea was still a common drink, but coffee was becoming more and more popular. It was usually roasted and ground in the home. A drink made from parched rye and other grain was sometimes used as a substitute. Soft drinks, such as root beer and birch beer, were becoming common beverages during the summer.

After 1850 came many changes in traditional food habits, and the gradual building up of the customs which hold today. One of the most startling changes was due to new types of food preservation. "Self-sealing" glass cans for home use were advertised in the mid-fifties, and within a decade were in great demand. Pint jars cost $2.50 per dozen and quarts $3.00 in 1865, and prices remained high until the original patents ran out. During the same period a market grew up, despite fears of poisoning, for foods canned in tin. In 1867 Hoagland and Company of Flemington advertised canned peaches, pineapple, blackberries, wortleberries, green corn, beans, peas, asparagus, tomatoes, spiced oysters, clams, canned lobsters, sardines, and salmon. Because of the increased use of canned goods, the drying of fruit and the making of old-fashioned fruit preserves and apple butter declined. As methods of commercial canning have improved, canned foods have become a usual part of the diet, so much so that their rationing in World War II has seemed a hardship. Home canning remained important, however, especially on the farms, and received a considerable impetus when commercially canned goods were rationed.

A change already alluded to in another connection was the decline in the home butchering and curing of meat late in the nineteenth century. A similar decline occurred in connection with home baking. For several decades bread, pies, and pastries formerly made in the old stone ovens were baked in kitchen ranges, as is occasionally done yet today; but as the number of bakeries grew apace, first villagers and then country people turned to the use of bread and other baked goods made by factory methods. Another gradual shift was that toward the eating of processed cereals. Stillwell, speaking of the year 1867, says, "When he first came to Readington the congregation were much interested in the fact that the new minister ate oats." A little later the use of other breakfast cereals under various fancy names presaged a trend which was to become universal. Other innovations were frozen dairy foods and cold drinks of various kinds, which were sold at "ice cream parlors," drug stores, and restaurants. The "oyster saloon," once so popular, was gradually driven from the field. Ice cream freezers holding from five to six quarts came into common home use by 1890.

Changes in methods of refrigeration have had a great influence also. During the greater part of the history of the county the usual method of keeping food from spoiling quickly during warm portions of the year was to put it in a spring house. Dairy products, eggs, meat, and fruit were kept in this small stone structure and cooled by water from a spring or a stream. Foods which spoiled most easily were often put into pails and let down into a well, and game was sometimes put there for three or four days to improve it as to flavor and tenderness. Storekeepers and farmers often harvested ice during the winter and packed it in sawdust or swamp hay to prevent melting. During the summer they sold to their neighbors the surpluses which they themselves did not need. The icebox became a common household convenience, until superseded in the present century by the electric refrigerator. Ice harvesting was hurt first by the competition of commercially made ice, then by the coming of the refrigerator, but only in recent years has it been entirely discontinued. The ability to keep foods has had a considerable influence on both eating and buying habits. One of the latest changes has been the quick freezing of meats, vegetables, and fruits. A freezing locker plant installed at Flemington in 1942 operates to capacity, and many farmers have bought freezing units for their own homes.

Every change in eating habits has had its bitter opponents. The whole gradual process has, for the most part, been away from the old self-contained home unit. One proponent of the old ways of doing things wrote as follows in 1921: "So they bought hams, many of which were painted over with a rank liquid in place of the old smoking, and pork took its place in the country store with pasteboard boxes of breakfast food, canned goods and evaporated fruit, baker's bread, factory made pies and cakes, coffee ground and bagged for months, mincemeat by the bucket and lard by the pound . . . so that a farmer today would starve to death in a blizzard if shut off from the country store in about two weeks." [11] Home canning and storing of food has continued, and at the present writing has been greatly stimulated by war conditions. As a result of this and the coming of quick-freezing equipment, many farm homes today are more self-contained than they have been for several decades.

The immigrants coming to the new land brought with them the clothing of their old homes. But frontier conditions and lack of touch with home soon caused divergences. Only in a few cases, such as that of the sunbonnet brought from England, did Old World dress survive in original form. Standard dress clothes for the male during the Colonial period were breeches fastened below the knee with silver or brass buckles, woolen stockings held up by fancy garters, cowhide shoes, linen shirt and

[11] F. K. Swain, "Passing Events."

stock, a "west" or waistcoat with brass buttons and skirts half way to the knee, a wig, a broad-brimmed beaver or woolen felt hat, and a greatcoat for use in winter. The women wore short gowns with petticoats, linen stockings, heavy shoes, hoods or sunbonnets according to season, and light or heavy coats.

Much of the cloth for dress clothes was imported from England, the favorite kinds during Colonial days going by the names of broadcloth, ozenbrig, buckram, calleminco, tamey, shaloon, drugget, camblet, calico, and muslin. Work clothes, on the other hand, were nearly always made from native cloth, most of it spun, woven, and dyed in the home from home-grown wool and linen. Linsey-woolsey, often used for outside clothes, had a warp of linen and a woof of wool. Work trousers and skirts were often made from tow, the heavier fibers of the flax plant. It has been said that a good pair of tow pants were able to stand without their owner being in them. Deerskin was made into jackets and trousers, and leather was used for making cowhide boots, shoes, and heavy work clothes. Furs and skins of wild animals were manufactured into caps, coats, gloves, mufflers, and moccasins. At his death in 1740 Adrian Hogeland of Amwell Township possessed the following apparel:

4 pair Leather Britches; 1 Light Colour'd Broad Cloth Coat; Dark Colour'd West and Britches; 1 Black West and Britches of Mixt Duroy; 2 Coat, 2 West home made Drugget; 1 Worstead Coat, 1 Linsey West Gray; 1 Old Duroy West; 2 Great Coats; 2 Strip'd Wests, 1 pair Britches; 2 pair Linnen Britches, 1 pair Trowsers; 7 Shirts; 2 Hats; 9 pair Stockings; 3 pair Mittings; 3 pair Leather Gloves; 3 pair Boots, 1 pair Shoes; Purse.

The ladies of the Capner household were surprised at but not critical of the difference between American and English dress. Hunterdon County natives seldom wore pattens or clogs for protection from the mud. Babies did not wear caps, children and even adults went barefoot in summer, and men sometimes had long hair which they braided into pigtails. Mrs. Capner noticed that the term "child in arms" was not used here, and commented, "Nor are they much in arms except when they suck or are dressed or clouted. They weare clouts and long petticoats nearly till they can walk alone in this Neighborhood except hot weather when they have no petticoats on." [12] Of ladies' hats her daughter wrote, "I hear my mother say those hats made without crowns was called Hives and was to let the Toppins of their caps be upon the outside that they might not be crushed. . . . [D]o but see how proud such a one is, she

[12] Mrs. Mary Capner to Mrs. Hugh Exton, Dec. 5, 1790, Capner Papers.

has wood heels to her shoes and goes without a stay in her cap." [13] She also commented upon the American habit of wearing aprons.

Soon after arriving at Flemington, Mary Choyce wrote: "In summer the women go without stays and wear only one peticoat [*sic*] and a short gown." From modesty she disliked the "short" and "long short gowns," but a Quaker neighbor took her in hand, saying, "My dear, thee must make thee some short gowns. Thee wilt find them much cooler and better. Besides, thee wilt save stuff. The quantity of cotton that makes one of these would almost make two short gowns and be much more convenient too." [14] Mary was soon convinced, especially after warm weather came, and remade some of her long gowns into short ones.

A Yorkshireman named Bachus, tarrying for a time at the Capners, said, "The beggars in this country ride on horseback and the farmers' wives go barelegged." [15] But in general the newcomers approved of American clothes as being sensible and even in good taste. For instance, Isaac Passand thought the custom of wearing flannel shirts in winter a good one. On May 18, 1788, Mrs. Capner described a neighbor woman thus: "She had now Done her work and sat Sewing in a clean flowered short gown, camblet coat, white Handkerchief, gauze cap, cotton Stockings, and Stuff shoes."

The day books of the general stores of the latter part of the eighteenth century show that there was justification for much of the criticism of American buying of foreign goods. By that time the number of types of English cottons and woolens had multiplied enormously, and French laces, Spanish silks, and other luxury goods had good markets. Clothes of that period were often gay and sometimes gaudy. Lists of clothing stolen by runaway slaves and apprentices in the last years of the eighteenth century included a broadcloth coat with yellow buttons, light-colored pantaloons and vests, nankeen and brown broadcloth coats and trousers, velvet vests, a blue sailor's jacket, green short coats, cross-barred and striped jackets and trousers, a deep blue surtout, castor and rowen hats, silver-buckled shoes, and red silk handkerchiefs. Peg, a runaway slave in 1799, stole from her mistress a calico short gown, a calico petticoat, and a durant skirt. The emphasis on "suitable clothing" was strong. Two men excommunicated from Hopewell Church for nonattendance in 1794 excused themselves on the grounds that their clothes were not good enough for such a solemn occasion as a church service.

13 Ann Capner to Mrs. Hugh Exton, Dec. 5, 1790, *ibid.*

14 Mary Choyce to Mrs. Hugh Exton, Aug. 6, 1793, *ibid.*

15 Mrs. Mary Capner to Mrs. Hugh Exton, Feb. 7, 1791, *ibid.*

During the early decades of the next century the tendency toward the buying of manufactured cloth and of clothing made outside the home continued. Tough homespun work clothes were the last to be displaced; and even in the 1830's and 1840's tow pants, linsey-woolsey jackets, and linen shirts were still found along with new types of cotton clothing. Men's dress clothes, gradually becoming like those of today—coat, vest, and trousers—were made by this time not only from manufactured cloth but were usually cut and assembled by professional tailors. In 1828 a Flemington shoemaker who alarmed his family by staying too long in New York wore a "brown coat, mixt pantaloons, [and] a light thin vest with a dark striped one over it." [16] Patent leather shoes became popular in the 1840's, and by the end of the decade "gum shoes" were made to protect them from the mud.

As for women, they were less inclined than ever to wear sober clothing. Among articles sold at a store in the Oldwick area in 1815 were "Yankee buttons," scarlet cloth, and pink ribbons. Cloth and finery of all kinds were usually available at the nearest general store. A Flemington firm advertised as follows in the *Hunterdon Gazette* of October 3, 1827:

> Veils, Laces, Ribbons, Bobinet,
> Fine Levantines and Gros des'ete;
> Dress handkerchieves of every kind
> That could be thought, to suit the mind;
> Chintzes rich, of fashion new,
> Of oddest style, of every hue;
> Gloves, stockings, belts, morocco shoes,
> And almost everything you choose.

On May 20, 1846, L. M. Prevost and Son of Frenchtown made an effort to catch the feminine eye by advertising thus in the same newspaper:

> For the Ladies—A choice assortment of calicoes, rich patterns, lawns of various style, Organdy plaid, ginghams, fancy mous de laines, lustres, Swiss and mull muslins, bishop's lawn, cambrics, India book, fancy bonnet and cap ribbons, Opera ties, fancy cravats, kid gloves, laces and edgings of almost every description, light gaiters, kid slippers, &c., mode col'd alpacas balzarines, light plaid and striped gros de Naples, fancy woolen carpetings, &c.

During these decades women became increasingly style conscious, and as a result there were some startling changes in female clothing. Godey's *Philadelphia Monthly Report of Ladies' Fashions* and other fashion magazines had their devoted followers in Hunterdon County. Another

[16] *Hunt. Gaz.*, Sept. 24, 1828.

mark of the times was the large number of well-patronized millinery establishments. An example of a style fad taking everybody by storm was the prevalence of puffed sleeves in the 1830's and of "jubilee skirts" in the 1840's. Perhaps the height of the ridiculous was reached with the coming of hoop skirts in the 1850's. On January 2, 1856, the editor of the *Hunterdon Gazette* wrathfully wrote:

Do any of our lady readers wear hoops? We hope not, for we believe no sensible woman would be so stupidly silly. Men are not such fools that they cannot see the effort made by anseriferous [anserous?] females to "stick out," and many a harty [*sic*] laugh do they have over the unnatural circumference achieved by cooperized women. . . . We would a thousand times rather see a Bloomer dress, much as we detest them. When will our females think and act for themselves despite the dictates of fashion?

He might well have spared his choler. Further remarks which he made on the subject were answered in a rival newspaper by a lady calling herself "Hoops," who angrily accused him of not minding his own business and of possessing an immodest curiosity. The *Gazette's* stand was said to have cost it a number of subscriptions. The fad increased until hoops for dressmaking and ready-made skirts with built-in hoops and springs were carried by every general store. The hoop skirt eventually went out of fashion about 1868, and its return in modified form in the 1880's was brief.

The long period of the supremacy of the hoop skirt was followed by that of the bustle, which in turn gave way to the skirt with a train, "which gathered up a good supply of dust and dirt, but made no apparent difference in the amount left on the street." [17] Along with these extremes in garments went other fads. For example, false hair was used a great deal. Just before 1870 the type known as the "waterfall" was in vogue, and later varieties were the switch, the "poodle friz," and a kind which covered the forehead. Queer hats, the "Grecian bend," tightly laced corsets, the use of powder and even paint, and extreme fashions in dress led to masculine taunts and to sermons on "absurd and extravagant fashions." Naturally, since women dress chiefly for the effect on other women, these things made little difference. To us today nothing seems more ridiculous than the wasp-waisted, flare-sleeved, pointed-lapeled, ankle-length clothes of the 1890's, the tight skirts of 1913 and 1914, or the short dresses, rolled-down stockings, bobbed hair, and ostentatious cigarettes of the "flapper" period from 1918 to 1922. Naturally, every-

[17] Stillwell, *op. cit.*, p. 19.

day clothes and the dress clothes of many women did not go to the extremes of fashion indicated above.

Masculine costumes did not keep pace with feminine ones, but there were, of course, fashion changes at times. A description of the fop of 1869 is very amusing:

> We like to see a young man with a choice cigar under his rapidly developing moustache, a closely fitting pair of pants, and a pair of boots so tight that his step resembles that of a long-legged Shanghai rooster in his attempts to straddle through a pile of brush. He shows taste and an aspiration for observation.[18]

Another description, dealing with the same period, shows a more moderate picture for the average citizen:

> As for clothing, men dressed pretty much as they do today. The negligee has completely supplanted the dicky and the stiff bosom shirt, and there is some difference in the style of the outer garments, but the changes have not been very great.[19]

Men had their fads in dress, too, such as the very tall shirt collars of the 1870's. In the 1880's the well-dressed man wore pointed button shoes, bags or wide pants, a white vest, and perhaps a cutaway coat, a Norfolk jacket, or a linen duster. But most men were too conservative to adopt changes in clothing quickly, and few of them accepted extreme changes at any time. The dress suit, often called "store clothes," gradually evolved into that of today. Sport clothes made more changes, but concerned very few men. For several decades after the Civil War a very important part of a man's costume was the covering of hair on his face. Mustaches of various types became stylish in the 1840's and remained so until about 1890. In the 1870's "mutton chops" vied with cutaway whiskers and chin whiskers as to popularity. Beard styles changed from time to time, and proponents of the clean face gained in numbers toward the end of the century.

A very significant development was that of ready-made clothes. A Flemington store in 1856 carried all types, and another in 1859 hired workmen to make clothing for sale. The cheapness of the new clothes soon restricted the custom work of tailors. In 1867 all-wool suits cost $12, and a good overcoat cost even less. The coming of the sewing machine not only simplified the making of clothes in the home, but also led to a greater and greater use of factory-made clothing of all types.

[18] *Hunt. Co. Dem.*, Jan. 13, 1869.
[19] Stillwell, *op. cit.*, p. 18.

The changes in the ways of living in Hunterdon County during the past two and a half centuries have been considerable, as has been the case in most other parts of the United States. Standards of comfort have changed, and many necessities of today outstrip the luxuries of an earlier period. Laborious, monotonous, and time-consuming tasks on the farm and in the home are now done by machine, and both the productivity and the leisure time of the average person have been greatly increased. Great changes have been made in the construction and furnishing of homes and in methods of heating, cooking, lighting, water supply, and sewage disposal. Food habits have been modified, sometimes for the better and sometimes for the worse nutritionally, by new types of food preservation, refrigeration, and processing. The evolution in clothing is even more pronounced, and it has been a long step from the homemade linen, woolen, and linsey-woolsey clothing of an earlier day to the machine-made dress and work clothes of today. If a farmer of the Colonial period or one of a century ago were permitted to return to this earth, he would be amazed at the material progress since his day and would find it hard to adjust himself to the machinery and gadgets and to the hustling and bustling which accompanies all activity today.

CHAPTER XIV

A New Day

SINCE AGRICULTURE is the chief occupation of the people of Hunterdon County, it may be well in closing to look more closely at some of the recent developments in that field and particularly at the forces which have made for change. Like most of the United States, this was a general farming area during the greater part of its history. Each farm had a variety of crops and domestic animals which made it largely self-sufficient, yet produced enough surplus and cash crops to give its owner purchasing power to buy the few things which he did not produce. As the nineteenth century drew to a close, the number of needs and wants multiplied, and home processing and domestic manufactures declined in competition with the growing factory system. A greater production of cash crops became necessary, and the agriculturists of the area groped about, searching for staples. The production of wheat, potatoes, beef, cattle, pork, and fruits were at one time or another expanded; but sooner or later the production of each declined because of inability to compete successfully with other regions. By 1900 the agriculture of Hunterdon County, like that of much of the East, was in a desperate plight. The day was saved when the lead of farmers located in regions nearer to large cities, who were already capitalizing on the enormous demand for dairy, poultry, and vegetable products, was followed. Increasing specialization in dairy cattle, poultry, and vegetables has been accompanied by thorough commercialization, and the welfare of the agriculture of the region has become dependent on continued markets with favorable price levels.

Paralleling these changes, there have been a gradual divorcement of industry from agriculture and a development of transportation, which have made possible the carrying of farm products rapidly to their ultimate markets. Under such conditions the development of efficient marketing devices has been of great importance. As already noted, there has been a decided trend toward the elimination of middlemen through the creation of farmers' cooperatives. The nine or ten "fruit exchanges" which were started as cooperative auction markets during the 1880's and 1890's were the first examples of this. The Dairymen's League, the

United Milk Producers' Association, the Interstate Milk Producers' Association, and the Consumers' Cooperative Association have acted as marketing cooperatives for milk. The Flemington Cooperative Auction Market has since 1930 brought dealers to Hunterdon County to bid against one another in buying eggs, live poultry, beef animals, and swine.

There has also been some development of cooperative buying, particularly of farm supplies. Such buying was tried by the local Granges of the county in the 1870's and occasionally later, by Farmers' Alliance groups in the 1890's, and by other farm organizations at one time or another; but in no case did it last very long. In 1921 a cooperative store was started by the Hunterdon County Farmers' Cooperative Association, which had the moral backing of the Hunterdon County Board of Agriculture. In 1928 the Association was bought out by the Cooperative Grange League Federation Exchange, commonly known as the "G.L.F." The G.L.F. is a large cooperative whose operations cover much of the Middle Atlantic area. Its operations in New Jersey are for the most part in the wholesale field, and the savings returned to customers each year are the result of large-scale purchasing and processing. In Hunterdon County, however, there are the retail outlet at Flemington already mentioned and a sub-store at Clinton. The member-consumers patronizing these two stores totaled approximately 2400 in 1944, and of these 922 did more than one hundred dollars worth of business each.[1]

A newer venture in cooperation is the Delaware Valley Cooperative organized at Frenchtown in 1938. This is a cooperative farm supply store owned by some two hundred members of the Farmers' Union located in Kingwood and adjoining townships. Except for the organizations mentioned cooperative purchasing has not progressed very rapidly as yet. However, the trend toward both cooperative buying and selling is noticeable. The agriculture of this and neighboring regions is faced with serious competition from lower-cost sections of the United States suddenly brought nearer by rapid transportation. Its success depends in part on the development of the most efficient marketing practices, so that farm products are delivered to ultimate consumers with a minimum of expense and labor, and on the working out of wholesale and retail practices which will lower the costs of farm supplies and perhaps the costs of living.

The farmers of the county have long felt the need for united action through effective organizations, and have made considerable progress,

[1] Information given by D. M. Babbitt, County Agricultural Agent, Flemington, N.J.

especially of late years, in that direction. In the period of dislocation in the 1870's, local farmers' clubs were sometimes organized to discuss agricultural ills and their remedies. The Union Farmers' Club of Delaware and West Amwell townships was a good example of such an organization. Meeting first in 1870, this group during the next two or three years discussed at bimonthly meetings many of the problems of the day, and often gave some well-reasoned answers. Their activities never went beyond the talking stage, however.

The same chaotic conditions which led to the formation of farmers' clubs resulted in the Grange movement. The Grangers or Patrons of Husbandry were militant in their early days and plunged into a course of action which included cooperative buying, nation-wide publicity, and lobbying at Washington, D.C. The first Grange in Hunterdon County and the twelfth in New Jersey was the one founded at Ringoes in 1873. It was soon followed by others at Readington, Mount Pleasant, Milford, Flemington, Locktown, Clinton, and Pittstown. A county or "Pomona" Grange was organized in 1875. For a time both it and the local Granges were very active in cooperative purchasing of farm supplies, but as times improved their business ventures declined. The old Granges which remained active and new ones started during the 1880's became almost purely social organizations.

Some of the Granges resumed their economic activities during the hard times of the 1890's, and Grange meetings became scenes of political ferment; but the most active groups of that period were the Farmers' Alliances founded at Jutland, Mount Pleasant, Clinton, Three Bridges, Sunnyside, Pleasant Run, and Sand Brook. Some of these units ran cooperative stores for handling farm supplies, and all of them did cooperative buying and gave their support to the fruit exchanges which were being formed at this time; but most of their energies went into political action. After the decline of the Populist party and the coming of better times in the late 1890's, the Alliance groups also turned to social activities. By 1900 all Alliance groups in Hunterdon County had merged with the Granges, so that by 1906 there were twelve local Granges. Since that time the Granges have been for the most part social organizations, yet they have actively championed many movements with both economic and political significance. The local and Pomona Granges have by resolution taken a stand in times past on taxation, cow-testing associations, prohibition, soil conservation, agriculture extension, artifical insemination of cattle, and other matters of direct and indirect concern to agriculture. Not the least of their services has been their willing cooperation with other groups also interested in the welfare of agriculture.

An organization which has been constantly and directly concerned with agricultural developments is the Hunterdon County Board of Agriculture, which numbers more than fourteen hundred members at the present writing and is the most powerful agricultural force in the county. The County Board of Agriculture was first organized in the late 1880's, and its original plan was like that of the farmers' clubs of an earlier time in that papers on various problems of agriculture were read and discussed at length at the meetings. For example, a paper given in 1891 was on a subject of special interest to many farmers of the time: "Can Poultry Raising Be Made Profitable?" The group met at different places in the county, often in connection with Grange meetings or Farmers' Institutes. The organization received considerable publicity, and many of its meetings were well attended; so it is probable that its programs over the years did much good in disseminating agricultural information. The Board of Agriculture operated according to its original plans without much alteration until the year 1920.

In 1920 the Board of Agriculture was completely reorganized. An executive committee was created, members of which were to represent the various townships. The group was to meet monthly and the Board of Agriculture itself only annually. The Hunterdon County organization joined the State Federation of County Boards of Agriculture, and through it became part of the National Farm Bureau Federation. In 1928 the custom of holding meetings in various sections of the county was revived. The membership of the executive committee has been increased from time to time, and "visiting members" representing the Hunterdon County Poultry Association, the Pomona Grange, the Dairymen's League, the Hunterdon County Bankers' Association, the Rural Ministers' Association and other county-wide organizations have been admitted. The Board of Agriculture, through its executive committee, has sponsored many movements and activities of benefit to Hunterdon County agriculture. Its "farmers' meetings" and tours of inspection of farms have led to much exchange of practical information, and in recent years its annual businessmen-farmers' picnic has promoted good relations with nonfarming groups. The Board gave moral backing for the Cooperative Auction Market and had much to do with the formation of the Artificial Breeding Association, the first commercial venture of its kind in the United States. It has acted as sponsor for the Agricultural

Extension Service, and has repeatedly and actively backed state legislation favorable to agriculture. The monthly meetings of the Board of Agriculture provide opportunities for the various agricultural interests of the county to voice their views. Members of the committee are for the most part farm leaders, and many of them have represented their communities for a long period. Clifford Snyder, still president of the Board, has served in that capacity for more than a quarter of a century.

A newer organization, the Farmers' Union, became active in Hunterdon County during the depression years of the 1930's. The first unit or local was organized at Ringoes in 1932 and a second in Round Valley in 1933. Although they envisioned the cooperative purchasing of farm supplies and marketing of milk, neither group lasted very long. Local Number 3, organized in Kingwood Township in 1936, has survived, however, and at present speaks for approximately two hundred farmers. The members of the local are also members of the Delaware Valley Cooperative, mentioned above. The Farmers' Union was militant in its early days and somewhat contemptuous of other farm organizations. Local members criticized the Board of Agriculture as being representative of only the more prosperous farmers, and one of them in 1936 wrote as follows concerning the Grange: "The Grange is a splendid social organization, but if we want to stay on our farms, we can't just go out and eat pie." [2] In 1940, however, Local Number 3 invited the executive committee of the County Board of Agriculture to meet with it, and there are other indications that Farmers' Union members are less critical of other local farm groups than was formerly the case.

The county, as well as New Jersey generally, has been favored by having very active agencies devoted to the finding of means of making its agriculture more efficient. The activities of the New Jersey Experiment Station at Rutgers University have been alluded to from time to time in this book. The many workers of that institution have by their experimentation and research done work of inestimable value to farmers of the state, including those of Hunterdon County. Leading farmers have learned to keep an eye open for the latest findings of their Experiment Station, for they know that they will continue to prosper only as long as they keep abreast of or slightly ahead of the times.

Equally important have been the various agencies for the dissemination of the latest agricultural information directly to farmers. Beginning in 1867, regular lectures by men from the newly-formed New Jersey College of Agriculture at Rutgers University were started. During the

[2] *Hunt. Co. Dem.,* June 18, 1936.

latter part of the century these lectures were usually given as part of Farmers' Institutes, at which there were a number of other speakers. The Institutes became increasingly popular, and the four meetings which were held in various parts of the county each winter drew large crowds. They continued until 1920, and at least one meeting was held as late as 1922. The Institute was an excellent educational device, and for many years kept thinking farmers informed as to the best practices known in agriculture.

The guiding hand in many of the recent changes in the agriculture of Hunterdon County has been the New Jersey Agricultural Extension Service, which was created in 1912. Hunterdon County was slow in adopting the program insofar as employing local extension workers was concerned, but interest was stirred up from the beginning by crop demonstrations on the farms of progressive farmers each year. For instance, five farm meetings for seed-corn selection demonstrations were conducted in the county by Extension Service representatives in 1917. The Ringoes Grange and the County Board of Agriculture both went on record in 1916 as favoring the hiring of a County Demonstrator, but the idea was apparently dropped for a time because of the war. The issue was again raised in 1918; and during the following years the Board of Agriculture, the various Granges, the cow-testing associations, the Lambertville Vigilant Society, and several other groups passed resolutions and sent committees to urge the Board of Chosen Freeholders to pay the county's share of such a plan. Letters and editorials on both sides of the issue appeared, and public meetings where the matter was discussed often became heated. In 1922 the proposition was decisively defeated by a referendum of the voters of the county, but its proponents set to work in even more determined fashion, bringing in Director H. J. Baker of the New Jersey Agricultural Extension Service and others to speak to groups in all parts of the county. The new editor of the *Hunterdon County Democrat,* D. H. Moreau, gave the project constant and favorable publicity, and other county papers also supported it. Finally, in 1927, the Board of Chosen Freeholders budgeted sufficient funds, and Hunterdon County gained a County Agricultural Agent, as county representatives of the Extension Service were called by this time.

Edwin A. Gauntt, the first County Agricultural Agent, took office on February 15, 1927, having been chosen by the executive committee of the Board of Agriculture from nominations submitted by the New Jersey Agricultural Extension Service. His term of office saw the disappearance of opposition to the Extension Service except from a few die-hards. Gauntt took an active part in organizing the Cooperative Auction Market, which started in 1930, and gave it his guidance during its period of infancy. He served until 1934, when he was succeeded by Dwight M. Babbitt, who is serving at the present writing. Notable achievements during the latter's decade in office have been the founding of the Artificial Breeding Association and the development of a promising cannery-tomato growing industry. In general, it should be emphasized that the period during which Hunterdon County has had an Agricultural Agent has seen it forge ahead from a somewhat backward to a leading position in New Jersey Agriculture. The New Jersey Agricultural Extension Service, largely through its county representatives, has furnished much of the leadership which brought this progress.

Since 1933 the County Agricultural Agent has had general supervision of the Agricultural Adjustment Administration program of the county. The "Triple A" in that year curbed wheat production through federal payments to cooperators, and in the following year that of corn and hogs. Further action was stopped for a time by the Supreme Court decision which declared unconstitutional the Congressional act which had established the program. The Agricultural Adjustment Administration was reestablished in new guise in 1936, however, by the Agricultural Conservation Act. Since that time between one thousand and twelve hundred farmers have received governmental benefits each year in return for restricting of acreage of certain crops, plowing under green manure, applying fertilizers, and in other ways cooperating. Opinions in the county differ as to the political philosophy back of the program, but all will agree that the soil resources of the region have been conserved and increased because of it.

The 4-H Club program is a phase of Agricultural Extension work which has accomplished a great deal in a short time and which is very promising so far as the future is concerned. Through it hundreds of boys and girls have learned better agricultural and homemaking methods, as well as the benefits to be gained by working together. The records are not specific, but the first clubs in Hunterdon County were apparently started about 1915 or 1916, supervision being given by various Extension workers from New Brunswick. By 1919 there were at least twenty-three clubs, whose members carried homemaking, gardening, poultry, and livestock projects. Livestock projects, especially those in dairying, were given impetus by the creation at Trenton in 1921 of the New Jersey Junior Breeders' Fund, a revolving fund from which loans were made to 4-H and Vocational Agriculture boys of the state to buy pure bred animals.

The "Calf Clubs" which were being founded at this time taught not only the boys and girls but also their parents the advantages of careful breeding, milk testing, and testing for bovine tuberculosis. Within fourteen months after the forming of the Mount Airy Purebred Calf Club in 1923, it was reported that the fathers of all members had acquired pure bred herd sires and that over one half had joined a cow-testing association. In 1937 the dairy club enrollment was fifty-six members, the poultry club enrollment twenty-eight, and the gardening enrollment eighty-five. Forestry clubs, which were discussed in an earlier chapter, have been in existence for over a decade, with approximately one thousand members enrolled annually in the elementary schools of the county.

In 1928 an assistant to the County Agricultural Agent took over supervision of the agricultural clubs for a time, and in 1930 a part-time club agent was appointed. Brandon H. Wright, the new appointee, during the next five years carried out a vigorous program in Hunterdon County and in Somerset and Warren counties, in which he also had the supervision of clubs. B. F. Ramsburg, the present County 4-H Club Agent, took charge of boys' club work in the three counties in 1935, and a few months later was made the first full-time 4-H Club Agent for Hunterdon County. Part-time home demonstration agents began work in the county in 1927, and a full-time appointee was named in 1938. The Home Demonstration Agent spends approximately one half of her time supervising girls' 4-H clubs in homemaking and the remainder on adult work in nutrition, clothing, child training, and other matters pertaining to homemaking. Today 4-H Club work in Hunterdon County is strongly entrenched and is without doubt one of the most effective means of disseminating widely the latest agricultural information.

More intensive training in agriculture has been given to a smaller number of boys in the Vocational Agriculture courses in the high schools during recent years. The first high school to offer a course was that of Flemington in 1918, probably as a result of the impetus given to agriculture by the war. In 1943 thirty-nine boys were enrolled for this course at this school. An agriculture department was established at the Clinton High School in 1925 and at Lambertville in 1926. An undetermined number of young men from the county have attended regular courses given by colleges of agriculture, particularly those given at the New Jersey College of Agriculture at Rutgers University in recent decades. There is no way of knowing how many have returned to Hunterdon County to engage in farming, but some have done so. A larger number have taken "short courses" offered during winter months. Most of these have been given at New Brunswick, but some have been offered at Flemington by teachers from Rutgers. A firm belief in technical training in agriculture has replaced the former distrust for "book learning."

Hunterdon County agriculture has in many ways entered a new day—a mechanized, specialized, commercialized era. Farmers here, as in all of New Jersey, have capitalized on their nearness to markets, and with the increasing improvement of transportation have had to become more and more effective in their methods. Scientific feeding, selective breeding, milk testing, artificial insemination of cattle, cooperative buying and selling, effective use of the newest machinery, and employment of technical advice have all helped to make agriculture of this area more efficient; but, of course, these things have been or are being adopted by rival regions, which means that new techniques and perhaps new types of production are needed here if present prosperity and present high standards of living are to be maintained. Dairymen and poultry raisers of Hunterdon County have been prosperous during recent years, but they realize the necessity for constantly increasing efficiency and lower costs. The commercial growing of tomatoes and other cannery crops is expanding, and time will prove whether truck gardening can take its place beside the production of milk and poultry products or even, to some extent, supersede them. The new day in agriculture has its dangers, but as never before the agriculturists of Hunterdon County are alive to their problems and willing to utilize individually and collectively the best knowledge of the day.

Bibliography

PRIMARY SOURCES

I. *Manuscripts*

Account Books

"Account of Tickets Sold, Flemington Station, commencing Decr. 4th 1854. F. C. Haward, Agt." Hunt. Co. Hist. Soc., Flemington, N.J.

ANDERSON and WELLES (?), general store, New Hampton, N.J., day books, 1865–68. 2 vols. Rutgers Univ. Library, New Brunswick, N.J.

ATKINSON, ASHER, family expense day book, Flemington, 1811–22. Hunt. Co. Hist. Soc.

ATKINSON, ASHER, general store, Flemington, day book, 1820–29. Hunt. Co. Hist. Soc.

ATKINSON and JOHNSON, general store, Flemington, day book, 1793–94. Hunt. Co. Hist. Soc.

BERKAW, BERGUN, Readington Twp., day book, 1812–51. Rutgers Univ. Library.

Bethlehem Presbyterian Church, accounts for building graveyard wall, 1795. Hunt. Co. Hist. Soc.

BOWNE, JOHN, day book, 1814–18; ledgers 1801–34. Includes personal and farm accounts as well as those of Dr. Bowne's medical practice. Hunt. Co. Hist. Soc.

Building accounts, Flemington courthouse, 1791–92. Ledger and folio. Hunt. Co. Hist. Soc.

———, 1828–30. 2 vols. Hunt. Co. Hist. Soc.

BURROUGH, JAMES, Hopewell, N.J., farm account book, 1833–50. Rutgers Univ. Library.

CAPNER, J. and J. Q., general store, Flemington, day books, 1834–39. Hunt. Co. Hist. Soc.

CAPNER, JOHN H., day book, personal accounts, 1840. Hunt. Co. Hist. Soc.

CAPNER, THOMAS, estate, cash book, 1832. Hunt. Co. Hist. Soc.

Case tannery, Flemington, account books, 1783–1851. Includes day books, ledgers, hide books, bark books, and other records in complete order. Personal and farm expenses are often included with those of the tannery. Hunt. Co. Hist. Soc.

Copper mine accounts, Flemington, 1853–55, 1858–61, 1864–65. The entries for the first period, though not so labeled, were probably for the Flemington Copper Company; those for the second are for the Hunterdon Copper Company; and those for the third period are for the Kent Copper Company. Hunt. Co. Hist. Soc.

DEATS, H. E., farm account books, 1910–30. Property of H. E. Deats, Flemington.

DEATS, HIRAM, machine shop and foundry, Quakertown, N.J., day books, 1831–40. Hunt. Co. Hist. Soc.

DEATS, CASE and COMPANY, machine shop and foundry, Pittstown, N.J., day book, 1869–74. Hunt. Co. Hist. Soc.

Delaware Valley Farmers' and Mechanics' Agricultural Society, account book, 1877–81. Hunt. Co. Hist. Soc.

HOLCOMBE, SAMUEL, general store, Mount Airy, N.J., day books, 1824–32, 1868–71. Hunt. Co. Hist. Soc.

JANEWAY and BROUGHTON, general store, Bound Brook, N.J. (?), day book, 1735–45, ledger, 1738–47. Rutgers Univ. Library.

LAUDERDALE, THOMAS, farm account books, 1886–98. Property of William Lauderdale, Mount Airy.

LAUDERDALE, WILLIAM, farm account books, 1914–40. Property of William Lauderdale, Mount Airy.

Miss Allen's school, Flemington, account book, 1813–48. Hunt. Co. Hist. Soc.

"Mowing and Harvesting," account book of unidentified Hunterdon County farmer, 1827. Hunt. Co. Hist. Soc.

SMITH, W. R., general store, Jacksonville (Lebanon), N.J., day books, 1828–29. Rutgers Univ. Library.

SMITH and BOEMAN, general store, Jacksonville (Lebanon), ledger, 1832–33. Rutgers Univ. Library.

Unidentified distillery, Tewksbury Twp., ledger, 1825–27. Hunt. Co. Hist. Soc.

Unidentified general store, Amwell Twp., day book, 1798–99. Hunt. Co. Hist. Soc.

———, Annandale, N.J. (?), day book, 1880–83. Rutgers Univ. Library.

———, Clinton, N.J., day book, 1865. Hunt. Co. Hist. Soc.

———, Flemington, day book, 1793–94. Hunt. Co. Hist. Soc.

———, Flemington, day book, 1855–56. Hunt. Co. Hist. Soc.

———, Jacksonville (Lebanon), N.J., day books, 1828–32. Hunt. Co. Hist. Soc.

———, New Germantown (Oldwick), N.J., day book, 1815–21. Rutgers Univ. Library.

———, Whitehouse, N.J. (?), ledger, 1771–72. Hunt. Co. Hist. Soc.

Unidentified memorandum book, 1824. Mostly personal accounts. Hunt. Co. Hist. Soc.

Unidentified shoemaker, located in turn at Cherryville, Raritan, and Rowland's Mills, N.J., day books, 1854–98. Hunt. Co. Hist. Soc.

"Work Book," labor accounts of unidentified person, Flemington, 1829–31. Rutgers Univ. Library.

Census

Hunterdon County local schedules, U.S. Census, 1850. Rutgers Univ. Library.

Correspondence and Personal Papers

Blackwell Papers. Correspondence, personal papers, and accounts of James H. Blackwell and family, Flemington, 1820–70. Hunt. Co. Hist. Soc.

Bonnell Papers. Correspondence and personal papers of Elmer E. Bonnell and family, Clinton, 1849–1902. Property of Miss Julia Haver, Clinton.

Capner Papers. Correspondence, personal papers, and accounts of Capner (Capnerhurst) family, Flemington, and various relatives and friends, 1776–1846. Hunt. Co. Hist. Soc.

[Choyce], Richard, accounts, *circa* 1810. Records of fulling and weaving shop and farm at Flemington, interspersed with narrative. Hunt. Co. Hist. Soc.

Emley Papers. Accounts, surveyor's records, agent's accounts, and personal papers of John Emley, Kingwood Twp., *circa* 1750–70. Hunt. Co. Hist. Soc.

Hunterdon County Historical Society, Flemington, miscellaneous correspondence and papers. Hunt. Co. Hist. Soc.

Leigh, I. N., "A Few Historical Facts Relating to Hunterdon County, Gleaned from Some Old Letters," 1887. Hunt. Co. Hist. Soc.

"Mementoes for the One Hundred and Seventy-Fifth Anniversary of the Reformed Church of Readington, N.J." (scrapbook, 1894). Contains old letters and records of various kinds donated by members of the congregation in honor of the occasion. Rutgers Univ. Library.

Race Papers. Correspondence and papers of Dr. Henry Race, Pittstown, N.J., *circa* 1820–60. Hunt. Co. Hist. Soc.

Diaries and Reminiscences

Bush, E. T., "Croton and Vicinity" (typed MS., 1896). Hunt. Co. Hist. Soc.

Blackwell, James H., Flemington, diary, 1843–66. Hunt. Co. Hist. Soc.

Bonnell, Elmer E., Clinton, N.J., diary, 1878–1895. Property of Miss Julia Haver, Clinton.

"Extracts from the Diary of Peter Haward, 1800–58." Excerpts copied from original manuscript, now lost. Hunt. Co. Hist. Soc.

Hall, John, "Jno. Halls Book, 1779 [–1807]." This diary is the work of two men, John Hall and his nephew, John Hall Capner. The first portion was written by John Hall while he was still in England, the second, following a gap of several years, by his nephew after their emigration to America. Hunt. Co. Hist. Soc.

Hankinson family, Amwell Twp., commonplace book, 1787–1842. Contains entries of many kinds, including personal and family records. Items are usually anonymous and seldom dated. Rutgers Univ. Library.

LAMBERT, JOHN, "John Lambert's Almanack or Account Book, 1774." Personal observations and mill and store records, entered in a copy of *Aitken's General American Register and Calendar, for the Year, 1774.* Hunt. Co. Hist. Soc.

LAIR, DAVID, "Log of a Journey to Ohio by Philip Gordon Family, 1839." Typed copy of original MS. Hunt. Co. Hist. Soc.

LARISON, G. H., "Last Primitive Houses Near Howell's Ferry," undated. Personal recollections. Hunt. Co. Hist. Soc.

LEQUEAR, J. W., "Flax Culture and Manufacture," 1889. Hunt. Co. Hist. Soc.

———, "Reminiscences," undated. Hunt. Co. Hist. Soc.

MOORE, JOHN, memorandum book, 1800–34. Hunt. Co. Hist. Soc.

THOMPSON, J. B., Readington, diary, 1848–52, 1857–59. Rutgers Univ. Library.

Minute Books and Records of Organizations

Alexandria Vigilant Society, minute book, 1818–27. Hunt. Co. Hist. Soc.

Flemington Soldiers Aid Society, minutes, 1862. Hunt. Co. Hist. Soc.

Flemington Vigilant Society, minutes, 1822–83. Hunt. Co. Hist. Soc.

Hunterdon County Board of Agriculture, minutes, 1915–22. Property of Hunt. Co. Board of Agriculture.

Hunterdon County Historical Society, minutes, 1885–1939. Hunt. Co. Hist. Soc.

Lambertville Vigilant Society, minutes, 1837–1922. Hunt. Co. Hist. Soc.

Woman's Relief Corps, Lambert Boeman Post, G.A.R., Flemington, record books. Hunt. Co. Hist. Soc.

Miscellaneous

"An Articel [*sic*] of Agreement between G. C. Sergeant and Joseph Sergeant about the Mine," 1835. Hunt. Co. Hist. Soc.

"Articles of Agreement . . . between . . . President and Directors of 'the Flemington Aqueduct Company' . . . and John Hughes," 1807. Hunt. Co. Hist. Soc.

HORDE, J. WILHELM, Raubach, Germany, certificate of character given to Pieter Jung, 1727. Tr. by D. A. Sinclair, New Brunswick. Hunt. Co. Hist. Soc.

Hunterdon County Agricultural Society, records, Hunt. Co. Hist. Soc.

Letters patent issued to John Deats, Apr. 26, 1828. Original patent for the Deats plow. Property of H. E. Deats, Flemington.

RACE, HENRY, teacher's subscription list and day book of Quakertown school, 1837.

Rental agreement between H. E. Deats and Henry Butcher, 1898. Property of H. E. Deats, Flemington.

Rental agreement between Rebeckah Compton and William Birch, 1811. Hunt. Co. Hist. Soc.

"Rules and Subscription List of Flemington Telephone Co., *circa* 1886. Hunt. Co. Hist. Soc.

Official Records

BRUNNER, PETER, justice of the peace, Readington Twp., dockets, 1780–88. Rutgers Univ. Library.

COLE, EZEKIEL, justice of the peace, Readington Twp., dockets, 1790–1804. Rutgers Univ. Library.

Deed record books, Hunterdon County, 1714–1944. Office of County Clerk, Flemington.

Deed record books. Office of Sec. of State, Trenton.

Hunterdon County court records, 1714–1944. Hall of Records, Flemington.

New Jersey Commissions. Office of Sec. of State, Trenton.

New Jersey Wills. Probate records, including wills, inventories, records of vendues, reports of executors, etc. Office of Secretary of State, Trenton.

Readington Twp., tax assessment book, 1846. Hunt. Co. Hist. Soc.

RODENBOCKS, ADAM, Inventory, personal estate of Adam Rodenbocks, Amwell Twp., 1809. Hunt. Co. Hist. Soc.

West Amwell Twp., Civil War draft record book. Property of West Amwell Twp.

———, record books, 1849–1940. Property of West Amwell Twp.

II. *Printed Sources*

Archives

Archives of the State of New Jersey (Newark, Paterson, Somerville, and Trenton, 1880–1944).

Documentary History of the State of New York (New York, 1933–37). 10 vols.

Documents Relating to the Colonial History of New York (Albany, 1853–87). 15 vols.

Town Records of Hopewell, New Jersey (New York, 1931). Records of Hopewell Township, 1721–99, and of Hopewell Baptist Church, 1749–1849.

Articles

ACTON, H. E., "Charles Kirk's Review of a Century," *Papers Bucks Co. Hist. Soc.*, VII, 95–103.

ANDERSON, J. A., "Silk Culture in Bucks County," *ibid.*, III, 579–85.

BELLIS, JOHN, "Recollections of Baptisttown and Vicinity," *Hunt. Ind.*, clippings, 1893–97. Hunt. Co. Hist. Soc.

BELLIS, THEODORE, "Recollections," *Hunt. Co. Dem.*, Aug. 28, 1924.

BELLIS, THEODORE, "Some Historical Notes of Local Interest," *Home Visitor*, Mar. 11, 1896.

BLACKFAN, F. K., "Old Household Industries," *Papers Bucks Co. Hist. Soc.,* V, 418–22.
BLANE, JOHN, "About Patenburgh," *Our Home,* I (1873), p. 189.
———, "Recollections of Hunterdon County Forty-Five Years Ago," *ibid.,* 470–75.
BOND, M. S., "Higgins School," *Jerseyman,* III (1895–97), 34–35.
BUCHANAN, JAMES, "Reminiscences," *ibid.,* VI (1900), 23–24, 34–35.
BURKETT, A. J., "Harmony School House," *Home Visitor,* Mar. 18, 1896.
BUSH, E. T., "The Records of an Early Physician," undated clipping, *Hunt. Co. Dem.* Hunt. Co. Hist. Soc.
CARPENTER, JOHN, "A Local Retrospect," *Clinton Democrat,* clippings, *circa* 1894. *Hunt. Co. Hist. Soc.*
CASE, E. R., "Growing, Treating and Drying Flax," *Papers Bucks Co. Hist. Soc.,* V, 282–84.
COLLINS, A. M., "Octagonal or So-called 'Eight Square' Schoolhouse," *ibid.* V, 251–60.
ELY, W. S., "Lime Burning Industry, Its Rise and Decay in Bucks County," *ibid.,* IV, 69–74.
FITZGERALD, J. H., "Peddlers and Other Itinerants," *ibid.,* VI, 282–89.
———, "Wells and Pumps in Bucks Co.," *ibid.,* V, 454–60.
"Frenchtown 40 Years Ago," *Hunt. Ind.,* clippings, 1879. Hunt. Co. Hist. Soc.
FRETZ, C. D., "Letters from Native Bucks Countians Living in Canada, 1815," *Papers Bucks Co. Hist. Soc.,* VI, 246–48.
GEHAM, DAVID, "The Last Days of Harness Making in Bucks Co.," *ibid.,* VI, 79–81.
GROSS, (MRS.) A. H., "Pie Baking," *ibid.,* IV, 325–27.
GROSS, H. W. "Old Fences in Bucks Co.," *ibid.,* V, 429–32.
HALL, M. H., "Notes at Random from My Life's Experience," *ibid.,* VI, 82–87.
"Instructions of Freeholders of Hunterdon to the Representatives of the County, 1771," *Proc. N.J.H.S.,* 1st ser., X, 81–82.
JAMES, I. M., "Soap Making of Old," *Papers Bucks Co. Hist. Soc.,* IV, 334–35.
"Journal of John Reading [1715–16]," *Proc. N.J.H.S.,* 3d ser., X, 35–46, 90–110, 128–33.
LARGE, JOHN K., "The Mines of Hunterdon County," *Our Home,* I, 141.
LEQUEAR, J. W., "My First Visit to Flemington," *Jerseyman,* III (1895–97), 6–7.
———, "Recollections of Pittstown General Training," *ibid.,* II (1893), 12–14.
———, "Some Recollections of Lambertville," *Lambertville Record,* clipping, 1891. Hunt. Co. Hist. Soc.
MERCER, H. C., "Notes of Forgotten Trades," *Papers Bucks Co. Hist. Soc.,* V, 207–11.

———, "Notes Taken at Random," *ibid.,* II, 122–31.

——— and H. M. Mann, "Random Notes on Forgotten Trades," *ibid.,* V, 740–45.

Parse, J. E., "An Interesting History," *Hunt. Co. Dem.,* May 31, 1916—Aug. 2, 1916.

———, "Interesting Reminiscences of 34 Years of Commuting to New York," *ibid.,* Apr. 5, 1916.

Paxson, (Mrs.) Henry D., "Roasting on the Spit," *Papers Bucks Co. Hist. Soc.,* IV, 320–22.

Preston, (Mrs.) E. K., "The Art of Frying as in Olden Times," *ibid.,* IV, 322–24.

Putnam, F. C., "An Address Delivered at Re-opening of St. Andrew's Church, Amwell, at Lambertville, on Sunday, August 4th, 1867," clipping, *Lambertville Beacon,* Aug. 16, 1867. Hunt. Co. Hist. Soc.

Race, Henry, ed., "Parish Register of the German Reformed Church of Alexandria, Hunterdon County, New Jersey, 1763–1802," *Jerseyman,* III (1895–97), 37–53.

———, "Contributions to Hunterdon County History," *Proc. N.J.H.S.,* 2d ser., XI, 157–61.

———, "Rev. William Frazer's Three Parishes,—St. Thomas's, St. Andrew's, and Musconetcong, N.J., 1768–70," *Pa. Mag. of Hist. and Biog.,* XII (1888), 212–32.

Runkle, Lewis, "A Chapter on Butter," *Hunt. Co. Dem.,* Mar. 4, 1890.

Rutherford, John, "Notes on the State of New Jersey," 1788, *Proc. N.J.H.S.,* 2d ser., I, 79–89.

Scott, J. E., "Old Shad Fisheries on the Delaware River," *Papers Bucks Co. Hist. Soc.,* III, 534–41.

Scheetz, Grier, "Broom Making By Hand," *ibid.,* V, 312–15.

———, "Flax and Its Culture," *ibid.,* III, 482–86.

———, *et al.,* "Flax Seed Mills," *ibid.,* IV, 724–29.

Scott, (Mrs.) J. E., "Cooking Shad by the Open Fire," *ibid.,* IV, 316.

Stevenson, J. R., "The Council of Proprietors of New Jersey," *Proc. N.J.H.S.,* ser. 3, VII, 129–36.

Swain, F. K., "Passing Events," *Papers Bucks Co. Hist. Soc.,* V, 324–39, 407–18.

Swain, G. G., "Former and Present Ways of Brick Making," *ibid.,* IV, 474–77.

Taylor, Kate, *Reminiscences of Shuster's Hollow,* clipping from unidentified newspaper, 1892. Hunt. Co. Hist. Soc.

Thomas, (Miss) E. M., "Old Dutch Ovens," *Papers Bucks Co. Hist. Soc.,* IV.

"Traditions of Our Ancestors," *Hunt. Rep.,* Dec. 30, 1869—Aug. 4, 1870. Based on interviews with old men and women of that day. Authorship usually credited to J. W. Lequear.

"Two Revolutionary Letters," *Jerseyman,* III (1895–97), 35–36.

VANDERBILT, G. O., "Recollections Along the Nesasacaway," *Hunt. Co. Dem.*, Jan. 23—May 7, 1912.

VOSSELLER, ELIAS, "An Interesting Find of Two Parchments Relating to Early New Jersey," *Jerseyman*, VII (1901), 31–33.

———, "Reminiscences of the Kase Family," *ibid.*, I (1891–92), 17–18.

WOODMAN, M. S., "Manners and Customs of Eighty Years Ago," *Papers Bucks Co. Hist. Soc.*, V, 90–95.

Books

ACRELIUS, ISRAEL, *A History of New Sweden*, tr. by W. M. Reynolds (Philadelphia, 1874; first published in Stockholm, 1759).

Aitken's General American Register and Calendar, For the Year 1774 (Philadelphia, 1774).

American Husbandry, ed. by H. J. Carman (New York, 1939; originally published in London, 1775).

ANBUREY, THOMAS, *Travels Through the Interior Parts of America* (London, 1791). 2 vols.

BARBER, J. W. and HENRY HOWE, *Historical Collections of the State of New Jersey . . . with Geographical Descriptions of Every Township in the State* (New York, 1844).

BARNUM, H. L., *The American Farrier* (Philadelphia, 1832).

BLOWE, DANIEL, *A Geographical, Historical, Commercial, and Agricultural View of the United States, forming a Complete Emigrant's Directory through Every Part of the Republic* (London, 1820).

BOYER, C. S., *Early Forges & Furnaces in New Jersey* (Philadelphia, 1931).

BRACHEN, HENRY, *Farriery Improved* (London, 1785).

BUDD, THOMAS, *Good Order Established in New Jersey in America* (New York, 1865; first published in 1685).

BURNABY, ANDREW, *Travels Through the Middle Settlements in North-America* (London, 1775).

CAZENOVE, THEOPHILE, *Journal, 1794*, tr. and ed. by R. W. Kelsey (Haverford, Pa., 1922).

COOK, G. H., *Geology of New Jersey* (Newark, 1868).

CRÉVECŒUR, J. H. ST. J. DE, *Letters from an American Farmer* (London, 1782).

———, *Sketches of Eighteenth Century America*, ed. by H. L. Bourdin *et al.* (New Haven, 1925).

DANCKAERTS, JASPAR and PETER SLUYTER, *Journal of a Voyage to New York and a Tour in Several of the American Colonies in 1679–80*, tr. and ed. by H. C. Murphy (Brooklyn, 1867).

DEATS, H. E., ed., *Marriage Records of Hunterdon County, New Jersey, 1795–1875* (Flemington, 1918).

[Furman, Moore], *Letters of Moore Furman* (New York, 1912).
Gordon, T. F., *A Gazeteer of the State of New Jersey* (Trenton, 1834).
Kalm, Peter, *The America of 1750; Travels in North America*, ed. by A. B. Benson (New York, 1937; text of English version of 1770, revised from original Swedish).
Messler, Abraham, *Forty Years at Raritan; Eight Memorial Sermons with Notes for a History of the Reformed Dutch Churches in Somerset County, New Jersey* (New York, 1873).
Morse, Jedidiah, *The American Universal Geography* (Boston, 1796).
The Pocket Farrier or Farmers' Receipt Book (Boston, 1840).
Roberts, Job, *The Pennsylvania Farmer* (Philadelphia, 1804).
Schoepf, J. D., *Travels in the Confederation, 1783–84*, tr. by A. J. Morrison (Philadelphia, 1911).
Schrabisch, Max, *Archaeology of Warren and Hunterdon Counties*, Bulletin 18, Geologic Series, Reports of the Department of Conservation and Development, State of New Jersey (Trenton, 1917).
Stillwell, J. L., *Old Readington, 1867–1876* (Somerville, N.J., 1935).
Thomas, Gabriel, *An Historical and Geographical Account of the Province and Country of Pensilvania and of West-New-Jersey in America* (New York, 1848; first published in London in 1698).
Varlo, Charles, *A New System of Husbandry* (Philadelphia, 1785).
Woodward, C. R., *Ploughs and Politicks* (New Brunswick, 1941).

Census Records

"Census of New Jersey, 1726, 1737, 1745," *Proc. N.J.H.S.*, 1st ser., IX (1860–64), 202–03.
State of New Jersey Census of 1915 (Union Hill, N.J., 1916).
State of New Jersey, Compendium of Censuses, 1726–1905, Together with the Tabulated Returns of 1905, and the Federal Census of 1910 (Trenton, 1910).
United States Census, 1790–1940.
Statistical Handbook of New Jersey Agriculture (Trenton, 1929).

Legislation

Allinson, Samuel, comp., *Acts of the General Assembly of the Province of New-Jersey, 1702–76* (Burlington, N.J., 1776).
Elmer, L. Q. C., *A Digest of the Laws of New Jersey* (Bridgeton, N.J., 1838).
Nevill, Samuel, comp., *Acts of the General Assembly of the Province of New-Jersey* (1752 and 1761), 2 vols.
New Jersey Session Laws, 1776–1944.
Paterson, William, comp., *Laws of the State of New-Jersey* (Newark, 1800).

Newspapers

Daily True American, New Brunswick, 1860–88. Rutgers Univ. Library.

Democrat-Advertiser, Flemington, 1881–1926. Incomplete files. Office of *Hunt. Co. Dem.,* Flemington.

Democrat, Flemington, 1866–67. Office of *Hunt. Co. Dem.* and Hall of Records, Flemington.

Federalist and New-Jersey Gazette, Trenton, 1796–1802. N.J. State Library, Trenton.

Home Visitor, Flemington, 1885–1902. Only occasional copies and clippings available, for the most part at the Hunt. Co. Hist. Soc.

Hunterdon County Democrat, Flemington, 1867–1944. Office of the *Hunt. Co. Dem.* and Hall of Records, Flemington.

Hunterdon Democrat, Flemington, 1838–66. Office of *Hunt. Co. Dem.* and Hall of Records, Flemington.

Hunterdon Farm Bureau News, Flemington, 1937–44. N.J. Agricultural Extension Service, Flemington.

Hunterdon Gazette, Flemington, 1825–67. Office of *Hunt. Co. Dem.,* and Hall of Records, Flemington.

Hunterdon Independent, Frenchtown, N.J., 1878–90, clippings 1893–97. Hunt. Co. Hist. Soc.

Hunterdon Republican, Flemington, 1856–1944. Incomplete files, Hunt. Co. Hist. Soc. and Hall of Records, Flemington.

Lambertville Press, Lambertville, N.J., 1858–61. Hunt. Co. Hist. Soc.

Lambertville Record, Lambertville, N.J., clipping, 1891. *Hunt. Co. Hist. Soc.*

Milford Leader, Milford, N.J., clippings, Dec. 2, 1897—Feb. 10, 1898. Hunt. Co. Hist. Soc.

New-Brunswick Daily Fredonian, New Brunswick, Jan. 22, Apr. 1, 1868. Rutgers Univ. Library.

New York Times, 1932–44.

Whitehouse Review, Whitehouse, N.J., Apr. 18, 1939. *Hunt. Co. Hist. Soc.*

Pamphlets

Address of the Committee Appointed by a Republican Meeting in the County of Hunterdon, Recommending Gen. Andrew Jackson, of Tennessee, to the People of New-Jersey, as President of the United States (Trenton, 1824).

An Alphabetical List of the Members of the Flemington, N.J., Baptist Church, June, 1888; Together with a Statistical Table of the Annual Reports of the Church from its Organization in 1798 to the Year 1887 (Clinton, 1888).

Anniversary Services, Presbyterian Church, Flemington, N.J., 1894 (Flemington, 1894).

Annual Statements of the Several Banks of the State of New Jersey (Trenton, 1849).

Appeal to the Legislative Council and General Assembly of the State of New Jersey on Behalf of the Religious Society of Friends (Philadelphia, 1836).

Business Review of the Counties of Hunterdon, Morris and Somerset, New Jersey (Philadelphia, 1891).

By-Laws of the Hunterdon County Historical Society (Flemington, 1900).

Charter and By-Laws of the Delaware Horse and Mule Mutual Insurance Company (Lambertville, 1887).

CLEMENT, JOHN, ed., *Notes and Memoranda Relating to the West New Jersey Society, of West New Jersey* (Camden, N.J., 1880).

Discourses Memorial of Abraham Lincoln . . . By the Pastors of the Different Churches [of Flemington] *on Wednesday, April 19th, 1865* (Lambertville, 1865).

HARDEN, J. S., *Life, Confession, and Letters of Courtship of Rev. Jacob S. Harden, of the M. E. Church, Mount Lebanon, Hunterdon County, N.J., Executed for the Murder of His Wife, on the 6th of July, 1860* (Hackettstown, N.J., 1860).

HOLCOMBE, JOHN, *The Will and Testament and Codicils of John Holcombe of Lambertville, Hunterdon Co., New Jersey, Who Departed This Life, September 29th, 1851* (Lambertville, N.J., n.d.)

LANDIS, ROBERT, *Bethlehem Church and Its Pastor; A Narrative of the Injurious Proceedings of the Rev. Messrs. Albert Barnes and H. W. Hunt, Senr., in Relation to the Pastor of the Presbyterian Church, Bethlehem, New Jersey* (n.p., 1851).

Memorial of Rev. Peter O. Studdiford, D.D., Late Pastor of the Presbyterian Church of Lambertville, New Jersey (Philadelphia, 1866).

MOORE, J. W., ed., *Records of the Kingwood Monthly Meeting of Friends, Hunterdon County, New Jersey* (Flemington, 1900).

MOTT, G. S., *Lessons of the Hour; A Sermon Delivered in the Presbyterian Church, Upon the Death of President Garfield* (Flemington, 1881).

———, *Two Sermons* (Flemington, 1889).

New Jersey; Its Cities, Towns and Railroads, Together with Descriptions of New Jersey Real Estate Offered for Sale by A. D. Melick, Jr. & Bro., No. 6 Pine Street, New York (New York, 1873).

OLMSTEAD, J. M., *The Voice of the Rod; A Sermon Preached at Flemington, N.J., Friday, May 14, 1841, on the Occasion of the National Fast* (Philadelphia, 1841).

PATRICK, A. L., *et al., Soil Survey of the Belvidere Area, New Jersey* (Washington, 1920).

———, *Soil Survey of the Bernardsville Area, New Jersey* (Washington, 1923).

Proposed Consolidated School and Community Center, East Amwell Township, Hunterdon County, New Jersey (n.p., 1938).

Regulations Ordained and Established by the Trustees for the Government of the "Flemington Library Company" Together with a Catalogue of Books Belonging to Said Library (Trenton, 1802).

RUSTON, W. O., *Sermon Delivered . . . in the Fairmount, N.J., Presbyterian Church* (n.p., 1876).

Stangl Pottery (Flemington, n.d.).

STILLWELL, J. L., *Old Readington, 1867–1876* (Somerville, 1935).

VASSAR, T. E., *An Address Delivered at the Funeral of Robert J. Killgore, Flemington, New Jersey, October 3, 1898* (Flemington, 1898).

———, *A Ministry of Fifty Years Reviewed* (Elizabeth, N.J., 1907).

———, *Three-Fourths of a Century Reviewed* (Flemington, 1873).

WILLIAMSON, J. G., *Commemorative Services, Fiftieth Year as Pastor* (n.p., 1899).

———, *Sermon Delivered August 7th, 1889, by Rev. J. G. Williamson, Pastor of the Bethlehem Presbyterian Church, of Hunterdon County, New Jersey, on the 40th Anniversary of His Installation, with Addresses by Rev. Messrs. Mott, Blauvelt, Ewing, Davis and Kline and Drs. Race and Hunt* (Clinton, 1889).

SECONDARY SOURCES

I. *Manuscripts*

"Account of the Minisink Indians," undated. Rutgers Univ. Library.

BARBER, E. A., "History of the Pottery Industry in New Jersey." Rutgers Univ. Library.

BARTOW, G. W., "Dr. George Andrew Viersellius, an Early Hunterdon County Physician, and His Environment," 1902. Hunt. Co. Hist. Soc.

BUSH, E. T., "The Delaware Guards," undated. Hunt. Co. Hist. Soc.

EICHES, O. P., "The Baptist Churches of New Jersey," 1897. Rutgers Univ. Library.

HAMMOND, D. S., "Early Land Ownership in Northern Hunterdon County," 1916. Hunt. Co. Hist. Soc.

HAYHURST, C. S., "Memorial of C. W. Larison, M.D.," 1910. Hunt. Co. Hist. Soc.

HOLCOMBE, M. J., "Sketch of Dr. Gershom Craven, Taken From Data Collected by the Late George H. Larison, M.D.," undated. Hunt. Co. Hist. Soc.

LARISON, G. H., "Coryell's Ferry in the Revolution," undated. Hunt. Co. Hist. Soc.

LEQUEAR, J. W., "A Biographical Sketch of Robert Rittenhouse," undated. Hunt. Co. Hist. Soc.

———, "Genealogical Notes," undated. Hunt. Co. Hist. Soc.

———, "History of the Baptisttown Baptist Church," *circa* 1887. Hunt. Co. Hist. Soc.

———, "Hunterdon County Sabbath School Association," 1887. Hunt. Co. Hist. Soc.

———, "A Paper on the Early Settlement and History of the Baptists in Hunterdon County," 1885. Hunt. Co. Hist. Soc.

———, "Rittenhouse Memorial," 1887. Hunt. Co. Hist. Soc.

MARSHAL, C. S., "An Ancient Mill," 1903. Hunt. Co. Hist. Soc.

SINCLAIR, JESSE, "Early Settlers of Holland Township," undated. Hunt. Co. Hist. Soc.

STAMETS, CORA W., "History of Ringoes Grange," 1933. Property of Cora W. Stamets, Ringoes, N.J.

"The Early German Lutherans," undated. Rutgers Univ. Library.

VOORHEES, O. M., "Chronicles Ancient and Modern Respecting the Iron Industry . . . At High Bridge, New Jersey," undated. Property of O. M. Voorhees, New Brunswick.

———, "An Unusual New Land Lottery," 1929. Property of O. M. Voorhees, New Brunswick.

———, "The Religious History of Hunterdon County." Property of O. M. Voorhees, New Brunswick.

———, "William Alexander, Earl of Stirling," 1902. Hunt. Co. Hist. Soc.

II. *Printed Sources*

Articles

ALLEN, A. B., "Joachim Hill, Clock Maker," *Jerseyman,* XI (1905), 7–9.

"An Ancient Homestead and Its Inhabitants," *Home Visitor,* Mar. 4, 1896.

BAILEY, WILLIAM, "More Local History—White House," *Our Home,* I (1873), 385–89.

CARVER, T. N., "Historical Sketch of American Agriculture," L. H. Bailey, *Cyclopedia of American Agriculture* (New York, 1909), IV, 39–71.

CONNOLLY, J. C., "Slavery in Colonial New Jersey, and the Causes Operating against Its Extension," *Proc. N.J.H.S.,* 3rd ser., XIV, 181–202.

"Customs of Our Early Indians, the Original Inhabitants of Pennsylvania and New Jersey," undated newspaper clipping. Rutgers Univ. Library.

DEATS, H. E., "An Historical Sketch of the Cherryville Baptist Church, 1849–1899," Jerseyman, VI (1900), 5–8, 15–16.

FACKENTHAL, B. F., "The Durham Iron Works in Durham Township," *Papers Bucks Co. Hist. Soc.,* VII, 59–93.

———, "Henry Quinn, Author of 'Temple of Reason,' " *ibid.,* III, 526–34.

———, "Improving Navigation on the Delaware River, with Some Account of Its Ferries, Bridges, Canals, and Floods," *ibid.,* VI, 104–230.

FOLSOM, J. H., "New Jersey's Part in the Revolution," *Proc. N.J.H.S.,* 3d ser., VII, 65–74.

GARDINER, D. H., "The Emancipation of Slaves in New Jersey," *ibid.*, 3rd ser., IX, 1–21.
GRIM, G. M., "Cupping and Bleeding," *Papers Bucks Co. Hist. Soc.*, V, 95–100.
GROFF, L. A., "Van Dolah's School," *Jerseyman*, III (1895–97), 23–26.
GROSS, H. W., "Horse Hopples," *Papers Bucks Co. Hist. Soc.*, V, 186–89.
HOLMES, G. K., "Aboriginal Agriculture—The American Indians," in L. H. Bailey, ed., *Cyclopedia of American Agriculture* (New York, 1909), IV, 24–39.
HOLCOMBE, R. C., "The Ferry Tract at New Hope, Penn., and Coryell's Ferry in New Jersey," *Papers Bucks Co. Hist. Soc.*, V, 584–612.
———, "The Old York Road," *ibid.*, V, 650–716.
HONEYMAN, J. C., "Zion, St. Paul and Other Early Lutheran Churches in Central New Jersey," *Proc. N.J.H.S.*, new ser., IX, 255–73, 347–71; X, 41–56, 294–306, 395–409; XI, 57–70, 191–98, 378–96, 532–42; XII, 67–78, 214–23, 326–35, 462–71; XIII, 209–23, 330–47, 433–46; XIV, 55–69, 336–50, 466–75; XV, 95–108, 250–61, 392–401, 503–8; XVI, 34–50, 180–85.
HONEYMAN, R. R., "More Local History—New Germantown," *Our Home*, I, 116–30.
HOPPOCK, J. M., "Delaware's Old-Time Mills," *Dem.-Adv.*, Sept. 13, 1906.
———, "Hunterdon's Oldest Schoolhouse," *ibid.*, Jan. 25, 1906.
———, "The Old Sergeant Mill," *ibid.*, July 20, 1905.
KEASBEY, A. Q., "Slavery in New Jersey," *Proc. N.J.H.S.*, 3d ser., IV, 90–96, 147–54, V, 12–20, 79–86.
KENDERDINE, T. S., "Hunting, Trapping and Fishing in Bucks Co.," *ibid.* V, 736–40.
———, "Lumbering Days on the Delaware River," *ibid.*, IV, 238–52.
———, "An Old Mowing Machine," *Papers Bucks Co. Hist. Soc.*, III, 373–78.
KLINE, J. A., "The Flemington Academy," *Jerseyman*, VII (1901), 17–19.
KUGLER, J. B., "The Beginning of the Presbyterian Church at Frenchtown, N.J.," *ibid.*, III (1895–97), 27–31.
LANING, A. J., "Settlement and Settlers of Ringoes," *Proc. N.J.H.S.*, 3d ser., VII, 29–30.
LARISON, C. W., "The Boss Family of Amwell," *Dem.-Adv.*, clipping, 1906. Hunt. Co. Hist. Soc.
———, "History of Ringos," *Ringos* [*sic*], I (1889–90).
———, "The Old Dawlis Mill," *Dem.-Adv.*, Sept. 21, 1905.
———, "The Old Landis House, Ringoes," *ibid.*, Aug. 17, 1905.
LARISON, G. H., "The Mode of Life in Our Early Settlements," *Papers Bucks Co. Hist. Soc.*, I, 459–66.
MCDOWELL, A. W., "Pluckemin One Hundred Years Ago," *Our Home*, I, 433–39, 481–87, 530–34.

MacReynolds, George, "Medical Uses of Plants by the Indians," *ibid.*, IV, 415–21.

Mann, H. M., "Shad Fishing on the Delaware River," *ibid.*, V, 279–82.

"Medical Survey of Hunterdon County," *Journal of the Medical Society of New Jersey*, XXXIV (1937).

Mercer, H. C., "Ancient Methods of Threshing in Bucks Co.," *Papers Bucks Co. Hist. Soc.*, V, 315–23.

———, "The Common Tinder Box of Colonial Days," *ibid.*, IV, 359–60.

———, "Open Fire Cooking in Bucks County," *ibid.*, IV, 312–15.

———, "Origin of Log Houses in the United States," *ibid.*, V, 569–83.

———, "The Tools of the Nation Maker," *ibid.*, II, 480–89; III, 469–81.

Messler, Abraham, "The Hollanders in New Jersey," *Proc. N.J.H.S.*, 1st ser., V, 69–89.

Mott, G. S., "The First Century of Hunterdon County, State of New Jersey," *ibid.*, 2d ser., V, 60–111.

Parker, R. W., "Taxes and Money in New Jersey before the Revolution," *ibid.*, 2d ser., VII, 143–51.

Parry, R. R., "Old New Hope, formerly Coryell's Ferry, Pa.," *Papers Bucks Co. Hist. Soc.*, III, 547–64.

Powell, J. W., "The North American Indians," in N. S. Shaler, ed., "The United States of America" (London, 1894), pp. 190–212.

Powell, R. P., Jr., "Transportation and Travel in Colonial New Jersey," *Proc. N.J.H.S.*, new ser., XVI, 284–310.

Race, Henry, "Bethlehem's Township: When Constituted and Why So Named," *Jerseyman*, III (1895–97), 31–33.

———, "The First Years of Frenchtown," *ibid.*, II (1893), 22–25.

———, "History of the Presbyterian Church at Kingwood, Hunterdon County, New Jersey," *ibid.*, 3rd ser., III, 13–23.

———, "St. Thomas' Church of Alexandria, Hunterdon County, N.J.," *Penn. Mag. of Hist. and Biog.*, X (1886), 256–62.

———, "A Sketch of the Pittstown Branch of the Rockhill Family," *Jerseyman*, II (1893), 14–16.

———, "The Two Colonels John Taylor," *ibid.*, I (1891–92), 33–35.

———, "The West Jersey Society's Great Tract in Hunterdon County," *ibid.*, pp. 1–6.

"Ringoes to Honor Founder," *New York Times*, Mar. 13, 1932.

Sassaman, H. D., "History of the First Presbyterian Church of Alexandria," *Milford Leader*, Dec. 2, 1897—Feb. 10, 1898.

Schenck, G. C., "Van Doren Ancestors," *Our Home*, I, 571–72.

"Sketch of the New Jersey Branch of the Ingham Family," *Hunt. Rep.*, June 20, 1896.

"Sketches from Local History," *Hunt. Ind.*, clipping, *circa* 1877. Hunt. Co. Hist. Soc.

SMITH, E. D., "The Smith Plow," *Papers Bucks Co. Hist. Soc.*, III, 11–17.

SMITH, R. A., "St. Thomas' Church, Alexandria," *Dem.-Adv.*, Aug. 24, 1905.

SPECK, F. G., and C. M. STORY, "Notes on Indian Life in Bergen County," *Papers and Proceedings of the Bergen County Historical Society*, III (1906–07).

SROPE, W. T., "Sheriffs of Hunterdon," *Dem.-Adv.*, Oct. 3, 10, 17, 1890.

STEVENSON, J. R., "The Councils of Proprietors of New Jersey," *Proc. N.J.H.S.*, 3d ser., VII, 129–36.

STRAWN, L. B., "Applebutter Making as Practiced by Our Ancestors," *Papers Bucks Co. Hist. Soc.*, IV, 331–33.

SUNDERLIN, J. C., "The Soldier's Monument at Flemington, New Jersey," *Jerseyman*, II (1893), 17–19.

THOMPSON, J. B., "Readington Negroes," in *"1719–1894: Historical Discourses and Addresses at the 175th Anniversary of the Reformed Church, Readington, N.J., October 17, 1894* (Somerville, N.J., 1894), pp. 52–60.

THOMPSON, R. T., "Transportation Combines and Pressure Politics in New Jersey, 1833–1836," *Proc. N.J.H.S.*, LVII, 1–15, 71–86.

TURNER, D. K., "John Fitch, The Inventor of Steam Navigation," *Papers Bucks Co. Hist. Soc.*, II, 22–34.

———, "Sketch of Log College," *ibid.*, III, 1–10.

VAIL, M. C., "A Historical Sketch of the Meeting and Meeting House of the Society of Friends at Quakertown, N.J.," *Jerseyman*, II (1893), 9–12.

VAN HORNE, B. G., "Old-Time Bergen County Doctors," *Papers and Proceedings of the Bergen County Historical Society*, III (1906–07), 29–36.

VIVIAN, C. H., "The Country's Oldest Iron and Steel Producer," reprint from *Compressed Air Magazine*, Jan., 1940.

VOORHEES, O. M., "Captain David Schamp of Readington," *Jerseyman*, VII (1901), 25–30.

VOORHEES, RALPH, "The Raritan and Its Early Holland Settlers," *Our Home*, I.

VOSSELLER, ELIAS, "James W. Marshall, the New Jersey Discoverer of Gold," *Proc. N.J.H.S.*, 3d ser., VII, 278–86.

———, "Revolutionary Monuments," *Jerseyman*, III (1895–97), 12.

———, "The Hunterdon County Historical Society," *ibid.*, II (1893), 19–21.

———, "Tucca-Ramma-Hacking," *ibid.*, VII (1901), 9–12.

WACK, C. P., "The German Reformed of Lebanon," *Our Home*, I, 389.

Books with Specific Local Application

ANDERSON, J. A., *Navigation of the Upper Delaware* (Trenton, 1913).

BITTINGER, L. F., *The Germans in Colonial Times* (Philadelphia, 1901).

BLANE, JOHN, *History of the District Medical Society for the County of Hunterdon from its Organization in 1821 to the Annual Meeting in 1871; Together with the Medical History of the County (As Its Boundaries Now*

Exist), from its First Settlement to the Present Time, 1872 (Newark, 1872).

Boer, L. P. de, ed., *The Sutphen Family* (New York, 1926).

Bond, B. W., Jr., *The Quit-Rent System in the American Colonies* (New Haven, 1919).

Bush, E. T., *In the Grip of the Expert* (New York, 1904).

———, *When Leaves Grow Old and Other Poems* (Boston, 1916).

Carpenter, W. H., and T. S. Arthur, *The History of New Jersey from its Earliest Settlement to the Present Time* (Philadelphia, 1853).

Chambers, T. F., *The Early Germans of New Jersey* (Dover, N.J., 1895).

Clement, John, *Historical Sketches Relating to Early Settlements in West New Jersey* (n.p., n.d.).

Converse, C. S., *History of the United First Presbyterian Church of Amwell,* N.J., (Trenton, 1881).

Conway, M. D., *The Life of Thomas Paine* (New York, 1892). Contains long quotations from the diary of John Hall, now lost.

Cooley, H. S., *Slavery in New Jersey* (Baltimore, 1896).

Ege, Ralph, *Pioneers of Old Hopewell* (Hopewell, N.J., 1908).

Fargo, C. B., *History of Frenchtown* (New York, 1933).

———, *The Story of the Delaware Valley* (Frenchtown, N.J., 1936).

Gibbons, H. A., *John Wanamaker* (New York, 1926).

Jones, E. A., *The Loyalists of New Jersey* (Newark, 1927).

Kugler, J. B., *History of the First English Presbyterian Church in Amwell* (Somerville, 1912).

Larison, C. W., *A Skech [sic] of the Fisher Family of Old Amwell Township in Hunterdon County, New Jersey* (Ringoes, N.J., 1890).

Leahy, W. T., *The Catholic Church of the Diocese of New Jersey* (Princeton, 1906).

Mellick, A. D., Jr., *Story of an Old Farm* (Somerville, 1889).

Mercer, H. C., *Light and Fire Making* (Philadelphia, 1898).

Mott, G. S., *History of the Presbyterian Church in Flemington, N.J., with Sketches of Local Matters for Two Hundred Years* (New York, 1894).

———, *The Resurrection of the Dead* (New York, 1867).

Nelson, William, *Indians of New Jersey* (Paterson, 1894).

Opdyke, C. W., *The op Dyck Genealogy* (Albany, N.Y., 1889).

Prevost, Severo Mallet, comp., *Historical Notes and Biographical Sketches Regarding the American Branch of the Mallet Family, 1794–1930* (New York, 1930).

Rockefeller, H. O., ed., *Transactions of the Rockefeller Family Association for the Five Years, 1905–1909, with Genealogy* (New York, 1910).

Schuyler, Hamilton, *A History of St. Michael's Church* (Princeton, 1926).

1719–1894; Historical Discourse and Addresses Delivered at the 175th Anniversary of the Reformed Church, Readington, N.J., October 17, 1894 (Somerville, 1894).

SMITH, SAMUEL, *A History of the Colony of Nova-Caesaria, or New-Jersey*, ed. by W. S. Sharp (Trenton, 1877; originally published at Burlington, 1765).

SNELL, J. P., comp., *History of Hunterdon and Somerset Counties, New Jersey* (Philadelphia, 1881).

TANNER, E. P., *The Province of New Jersey, 1664–1738* (New York, 1908).

THOMPSON, H. P., *History of the Reformed Church at Readington, N.J., 1719–1881* (New York, 1882).

VAN SYCKLE, EMOGENE, *The Old York Road and Its Stage Coach Days* (Flemington, 1936).

WOODWARD, C. R., *Agriculture in New Jersey* (New York, 1930).

———, *The Development of Agriculture in New Jersey, 1640–1880* (New Brunswick, 1927).

——— and I. N. WALLER, *New Jersey's Agricultural Experiment Station, 1880–1930* (New Brunswick, 1932).

Books—General

ABBOTT, EDITH, *Historical Aspects of the Immigration Problem* (Chicago, 1926).

———, *Women in Industry* (New York, 1926).

ADAMS, J. T., *Provincial Society, 1690–1763* (New York, 1924).

ADAMS, W. F., *Ireland and Irish Emigration to the New World from 1815 to the Famine* (New Haven, 1932).

ANDREWS, C. M., *The Colonial Period of American History* (New Haven, 1934–38), 4 vols.

ARCHER, G. L., *Big Business and Radio* (New York, 1939).

———, *History of Radio to 1926* (New York, 1938).

BAILEY, L. H., ed., *Cyclopedia of American Agriculture* (New York, 1909). 4 vols.

BAIRD, C. W., *History of the Huguenot Emigration to America* (New York, 1885). 2 vols.

BALCH, E. G., *Our Slavic Fellow Citizens* (New York, 1910).

BASSETT, J. S., *Short History of the United States* (New York, 1939).

BAYER, H. G., *The Belgians First Settlers in New York and in the Middle States* (New York, 1925).

BEER, G. L., *British Colonial Policy, 1754–65* (New York, 1907).

BERNHEIMER, C. S., *The Russian Jew in the United States* (Philadelphia, 1905).

BIDWELL, P. W., and J. I. FALCONER, *History of Agriculture in the Northern United States, 1620–1860* (Washington, 1925).

BISHOP, J. L., *A History of American Manufactures from 1608 to 1860* (Philadelphia, 1866). 3 vols.

BOGART, E. L., *Economic History of the American People* (New York, 1935).

——— and C. M. THOMPSON, *Readings in the Economic History of the United States* (New York, 1916).

BOLTON, C. K., *Scotch-Irish Pioneers in Ulster and America* (Boston, 1910).

BRUNNER, E. D., *Immigrant Farmers and Their Children* (Garden City, N.Y., 1929).

——— *et al., Radio and the Farmer* (New York, n.d.).

BUCK, S. J., *The Agrarian Crusade* (New Haven, 1920).

———, *The Granger Movement* (Cambridge, 1913).

BURTT-DAVY, JOSEPH, *Maize* (London, 1914).

CAPEK, THOMAS, *The Cechs in America* (New York, 1920).

CARRIER, LYMAN, *The Beginnings of Agriculture in America* (New York, 1923).

CASSON, H. N., *Romance of the Reaper* (New York, 1902).

CHANNING, EDWARD, *A History of the United States* (New York, 1905–25). 6 vols.

CHASTELLUX, FRANÇOIS JEAN, MARQUIS DE, *Travels in North America in the Years 1780, 1781, and 1782*, tr. by G. Grieve (London, 1787).

CLARK, T. D., *Pills, Petticoats and Plows* (Indianapolis, 1944).

CLARK, V. S., *History of Manufactures in the United States* (New York, 1929). 3 vols.

COMMONS, J. R., *et al., History of Labour in the United States* (New York, 1918–35). 4 vols.

———, ed., *Labor Movement, 1840–1860* (Cleveland, 1910). 2 vols.

———, *Races and Immigrants in America* (New York, 1927).

CRESSY, EDWARD, *Discoveries and Inventions of the Twentieth Century* (London, 1923).

CURTLER, W. H. R., *A Short History of English Agriculture* (Oxford, 1909).

CUSHING, MARSHALL, *Story of Our Post Office* (Boston, 1893).

DAVIS, JEROME, *The Russian Immigrant* (New York, 1922).

DEFEBAUGH, J. E., *History of the Lumber Industry of America* (New York, 1906–07). 2 vols.

DOUGLAS, P. H., *Real Wages in the United States, 1890–1926* (Cambridge, 1930).

DUNBAR, SEYMOUR, *A History of Travel in America* (New York, 1937).

DUNLAP, O. E., *Story of Radio* (New York, 1935).

EARLE, A. M., *Home Life in Colonial Days* (New York, 1891).

EGERTON, H. E., *Short History of British Colonial Policy* (London, 1897).

FAIRCHILD, H. P., *Immigration, a World Movement and Its American Significance* (New York, 1926).

FAULKNER, H. U., *American Economic History* (New York, 1938).

FAUST, A. B., *The German Element in the United States* (New York, 1909). 2 vols.

FISH, C. R., *The Rise of the Common Man, 1830–1850* (New York, 1929).

FISHER, JOHN, *The Dutch and Quaker Colonies in America* (New York, 1899). 2 vols.

FOERSTER, R. F., *The Italian Emigration of Our Time* (Cambridge, 1919).

FORD, H. J., *The Scotch-Irish in America* (Princeton, 1915).

GILBERT, A. W., *The Potato* (New York, 1917).

GOLDENWEISER, E. A., and L. E. TRUESDELL, *Farm Tenancy in the United States,* Census Monograph No. 4, U.S. Census, 1920 (Washington, 1924).

GREENWOOD, ERNEST, *Amber to Amperes* (New York, 1931).

———, *Prometheus, U.S.A.* (New York, 1929).

HANNA, C. A., *The Scotch-Irish* (New York, 1902).

HANSEN, M. L., *The Atlantic Migration, 1607–1860* (Cambridge, 1940).

HARLOW, A. F., *Old Wires and New Waves* (New York, 1936).

HAYWARD, A. H., *Colonial Lighting* (Boston, 1923).

HICKS, J. D., *The Populist Revolt* (Minneapolis, 1931).

HILL, J. A., *Women in Gainful Occupations 1870 to 1920,* Census Monograph No. IX (Washington, 1929).

History of Wages in the United States from Colonial Times to 1928, Bulletin of the United States Bureau of Labor Statistics No. 499 (Washington, 1929).

HOURWICK, I. A., *Immigration and Labor* (New York, 1912).

Huguenot-Walloon New Netherland Commission, *Huguenot-Walloon Tercentenary, 1624–1924* (New York, 1924).

HUNT, T. F., *Cereals in America* (New York, 1914).

HUTCHINSON, W. T., *Cyrus Hall McCormick* (New York, 1930–35). 2 vols.

JAMES, E., *et al., The Immigrant Jew in America* (New York, 1907).

JENKS, J. W., and W. J. LAUCK, *The Immigration Problem* (New York, 1926).

JOHNSON, S. C., *History of Emigration from the United Kingdom to North America, 1763–1912* (New York, 1914).

JONES, R. M., *The Quakers in the American Colonies* (London, 1911).

JOSEPH, SAMUEL, *Jewish Immigration to the United States from 1881 to 1910* (New York, 1914).

KAEMPFFERT, WALDEMAR, ed., *Popular History of American Invention* (New York, 1924). 2 vols.

KEMMERER, D. L., *Path to Freedom* (Princeton, N.J., 1940).

KETTERING, C. F., and A. ORTH, *The New Necessity* (Baltimore, 1932).

KUHLMANN, C. B., *Development of the Flour-Milling Industry in the United States* (New York, 1929).

KUHN, OSCAR, *The German and Swiss Settlements of Colonial Pennsylvania* (New York, 1901).

KULL, I. S., ed., *New Jersey, a History* (New York, 1930–32). 6 vols.

LANE, W. J., *From Indian Trail to Iron Horse* (Princeton, N.J., 1939).

LEE, F. B., *New Jersey as a Colony and as a State* (New York, 1903). 4 vols.

LIVELY, C. E., and C. TAEUBER, *Rural Migration in the United States* (Washington, 1939).
LUCKEISH, M., *Artificial Light* (New York, 1920).
LORD, E., *et al., The Italian in America* (New York, 1906).
LUNDIN, LEONARD, *Cockpit of the Revolution* (Princeton, N.J., 1940).
MACKENZIE, CATHERINE, *Alexander Graham Bell* (New York, 1928).
MADEIRA, C. C., JR., *The Delaware and Raritan Canal* (East Orange, N.J., 1941).
MILES, MANLY, *Silos, Ensilage and Silage* (New York, 1889).
McGEE, T. D., *A History of the Irish Settlers in North America from the Earliest Period to the Census of 1850* (Boston, 1852).
McGRANE, R. C., *The Panic of 1837* (New York, 1924).
MERCER, H. C., *Tools of the Nation Maker* (Doylestown, Pa., 1897).
MILLER, M. D., *Evolution of the Reaping Machine,* Dept. of Agri., Office of Experiment Station, Bulletin #103 (Washington, 1902).
National Industrial Conference Board, *The Immigration Problem in the United States* (New York, 1923).
OSGOOD, H. L., *The American Colonies in the Eighteenth Century* (New York, 1924).
PHILLIPS, U. B., *American Negro Slavery* (New York, 1928).
———, ed., *Plantation and Frontier* (Cleveland, 1910). 2 vols.
PLUMB, C. S., *Indian Corn Culture* (Chicago, 1908).
———, *Types and Breeds of Farm Animals* (Boston, 1920).
PROTHERO, R. E. (Lord Ernle), *English Farming Past and Present* (New York, 1927).
QUINTANCE, H. W., *Influence of Farm Machinery on Production and Labor* (New York, 1904).
REID, J. D., *The Telegraph in America* (New York, 1879).
REUTER, E. B., *The American Race Problem* (New York, 1927).
RICH, W. E., *History of the United States Post Office to the Year 1829* (Cambridge, 1924).
ROBERTS, E. F., *Ireland in America* (New York, 1931).
ROGIN, LEO, *The Introduction of Farm Machinery in Its Relation to the Productivity of Labor in the Agriculture of the United States* (Berkeley, Calif., 1931).
ROSENBERGER, J. L., *The Pennsylvania Germans* (Chicago, 1923).
SAMPSON, A. W., *Native American Forage Plants* (New York, 1924).
SANFORD, A. H., *The Story of Agriculture in the United States* (New York, 1916).
SCHLESINGER, A. M., *Political and Social Growth of the United States, 1852–1933* (New York, 1936).
———, *The Rise of the City* (New York, 1938).

SHALER, U. S., ed., *The United States of America,* I (London, 1894).

SHAW, THOMAS, *Soiling Crops and the Silo* (New York, 1904).

SHULTZ, W. J., and M. R. CAINE, *Financial Development of the United States* (New York, 1937).

STEPHENSON, G. M., *A History of American Immigration, 1820–1924* (Boston, 1926).

SUMNER, H. L., *History of Women in Industry in the United States,* 61st Congress 2d Session, Senate Document No. 645, Vol. IX (Washington, 1911).

THOMAS, W., and F. ZNANIECKI, *The Polish Peasant in America* (New York, 1927).

THOMPSON, R. T., *Colonel James Neilson* (New Brunswick, 1940).

TOWERS, W. K., *From Beacon Fire to Radio* (New York, 1924).

WEATHERWAX, PAUL, *The Story of the Maize Plant* (Chicago, 1923).

WESLEY, C. H., *Negro Labor in the United States* (New York, 1927).

WHITMAN, C. A., *Flax Culture* (Boston, 1888).

WING, J. E., *Alfalfa Farming in America* (Chicago, 1912).

Pamphlets

ANDERSON, H. C., *Gen. Washington at Coryell's Ferry* (n.p., 1928).

ARMSTRONG, H. W., *St. Andrew's,* Amwell (n.p., 1914).

The Bazaar Souvenir Illustrated Published in the Interests of the Reformed Church of High Bridge, New Jersey (n.p., 1906).

Brief History of the Cokesbury Methodist Episcopal Church and Vicinity (Newark, 1890).

BROWN, A. H., *Historical Sketch of the Synod of New Jersey . . . from 1861 to 1886* (Philadelphia, 1888).

———, *An Outline History of the Presbyterian Church in West or South Jersey, from 1700 to 1865, with an Appendix from 1865 to 1869* (Philadelphia, 1869).

CONNET, J. L., *Flemington, N.J.* (Flemington, 1898).

DEATS, H. E., *Colonel Stewart's Revolutionary War Service Recalled* (Flemington, 1932).

The Delaware and Raritan Canal (n.p., 1933).

DESHLER, C. D., "Early Roads in New Jersey" (New Brunswick, 1880). Rutgers Univ. Library.

DOHERTY, D. L., *Centennial Celebration, the Presbyterian Church, Milford, New Jersey,* March 19–26, 1933 (n.p., 1933).

EWAIN, N. R., *Early Brickmaking in the Colonies* (Camden, N.J., 1938).

FLEMING, JOHN, *History of Readington School, formerly called Holland Brook School, 1804–97* (Flemington, 1898).

History of the Baptist Church at Kingwood, N.J. (Trenton, 1833).

Idle Farms in Hunterdon County, New Jersey (Trenton, 1932). N.J. Dept. of Agri., Circular No. 227.

KIRKPATRICK, JAMES, *An Historical Discourse Delivered by Request of the Presbytery of Raritan in the United First Presbyterian Church of Amwell, N.J., June 20, 1860* (Philadelphia, 1860).

LARISON, C. W., *The Ancient Village of Amwell* (Flemington, 1916).

MATTHEWS, D. M., *Directory of the Frenchtown, M.E. Church, with Brief History* (Frenchtown, N.J., 1885).

MESSER, CHARLES, *et al., Progress in Achieving Soil and Water Conservation and Its Economic Aspects on Dairy Farms in the Flemington Area, New Jersey* (New Brunswick, 1941).

NAYLOR, M. P., *History of the Dutch Reformed Church of White House, New Jersey, and Its Forerunners* (n.p., 1938).

One Hundredth Anniversary of the Flemington Methodist Episcopal Church, Flemington, New Jersey, 1823–1923 (Flemington, 1923).

Road Map of Hunterdon County, New Jersey (New York, 1943).

RACE, HENRY, *Historico-Genealogical Sketch of Col. Thomas Lowrey, and Esther Fleming, His Wife* (Flemington, 1892).

ROBERSON, ELMER, *The Centre Bridge* (Flemington, 1928).

SCHMIDT, HUBERT, *Flax Culture in Hunterdon County, N.J.* (Flemington, 1939).

———, *Slavery and Attitudes on Slavery in Hunterdon County, New Jersey* (Flemington, 1941).

SCHULTZ, J. H. E., *History of the Methodist Episcopal Church of Flemington, N.J., 1822–1914* (Flemington, 1915).

1738–1938; 200th Anniversary, Old Amwell First Presbyterian Church of Reaville, New Jersey (n.p., 1938).

SIM, R. J., *Some Vanishing Phases of Rural Life in New Jersey* (Trenton, 1941). N.J. Dept. of Agri., Circular 327.

SMITH, R. W., *One Hundredth Anniversary of the Founding of the Methodist Episcopal Church, Quakertown, New Jersey, 1836–1936* (n.p., 1936).

STACKHOUSE, A. M., *Maladies, Remedies and Physicians of Colonial Days in Burlington County, and Other Kindred Topics* (Moorestown, N.J., 1908).

SYMMES, J. G., *Historical Sketch of the Monmouth Presbytery and Its Churches* (Trenton, 1877).

TAYLOR, KNOX, *et al., Historically Speaking* (n.p., 1917).

THOMPSON, C. O., *225 Years—a History of Zion Evangelical Lutheran Church* (Oldwick, N.J., 1939).

VAIL, M. C., *History of Land Titles in the Vicinity of Quakertown, New Jersey* (Flemington, 1915).

VOORHEES, O. M., *The Exterior and Interior Bounds of Hunterdon County* (Flemington, 1929).

VOSSELLER, ELIAS, *History of the Hunterdon County Historical Society* (Flemington, 1894).

———, *A Sketch of the Copper Mining Enterprise Near Flemington, New Jersey* (Flemington, 1891).

Periodicals

American Agriculturist, Feb. 4, 1939.
Compressed Air Magazine, New York, Jan., 1940.
Jerseyman, Hightstown, N.J., and Flemington, 1891–1905.
Journal of the Medical Society of New Jersey, XXXIV (1937).
New Jersey Agriculture, New Brunswick, 1925–44.
Our Home, Somerville, I (1873). Only one volume was published.
Papers and Proceedings of the Bergen County Historical Society, 1902–23.
Papers of the Bucks County (Pa.) *Historical Society,* Doylestown, Pa., 1908–44.
Pennsylvania Magazine of History and Biography, Philadelphia, 1877–1944.
Proceedings of the New Jersey Historical Society, Newark, 1847–1944.
Ringos [*sic*] (*A Monthly Magazine Devoted to the History of Ringos—Past and Present—and to the Current News of the Village and Vicinity*), vol. I, nos. 5–12, Ringoes, N.J., 1889–90.

Index